Extending the Evolutionary Synthesis

Darwin's Legacy Redesigned

Axel Lange

CRC Press is an imprint of the
Taylor & Francis Group, an **informa** business

Designed cover image: iStock (ID: 468462143)

First edition published 2023
by CRC Press
6000 Broken Sound Parkway NW, Suite 300, Boca Raton, FL 33487-2742

and by CRC Press
4 Park Square, Milton Park, Abingdon, Oxon, OX14 4RN

CRC Press is an imprint of Taylor & Francis Group, LLC

© 2023 Taylor & Francis Group, LLC

Originally published in German: Evolutionstheorie im Wandel – Ist Darwin überholt? from Dr. Axel Lange. Copyright © Springer-Verlag GmbH Germany, part of Springer Nature 2020. All rights reserved.

The English edition is corrected and updated. Proof reading: https://enago.com, Crimson Interactive Inc.

Reasonable efforts have been made to publish reliable data and information, but the author and publisher cannot assume responsibility for the validity of all materials or the consequences of their use. The authors and publishers have attempted to trace the copyright holders of all material reproduced in this publication and apologize to copyright holders if permission to publish in this form has not been obtained. If any copyright material has not been acknowledged please write and let us know so we may rectify in any future reprint.

Except as permitted under U.S. Copyright Law, no part of this book may be reprinted, reproduced, transmitted, or utilized in any form by any electronic, mechanical, or other means, now known or hereafter invented, including photocopying, microfilming, and recording, or in any information storage or retrieval system, without written permission from the publishers.

For permission to photocopy or use material electronically from this work, access www.copyright.com or contact the Copyright Clearance Center, Inc. (CCC), 222 Rosewood Drive, Danvers, MA 01923, 978-750-8400. For works that are not available on CCC please contact mpkbookspermissions@tandf.co.uk

Trademark notice: Product or corporate names may be trademarks or registered trademarks and are used only for identification and explanation without intent to infringe.

ISBN: 9781032376899 (hbk)
ISBN: 9781032376905 (pbk)
ISBN: 9781003341413 (ebk)

DOI: 10.1201/9781003341413

Typeset in Joanna
by codeMantra

It would be very strange indeed to believe that everything in the living world is the product of evolution except one thing — the process of generating [genetic] variation!

Eva Jablonka und Marion J. Lamb

We have to remember that what we observe is not nature herself, but nature exposed to our method of questioning.

Werner Heisenberg

An insect is more complex than a star.

Martin Rees

This book is dedicated to Olivia, Lemonia, Simon, and all future grandchildren. May their generations share nature's deep wisdom in adapting life to its home on Earth.

CONTENTS

Author / xv
Introduction / xvii

1 • DARWIN'S MILLENNIUM THEORY AND BATESON'S COUNTER MODEL / 1

1.1 Charles Darwin's Theory and Its Significance / 1
    1.1.1 At Least Three Theories in One / 2
    1.1.2 The Relationship of All Life / 3
    1.1.3 Darwin's Openness as a Scientist / 4
    1.1.4 Alfred Russel Wallace: The Co-Discoverer / 5
    1.1.5 Evolution Is a Fact / 6
1.2 William Bateson's Counter Model / 7
    1.2.1 Discontinuous Variation / 8
    1.2.2 A Hundred Years of Dispute / 9
1.3 The Time after Darwin / 10
    1.3.1 August Weismann: A Stubborn Doctrine / 10
    1.3.2 How Are Genes Defined? Where Are They Located? / 11
1.4 Summary / 12

References / 13
Tips and Resources for Further Reading and Clicking / 13

2 • THE MODERN SYNTHESIS: THE STANDARD MODEL OF EVOLUTION / 15

2.1 Emergence and Key Statements / 15
    2.1.1 Thomas Hunt Morgan and Thousands of Flies / 16
    2.1.2 Statisticians, Zoologists, and Others at the Same Table / 17
    2.1.3 Dobzhansky and Mayr: The Practitioners / 18
    2.1.4 Role of Dinosaurs and Plants in the Modern Synthesis / 20
2.2 Variation–Selection–Adaptation: The Practice / 21
    2.2.1 An Ingenious Experiment Confirms the Modern Synthesis / 21
    2.2.2 Guppies with and without an Enemy / 21

- 2.2.3 Perfect, Less Perfect, or No Adaptation / 22
- 2.3 Further Findings Up to 1980 / 24
  - 2.3.1 Punctualism / 25
  - 2.3.2 More Criticism of Adaptationism / 25
  - 2.3.3 New Worlds: Multilevel Selection and Sociobiology / 26
  - 2.3.4 The Neutral Theory of Molecular Evolution / 27
  - 2.3.5 Molecular Study of Lineages / 28
  - 2.3.6 Criticisms Remaining of the Modern Synthesis after 1980 / 28
- 2.4 Summary / 28
- References / 29
- Tips and Resources for Further Reading and Clicking / 30

3 • EVO-DEVO—THE BEST OF BOTH WORLDS / 31

- 3.1 Conrad Hal Waddington—Epigeneticist and Forefather of Evo-Devo / 31
  - 3.1.1 Veins in Fly Wings—An Attempt at a Proof / 31
  - 3.1.2 Buffers and Canalization in Development / 32
  - 3.1.3 Genetically Assimilated / 34
  - 3.1.4 Waddington Was Looking Forward / 34
- 3.2 Evo-Devo—History and Early Priorities / 34
  - 3.2.1 Embryo Was Already on the Radar in the 19th century / 34
  - 3.2.2 The Embryo Does Not Fit into the Picture of the Modern Synthesis / 35
  - 3.2.3 Developmental Genes—The Embryo Is Rediscovered / 36
  - 3.2.4 Ontogenesis and Phylogenesis—How a New Research Discipline Emerged / 39
- 3.3 Gene Regulatory Networks / 41
  - 3.3.1 Genetic Toolkit / 41
  - 3.3.2 The Arrival of the Fittest / 43
- 3.4 Systemic-Interdisciplinary View—A Research Discipline Acquires Order / 46
- 3.5 Facilitated Variation—The Perspective of Cells / 53
  - 3.5.1 Preserved for Hundreds of Millions of Years / 53
  - 3.5.2 And Yet Change Is Possible / 55
  - 3.5.3 The Exploratory Behavior of Cells / 55
  - 3.5.4 Weak Regulatory Couplings–Cells in Loose Conversation / 57
  - 3.5.5 Compartmentation–The Modular Solution / 58
  - 3.5.6 A Theory with New Informational Content / 59
- 3.6 Inheritance Is Much More than Mendel and Genes: Inclusive Inheritance / 60
  - 3.6.1 Genetic Mutation Does Not Always Occur by Chance / 61
  - 3.6.2 Epigenetic Inheritance–A Widely Discussed Topic in Scientific Literature and Media / 63
  - 3.6.3 Epigenetic Inheritance by

    Learning from the Ancestors / 65
  3.6.4 Decoupling of Evolution from Biology / 66
 3.7 The Music of Life / 67
  3.7.1 Questionable Determinacy / 67
  3.7.2 The Genetic Program Is an Illusion / 68
  3.7.3 We Inherit Cellular Machinery from Our Mother / 68
  3.7.4 Against Reductionism? / 69
  3.7.5 Biological Function from a System Perspective / 70
 3.8 Phenotype Plasticity and Genetic Assimilation: *Genes Are Followers* / 75
  3.8.1 The Environment in a New Role / 76
  3.8.2 Developmental Plasticity / 77
  3.8.3 Concrete Plasticity—A Goat on Two Legs and Tunnel-Digging Beetles / 78
  3.8.4 Genetic Accommodation / 79
  3.8.5 Environmental Influences Or Mutations—Which Is More Likely? / 80
  3.8.6 Heat Shock Proteins Buffer Mutations in Drosophila, Corals, and Darwin's Finches / 80
  3.8.7 Molecular Mechanisms / 81
 3.9 Innovations in Evolution / 87
  3.9.1 What Do We Mean by "New" in Evolution? / 87
  3.9.2 Conditions of Innovation Initiation / 91
  3.9.3 Conditions of Innovation Realization / 92
  3.9.4 Genetic Integration of Innovations / 94
 3.10 Summary / 95
References / 96
Tips and Resources for Further Reading and Clicking / 99

4 • SELECTED EVO-DEVO RESEARCH RESULTS / 101
 4.1 Evolution of Beak Shape in Darwin's Finches / 101
 4.2 Experiment 1: Fish Can Learn to Walk / 103
 4.3 Cichlids with Thick Lips and Large Bumps / 104
 4.4 Experiment 2: Butterflies with Eyes on Their Wings / 105
 4.5 Experiment 3: Genetic Assimilation in Tobacco Hornworm / 109
 4.6 The Almost Nonconstructable Turtle Shell / 109
 4.7 How Many Legs Do Millipedes Have? / 112
 4.8 Supernumerary Digits / 113
  4.8.1 The Uniqueness of the Human Hand in the Animal Kingdom / 115
  4.8.2 Why Is Five the Default Number of Fingers? / 115
  4.8.3 The Ancient History of Polydactyly / 117
  4.8.4 Polydactyly and Genetics—Only Half the Story / 119

- 4.8.5 Hemingway's Cats / 120
- 4.8.6 Mysterious Numbers of Toes / 121
- 4.8.7 Computer Modeling of Polydactyly / 123
- 4.8.8 How Toe Numbers Challenge Evolutionary Theory? / 125
- 4.9 Summary and Outlook / 126
- References / 127
- Tips and Resources for Further Reading and Clicking / 129

5 • THE NICHE CONSTRUCTION THEORY / 131

- 5.1 An Evolutionary Mechanism of Its Own / 131
- 5.2 Developmental Niche Construction—Castles and Palaces for the Descendants / 137
- 5.3 Niche Constructions of the Human Being / 138
- 5.4 Is There a Fault Line in the Modern Synthesis? / 142
- 5.5 What Constitutes a New Theory? / 143
- 5.6 Summary / 145
- References / 145
- Tips and Resources for Further Reading and Clicking / 146

6 • EXTENDED EVOLUTIONARY SYNTHESIS / 147

- 6.1 Emergence of the EES Project / 148
- 6.2 Objectives of the EES Project / 148
- 6.3 New Predictions about Evolution / 149
- 6.4 Brief Description of the Individual Research Projects / 149
- 6.5 A Project beyond EES: Instances of Agency in Living Systems / 155
- 6.6 Summary / 160
- References / 162
- Tips and Resources for Further Reading / 164

7 • THEORIES ON THE EVOLUTION OF THINKING / 167

- 7.1 Darwin's View of Thinking / 167
- 7.2 Theory of the Social Brain / 168
- 7.3 Tomasello's Natural History of Thinking / 172
- 7.4 Consciousness—A Questionable Scientific Object / 174
- 7.5 Summary / 181
- References / 181
- Tips and Resources for Further Reading / 182

8 • THE EVOLUTION OF HUMANKIND IN OUR (NON)BIOLOGICAL FUTURE / 183

- 8.1 Humankind Takes Control of Its Own Evolution / 184
  - 8.1.1 Synthetic Biology and Artificial Life / 185
  - 8.1.2 Genome Editing—Interventions in the Germ Line / 187
  - 8.1.3 Nanobot Technologies / 190
  - 8.1.4 Slowing Aging and Immortality / 190
  - 8.1.5 Human–Machine Combinations / 192
  - 8.1.6 The Future of Genetic Engineering and Transformation of Life / 193
- 8.2 Artificial Intelligence and Transhumanism in Evolution / 197
  - 8.2.1 AI and Humanoid Robots / 197
  - 8.2.2 Superintelligence—The Last Invention of Mankind? / 201
  - 8.2.3 Intelligence Explosion and Singularity / 205

8.3 The Theory of Evolution in the Technosphere / 207
8.4 Summary / 214
References / 215
Tips and Resources for Further Reading / 217

9 • MORE THAN ONE THEORY OF EVOLUTION–A PLURALISTIC APPROACH / 219

9.1 From Old to New Shores—Obstacles and Opportunities / 219
9.2 Evolution from Two Different Perspectives / 220
9.3 Summary and Outlook— For a Theoretical Pluralism / 221
References / 223

10 • THE PLAYERS OF THE NEW THINKING IN EVOLUTIONARY THEORY / 225

Glossary / 237

Index / 261

AUTHOR

Axel Lange graduated from the Jesuit College of St. Blasien in the Black Forest, Germany and proceeded to study economics and philosophy at the University of Freiburg. He worked in sales and marketing management in IT before his intense interest in evolutionary theory led him to completely redirect his efforts toward biology. Lange's first book on evo-devo and the extension of Modern Synthesis, published in 2012, provided the basis for his dissertation contract with the University of Vienna. There, in the Department of Theoretical Biology, Lange studied biology and conducted research on the evolutionary history of limb development in vertebrates and polydactyly—the formation of supernumerary fingers and toes in newborns. A new finger has thousands of cells with the potential to become bones, nerves, muscles, skin, and blood vessels; nevertheless, many of such fingers and toes commonly appear as new functional entities in a single generation in the embryo and are even inherited further in variable numbers. The standard theory of evolution cannot explain the mechanisms for this, which begs the following question: "How does this phenotype arise?" This is the question that preoccupied Lange. His article with Gerd. B. Müller on mankind's knowledge of polydactyly in development, inheritance, and evolution, from antiquity to the present, was published in March 2017 in the tradition-rich American journal *The Quarterly Review of Biology*. His other publications focused on the self-organizing capacity of the limb with simultaneous variation in finger numbers in the model.

Now a working biologist, Lange earned his doctorate with honors in 2018 and now delivers lectures on complex, epigenetic evo-devo processes in Germany and abroad. The (non)biological future of human evolution is also one of his favorite pretract topics. During his studies on post-Modern Synthesis in evolutionary biology, Lange formed lasting friendships with internationally renowned researchers. He loves the mountains, is passionate about playing romantic piano music, and makes his living as an author and science journalist in the south of Munich. Lange has three adult children.

INTRODUCTION

The theory of evolution is advancing, overcoming entrenched notions, and breaking new ground more than 160 years after Darwin's seminal work on the evolution of all forms of life on earth. This book will guide the reader through a series of fascinating theoretical developments in the field.

As a reader, you will be introduced to the history of evolutionary theory and confronted with questions left unanswered by the standard synthetic theory, originally known as the *Modern Synthesis*. As you advance from one chapter to the next, you will visualize the essence of a new theory: the *Extended Evolutionary Synthesis*, which critically re-evaluates the basic assumptions of *Modern Synthesis*, including its central focus on genetics, with random mutations as the only cause of phenotypic variation, and the narrowing of the scope of evolution to additive, minute changes. If you are of the opinion that genes are the sole carriers of heredity, your view will be shaken.

You will learn how evolutionary developmental biology (evo-devo) determines the constructive mechanisms of embryonic development that can provide for evolutionary changes at all biological levels. Evolutionary events acquire a new appearance—that of the inherent abilities of the embryo, which determines its own form and generates new variations. What are the consequences for the theory, that species shape their own environments, for example, beavers or termites with their constructions and birds with elaborate nests in which their young grow up? First and foremost, we humans are transforming the earth. The transformation of nature by humankind determines our own evolution. Moreover, we direct our own evolution through medicine and high-tech, artificial intelligence, and robotics. This was not on the radar of evolutionary theory until now.

I wrote this book for biologists and nonbiologists alike, students and adults, educators, and the curious. If you are interested in thinking outside the box and open to new ideas, then I invite you to dive into the modern science of evolution. I will (hopefully) present some challenging themes in an understandable manner and explain unavoidable technical terms. Important terms are also listed at the beginning of each chapter so that, if you wish, you can familiarize yourself with them a little in advance by looking them up in the glossary. Of course, some sections of the book deal with the concept of genetic interaction. Do not let this discourage you from reading on; however, you are welcome to skip forward a bit. I have also intentionally broken up the factual description in some places, where I bring my experiences into play or address you, the reader, in the text. These will serve as breathing spaces in your reading.

My own opinion on the state of the theory of evolution is subordinate. I will allow evolutionary biologists from all over the world to speak to you. In the end, you will be able to formulate your own opinion about how evolution proceeds from a 21st century perspective. Nevertheless, I hope to infect you with my enthusiasm for studying newly discovered mechanisms of evolution! I witnessed the emergence of the post-Modern Synthesis, met and had discussions with the world's leading researchers, and wrote a dissertation on a truly breathtaking evo-devo topic. Are you familiar with how additional fingers arise? Darwin was not, and he struggled to understand the process. In this book, you will discover the origins of polydactyly and other biological phenomena. You will look at evolution from a novel perspective, through the eyes of people of the 21st century.

AXEL LANGE
Taufkirchen (Munich), Germany
axel-lange@web.de

CHAPTER ONE

Darwin's Millennium Theory and Bateson's Counter Model

Charles Darwin was not the first to hypothesize that life on earth undergoes the processes of change. However, he was the first to introduce a mechanism for evolutionary change, natural selection. He was also the first to use several empirical examples, particularly those of breeding animals. Initially ridiculed for his work, Darwin eventually gained widespread acclaim. At present, Darwin (1809–1882) is rightly considered one of the foremost thinkers in human history. At present, 160 years have passed since the presentation of his seminal theory of the evolution of all life. Nevertheless, although one constantly reads or hears of new developments in the modern media regarding astronomy, genetics, or quantum physics, one tends to get the impression that no new developments have occurred in the theory of evolution since the convergence of Darwinism, Mendelism, and genetics. For the most part, only Darwin's basic ideas of random mutation and natural selection are reproduced in the public sphere, seemingly indicating that his theory has not developed significantly since. However, this book aims to show that nothing could be further from the truth.

Important technical terms in this chapter (see glossary): Adaptation, continuous and discontinuous variation, fitness, gene, gradualism, natural selection, phenotype, saltationism, Survival of the Fittest, and Weismann barrier.

1.1 CHARLES DARWIN'S THEORY AND ITS SIGNIFICANCE

Darwin's famous book *On the Origin of Species by Means of Natural Selection* was sold out a mere one day after its publication in 1859. His subsequent book, *The Descent of Man, and Selection in Relation to Sex* (Darwin 1871), was published 12 years later and received little acclaim. The man who derived our descent from the common origins of life with the apes was mocked for his ideas.

Darwin's magnum opus is a long and tedious read. Likely, few who write about evolutionary theory have read the book in its entirety. The reader literally feels the author's effort to set his hypothesis of natural selection on stable ground. For this purpose, he uses breeding animals, pigeons, dogs, cats, and ducks. By their example, he shows impressively how man succeeds in generating species variation via artificial selection. After illustrating the principle of artificial selection by man, Darwin turns to the process known universally as *natural selection*. Nature selects without the intervention of man or any other entity. The discovery of natural selection is thus the central mechanism of evolution and can be considered the center of Darwin's theory of evolution. We therefore address his theory as the selection theory. This book takes a critical approach to selection theory and its mode of operation.

However, the extent of Darwin's discoveries did not end there. Let us, therefore, briefly and sequentially consider the central theses of his theory. We will condense the 490 pages of the first edition of 1859 to a single page. It is not necessary to read a book that describes Darwin comprehensively (there are already quite a few of those) to discuss a few additions and unanswered questions that remain at the end. Rather, it is instructive to discuss how Darwin's theory and, above all, the theory of his successors—today's standard theory—can be extended by important arguments while overcoming the basic assumptions.

1.1.1 At Least Three Theories in One

To summarize, Darwin observed (or extended the observation of earlier researchers) that many species possess great fecundity; however, concurrent with a limited food supply, their population sizes do not explode, instead remain stable over time. His first theoretical conclusion was that competition among individuals in a population must occur to maintain their existence. Before Darwin, others had spoken of struggles between different species but not of those between individuals within a species. For example, Darwin's friend, the geologist Charles Lyell, imagined that species not only become extinct but also displace each other. Accordingly, one species could gain space only at the expense of another. Equilibria in nature varied unstably in this way.

Darwin extended the analysis astutely as follows. There are innumerable individual differences in a population, and no two individuals are identical. Such individual differences—variations, as he called them—are both hereditary and small. They accumulate slowly in small increments over many generations to form larger, visible variations. This is how, for example, the highly diverse limb shapes of vertebrates evolved. In this context, we speak of gradualism—an important term that we will revisit frequently in subsequent chapters and that has been instrumental in the modern criticism of evolutionary theory. Darwin's second central theoretical conclusion was drawn from the following observations: a natural selection process exists for all species because of their high fecundity. The survival of individual members of a species, according to this process of natural selection, depends on the variations they inherit.

Finally, Darwin's third core theoretical statement was that natural selection leads to an increased proportion of offspring of the best adapted member of a species in subsequent generations. This is the famous concept *Survival of the Fittest*, as he later called it. If the thesis holds true that individuals of a species *struggle to exist*, then it also holds true that favorable modifications tend to persist and unfavorable ones tend to disappear. Fitness is then the ability to pass on one's own advantageous traits to subsequent generations and, moreover, to do so under the particular living conditions of a population, not absolutely. These are Darwin's discoveries. There are others by him.

According to this theory, the complex structure and functionality of living things are solely a result of natural processes. Natural selection alone is the lever, the mechanism or process that determines evolutionary events on earth. This was the "bombshell" that Darwin detonated. Later, he added sexual selection as a complementary factor or a subtype of natural selection, a captivating topic in its own right. Imagine, for example, the magnificent birds of paradise. For our purposes, however, sexual selection is not a foremost concern (Figure 1.1).

Of course, Darwin's work leaves some essential questions unanswered. We may stipulate that his book contains two large gaps, which drew attention very early. We miss the *how* of heredity and the *whence* of variation. Of course, Darwin gave much thought to heredity. However, his theory was contradictory and, ultimately, incorrect. The mechanisms

Figure 1.1 Show-off fish. Darwin was intensely focused on breeding when he developed his theory. From artificial selection, he deduced the effect of natural selection. Here is a magnificent specimen of a molly (*Poecilica sphenops*). The exceptionally enlarged caudal and dorsal fins are produced in today's breeding populations by natural spontaneous mutations and human-induced outcrossing. In wild populations, the tail has a constant, adapted size and shape due to strong natural selective forces acting on its function as the main propulsive fin. An exception are the males of the related swordtail (*Xiphophorus hellerii*), which Darwin already described. In the aquarium, where natural selection is absent, sexual selection can have a stronger effect. Thus—as in the picture—striking secondary sexual characteristics may develop in a male and be well received by females.

of heredity were finally addressed and elucidated by Gregor Mendel and other early geneticists. Darwin, however, gave no explanation of why or how variations arise at all. The questions and the answers to them will run like a thread through this book.

1.1.2 The Relationship of All Life

From Darwin's groundbreaking view, it is unequivocal that all species are related to one another by virtue of *descent with modification*. All species have a history of descent. These lineages permit a chronological sequence of the splitting of the ancestral lines and the modifications that occurred in the process. Today, we know the following: humans and all mammals are descended from mouse-like species that lived on earth more than 150 million years ago and were inconspicuous contemporaries of the dinosaurs. Mammals, birds, reptiles, amphibians, and fish all trace their ancestry to small, worm-like animals that lived in the sea 600 million years ago. Finally, all animals and plants trace their ancestry to bacteria-like, single-celled organisms that lived 3,000 million years ago. Darwin's early assumption of a common descent of all life forms from simple organisms was a courageous thesis at that time. It was only proven correct much later, once it became possible at the beginning of this millennium to create genetic family trees that, for example, reveal human and mouse genes with a high degree of homology and thus reveal the relatedness of these organisms. Thus, even gene comparisons between humans and fruit flies, which are much older in evolutionary terms, are possible. Finally, the fact that the genetic code—the blueprint for amino acids and proteins—is almost identical in all living organisms points emphatically to a common ancestry of all life forms.

The theory of evolution lays out this spectacular scenario before us. The theory can scientifically justify the enormous number of different species, genera, and families on earth. Moreover, it can explain how they are all related to each other. It can explain why humans, cats, elephants, horses, bats, whales, and even dinosaurs share the same basic skeletal shape, the same limb construction—why they all have two eyes, one nose, one mouth, two lungs, and one heart. The evolutionist speaks of homologous structures. The design principle of the human hand applies to appendages with entirely different sizes and functions such as walking, swimming, flying, or writing. There must inevitably have been a common ancestor of all these homologous structures (Figure 1.2).

All this is indeed a grandiose show of life on earth, built on the pillars of Darwin's theory. Whether these pillars remain standing exactly as they were in Darwin's time, especially with selection as the sole driver of evolution, is a question we will explore critically, step by step, in this book.

1.1.3 Darwin's Openness as a Scientist

Darwin's most admirable quality was his openness in expressing the strengths and weaknesses of his own hypotheses. Preemptively, he discussed the most diverse possible objections that could be raised against his theory of natural selection. He did not categorically exclude alternatives but instead examined concepts and problems from various perspectives. Famously, Darwin encountered difficulties applying his theory to explain the evolution of complex structures such as the human eye; this example has been cited countless times. Speaking bluntly, he went so far as to confess

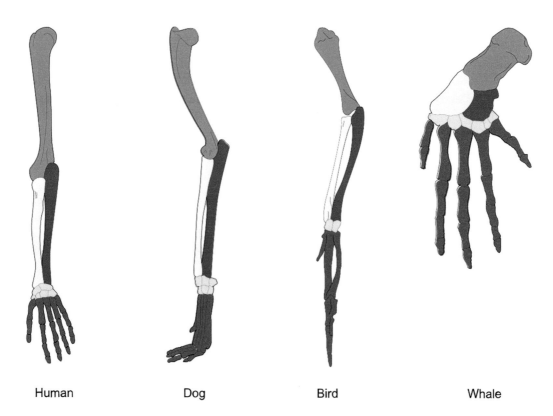

Figure 1.2 Homologous vertebrate extremities. Vertebrate extremities are homologous with respect to all bones, meaning that they share a common ancestor with the same bone elements. However, they may differ in function. The range of variation in size and shape is remarkable. In contrast, the wings of birds and insects developed not homologously but analogously, i.e., independently of each other.

that it seemed to him "absurd in the highest possible degree" that the eye could have developed by means of natural selection (Darwin 1859). Even to Darwin, the eye seemed like a system of independent parts: the retina with its photoreceptors to receive images, nerves to transmit signals from the retina to specific brain areas, and the lens to focus the image. On their own, none of these parts have any obvious use. What awareness could a lens have that it needs a retina to process light signals? Darwin conceded the possibility that the eye could only function in its current, apparently perfect, form if all its individual parts worked together ideally. Its path of origin was thus an open question.

In the same breath, Darwin formulated the conviction that subsequent generations of researchers could certainly uncover how the eye evolved from "an imperfect and simple" organ of sight that, in its initial form, functioned well enough to be useful to its owner and has since developed in no other way than by natural selection into the "perfect and complex eye" as we know it. By contrast, for an animal such as a snail, the eye structure is sufficient if it can distinguish light from dark. A slow-moving animal has no need of a more complex visual receptor to explore its proximity.

Darwin also places a restriction, for example, when he writes the often quoted phrase *Natura non facit saltum* (nature does not make a leap). He immediately adds that this is an "old, but somewhat exaggerated, canon in natural history" (Darwin 1859). Let us consider the following: Darwin's entire theoretical basis is that selection acts on accumulated, small variations in their order of appearance in the population. Darwin's theory offers no explanation for jumps. On how such jumps could arise, more discussion will follow. Nowadays, no scientist restricts his own hypotheses in this way—to do so would be far too risky. Who wants to imply that he has even the slightest doubt about his own work? At most, the readers should have doubts and criticisms but preferably no one does.

1.1.4 Alfred Russel Wallace: The Co-Discoverer

Here, it should be mentioned that Darwin also drew on the knowledge of other scientists, most notably, Robert Malthus. Malthus (1766–1834) was a national economist and founder of the modern economic discipline. He proposed the theory that food resources or soil yields on earth could be multiplied linearly, whereas living beings, including humans, multiply in a nonlinear, even exponential, manner. For example, overpopulation would inevitably occur over many generations when each family has four children, all of whom in turn have four children and so on, if it were not for food limitations. Darwin considered this doctrine as fundamental to the fact that there must be a struggle for limited resources. Members of a species that are well equipped for this struggle have a selective advantage over others that are not.

At this juncture, I would like to pay further tribute to one of Darwin's contemporaries, Alfred Russel Wallace (1823–1913; Figure 1.3). Like Darwin, he was an Englishman who embarked on a journey to the other side of the world, although somewhat later than Darwin himself. Wallace also conceived the idea of natural selection; however, his contribution was essentially forgotten. In a letter to Darwin, Wallace presented his theory, which resembled Darwin's own, as if it were written in his own hand. Darwin became considerably excited over this. In a gentlemen's agreement, the ideas of both researchers could finally be presented to the Linnean Society in London in 1858, before Darwin published his main work, long in preparation, in 1859. Wallace—generous as he was—allowed Darwin to publish first. He may not have had Darwin's foresight, but he was certainly a great naturalist. When referring to the mechanism of natural selection, it should be correctly called the Darwin–Wallace theory of natural selection.

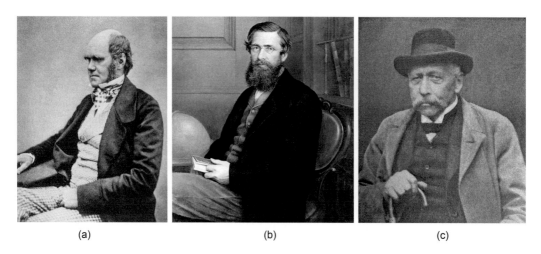

Figure 1.3 (a) Charles Darwin, (b) Alfred Russel Wallace, and (c) William Bateson. Three great British evolutionary theorists and biologists of the 19th century.

1.1.5 Evolution Is a Fact

Evolution is a fact. If we disregard complicated, little-used philosophical contortions, evolution is a theory in the sense of a system of scientifically based statements that are confirmed by observation. This theory can explain sections of reality with underlying regularities and derive predictions from them. One such prediction of evolutionary theory is that species diversity on earth will continue to decline in the coming decades if the average global temperature continues to increase. We will learn about several other predictions in subsequent chapters (4–6). At the same time, however, evolution is a fact in the sense that it is real, demonstrable, existing, and recognized. It is, according to the famous paleoanthropologist Donald Johanson, a fact, just as gravity is a fact (Leakey and Johanson 2011, YouTube). It has not required any further proof for a long time. Today, 160 years after Darwin, many gaps in the succession of species and between species and families have been clarified. Missing links are no longer a real issue because molecular biology, can be used for their detailed elucidation. We now know why the human embryo develops a tail in the womb only to resorb it long before the baby is born. Sometimes, however, this process goes awry, and the newborn has a bony tail as an extension of the spine. Furthermore, we know via molecular genetics how closely we are related to chimpanzees or bonobos. (Figure 1.4) and how distantly to mice, fruit flies, or annelids.

Nevertheless, there certainly remain phenomena that elude our understanding, even today. In this sense, we are like Darwin with the eye, and it is difficult for us to imagine how the genetic code could have arisen evolutionarily. It is the scheme, almost identical for all living things, according to which the smallest building blocks are produced that combine to form proteins, the complex substances of life that are present in all cells. Only a few years ago, I had difficulty imagining a good answer to the question of how mammals evolved. The evolutionarily older amphibians and lizards lay eggs that contain an adequate deposit of nutrients. The transition to embryonic development in the womb appears to be a huge evolutionary leap.

This is because the placenta, i.e., the interface that connects the mother with the embryo, must have developed first. It supplies the young life form with nutrients and oxygen from the maternal bloodstream. One imagines that such a placenta can be either present or absent, given that intermediate forms are

Figure 1.4 Relative. A young bonobo (*Pan paniscus*). The genetic relatedness of chimpanzees, bonobos, and gorillas to humans is very high. But with nearly 2% divergence, there are still approximately 60 million genetic differences. In addition, there are even more differences in gene regulation and development.

not known to exist. What could a step-by-step cumulative process for its emergence be like? Did a hundred or more individual steps precede the emergence of a functional placenta? In fact, more than one theory exists today as to how the placenta might have evolved in mammals; one such theory suggests that the emergence of the placenta resulted from viral invasion (by retroviruses) of the germ cells (Onoa et al. 2001). It cannot have been too complicated as there are several fish, including sharks, that give birth alive and others that do not. But did the maternal mammary glands not have to develop simultaneously with the placenta in mammals? Likewise, we can imagine these only as present or not present: the milk flows or it does not flow. Indeed, lactation is a rather complex task to coordinate for an evolutionary process that knows no goal and proceeds in small steps. I will not even attempt to explain the leap from a heart with three separate chambers (amphibians and reptiles) to one with four chambers (birds and mammals) when there can be no intermediary: 3.5 or even 3.6 chambers, with the appropriate inflows and outflows of blood, cannot exist; can it? Just as one cannot be "a little bit pregnant," on cannot have a nonwhole number of heart chambers.

However, serious doubt no longer exists regarding the facts that the "ingenious" genetic code, caring behavior of the mammalian mother, and other complex named and unnamed characteristics are one hundred percent the products of evolution. The remaining questions will eventually be answered. To conclude, evolution is built on a foundation of knowledge, about which only the eternal unbelievers retain doubts.

1.2 WILLIAM BATESON'S COUNTER MODEL

Like Darwin, William Bateson (1861–1926) was British, but the latter was 52 years younger than the former. While still alive, Bateson was famous in expert circles in both Europe and the United States. He was awarded the highest honors. Bateson (Figure 1.3) coined the term *genetics* and officially introduced this term to scientific circles at a meeting in London in 1906. He was one of the rediscoverers of Gregor Mendel's writing and was instrumental in spreading the "new" theory of heredity.

Above all, however, Bateson was known as a saltationist (from the Latin *saltare*, "to leap"). As I have mentioned above, nature allegedly does not make jumps; evolution, according to Darwin, proceeds gradually. Darwin himself stated that his theory would completely collapse if even one complex organ could be shown to not have arisen via successive, small modifications (Darwin 1859). This is precisely where Bateson critically steps in, and his contributions to evolutionary theory are of interest here.

1.2.1 Discontinuous Variation

It is better to work with the conceptual pair of continuous and discontinuous variations than to consider each one individually. Examples of continuous traits are height and weight. They assume a continuum of values. Bateson, on the other hand, was a vehement advocate of discontinuous variation. This type of variation has two or more clearly distinguishable discrete features, such as four or five toes. Bateson by no means flatly denied Darwin's teaching that variation is continuous and can accumulate in an organism through repeated selection processes. However, he flatly refused to elevate natural selection to a doctrine for evolution and ascribe to it an exclusive claim. Bateson saw both continuous and discontinuous changes in lineages. However, he noted a fundamental difficulty arising from Darwin's view: how can clearly delineated, i.e., discontinuous, species come about when the differences in environmental conditions that constitute Darwinian selection conditions are fluid over time? On the one hand, Bateson speaks of a specific diversity of the form of living beings, and on the other hand, a diversity of environments. Changes in the latter, however, imperceptibly merge into a continuous series (Bateson 1894). For example, the temperature and salinity of sea water do not rise or fall abruptly. Volcanic eruptions, which would represent an abrupt environmental change, are rare and usually do not affect an entire population. Thus, there is continuity everywhere, and yet we have clearly distinct species. How can this be possible?

Bateson turned to evolutionary theory for the solution to this cardinal problem. According to his logic, if environmental diversity is the ultimate determinant of species diversity, there is a wide range of environmental and structural differences within which no discernible result emerges. Everything remains as it is, in other words, "the relationship between environment and structure is not fine-tuned." In this case, Bateson argues, selection cannot be the sole directing or limiting cause of specific differences between species. It would then be problematic to accept natural selection as a doctrine (Bateson 1894). To better understand this inconsistency, he argued for a closer look at variation. "Variation is the essential phenomenon of evolution. Variation in fact, is evolution" (Bateson 1894).

To fully refine his idea of discontinuity, Bateson spared no effort. His main work, *Materials for the Study of Variation: Treated with Especial Regard to Discontinuity in the Origin of Species*, published in 1894, is 600 pages long. Its core statement is unmistakable: "The discontinuity of species depends on the discontinuity of variation." Bateson provides countless examples of discontinuous variation in the animal world, such as bees with legs instead of antennae or crabs with additional oviducts. In humans, he devoted himself to supernumerary fingers (polydactyly), extra ribs, and men with extra nipples. Everywhere, he found discontinuities (including, of course, in colors), to which he ascribed a role in the origin of species. His position was strengthened by the fact that traits do not disappear in crossbreeding experiments, for example, by being absorbed into continuous, mixed forms; rather, variety always remains.

In 1897, Bateson reported regarding a series of breeding experiments performed with a dainty flowering plantlet, the bucklermustard (*Biscutella laevigata*), in the Cambridge botanical gardens (Bateson 1897). In the wild, hairy and smooth forms of otherwise identical plants are known. Experimental crossing of the two forms, as expected, produced well-bred hybrid plants that still showed in their appearance either one or the other characteristic of the wild plant, with no fusion or regression of the trait to an intermediate form. Dimorphism, that is, two clearly distinguishable forms, remained.

1.2.2 A Hundred Years of Dispute

The fuse was lit for a dispute about the relevance of discontinuous traits in evolution. This dispute continued for a long time in the 20th century and was at times fervent and uncompromising. Bateson himself clarified in 1894 that his considerations on discontinuous variation were not, in principle, incompatible with the selection mechanism. However, as stated earlier, he resolutely rejected doctrinaire teachings claiming sole responsibility for natural selection. It is clear, however, that when selection operates on a fine scale, it has a different theoretical standing than it does in an environment of discontinuous variation. Intuitively, the attentive reader will correctly surmise that the scale and complexity of a discontinuity determine how easy or difficult it is to enforce throughout the population. Hopeful monsters, i.e., suddenly appearing mutants, later brought into play by the German geneticist Richard Goldschmidt (1878–1958), certainly do not have an easy time of it. Darwin's perspective seems incompatible with the relevance of such hopeful monsters to evolution, whereas in Bateson's view, they could be the key to rapid transitions, for example, from a three- to a four-chambered heart.

One need not envy the negative reputation that Richard Goldschmidt enjoyed during his lifetime and that reverberated after his death. Goldschmidt is something of a poster-boy villain in evolutionary biology, and the hardline proponents of the Modern Synthesis must uphold his views of how things do not work. At best, they permit him to assert that cross-species saltationism is very unlikely and, therefore, not worth addressing. Only much later, after the turn of the millennium, was a German geneticist, Günter Theißen from the University of Jena, able to correct the erroneous picture painted by Goldschmidt and draw a new picture of saltation incorporating numerous examples. The evolution of flowering plants (angiosperms) is one of the most remarkable cases, for which no gradual evolutionary line is recognizable. In addition, there are several other scenarios for which gradualist paths of continuous change do not seem plausible.

As Theißen makes clear, Goldschmidt was well aware that macromutations, in the vast majority of cases, do not end well for an organism; in other words, macromutations are lethal. The rejection of Goldschmidt's theory was based precisely on this point: jumping phenotype variations are unlikely to have a positive effect on fitness because affected individuals do not have a good chance of survival. Therefore, they were always downplayed. However, their improbability or presumed rarity says nothing logically about the possibility of their occurrence. This is precisely where Theißen focuses his attention (Theißen 2009). He clarifies in his article (as others have before him) that many paleobiological findings do not show gradualistic transitions between species but rather, abrupt changes. Although they are rare, so-called macroevolutionary changes could arise precisely because of innovations. Perhaps something similar to this occurs with a change in the blueprint only once in a million years. Thus, exactly at the points where we have no evidence from the fossil record for gradualistic, flowing evolutionary processes, such events would seem plausible, provided that one can explain how they come about.

Of interest at this point is a reference to the mechanisms cited by Goldschmidt for increasing phenotypic variation. Even before him, biologists other than Bateson, some of them famous, had repeatedly pointed out the possibility of evolutionary jumps; among them, for example, was the botanist Hugo de Vries (1884–1935), one of the rediscoverers of Mendel's writing. However, to assert something and to explain something are two different things. In fact, Goldschmidt named two possible mechanisms for macromutation. The first one may be forgotten right away; it refers to systematic rearrangements of chromosomes. The second mechanism, however, according to Theißen, points in a modern direction—that of embryonic development. Goldschmidt believed that important genes (control genes) could alter the early course of development and thus exert considerable effects on the adult phenotype (Theißen 2009). In light of what we know today about evolutionary developmental processes, about which we will learn in detail here, these were far-sighted thoughts. Goldschmidt deserves to be taken seriously. One who has read Chapters 3 and 4 of this book will probably make a similar judgment.

Today, discontinuities in evolutionary theory appear in a completely new light. Against this background, I will discuss what occurs in the embryo and how embryonic development brings about discontinuity. William Bateson was still denied deeper insights into the fascinating events in the embryo. Probably because of this, he did not highly value the potential of embryology to give new insights into evolution. Nevertheless, with his focus on discontinuous variation, this tireless researcher laid a distinctive foundation for today's research discipline, evolutionary developmental biology, or evo-devo for short.

1.3 THE TIME AFTER DARWIN

By 1900, it had become quieter around Darwin; one could also say his theory was as good as dead. Alternative views of evolution were increasingly in circulation. Even the thesis of the Frenchman Jean-Baptiste Lamarck (1744–1829), according to which characteristics acquired once during life are heritable (a view that had already not been completely rejected by Darwin), rose at that time like a phoenix from the ashes. Only the discovery that Gregor Mendel's theory of heredity could be combined with Darwin's ideas, i.e., that Modern Synthesis could be formed from it, revived the flagging discussion. For this, however, the writings of Mendel had first to be rediscovered in 1900, as mentioned above.

1.3.1 August Weismann: A Stubborn Doctrine

The steadfast hypothesis of the Freiburg zoologist August Weismann from 1883 proved to be a port in the storm. Weismann (1834–1914) argued that the transfer of information on acquired characteristics to germ cells (i.e., sperm and egg cells) was impossible. In other words, Weismann said that changes due to environmental influences on the body of an individual cannot have any effect on the phenotype, i.e., the totality of recognizable characteristics, of the following generation. For this to occur, environmental influences would have to be able to act on the germ cells. But this is exactly what he ruled out. Body and germ cells develop separately. Once a body cell has differentiated, there is no way back into the germ line. This view set thoughts on evolution on a fixed track for a 100 years, and the so-called Weismann barrier was to be overcome with great difficulty. Weismann's germline theory heralded the end of Lamarckism. But Lamarck's idea persisted and will occupy us again when I present epigenetic and cultural inheritance (Section 3.6).

Not to be overlooked is another contribution of Weismann's that remains unequivocally valid today. He discovered the importance of sexuality in evolution. Sexuality creates a significant widening of variation in heredity. In today's terms, this occurs because the child inherits only one haploid DNA strand from each parent, i.e., only one of the original two. Only on the diploid (complete) DNA strand newly created in the child is it then determined which copy of each gene is active, that of the father or that of the mother. This mechanism creates "an inexhaustible abundance of ever new combinations of individual variations, as is essential for selection processes" (Weismann 1892). An irreversible mixing of the paternal and maternal components does not take place. They remain as units but are recombined. With these insights into the workings of sexual reproduction, Weismann provided fundamental insights into the driving forces of species change.

1.3.2 How Are Genes Defined? Where Are They Located?

Meanwhile, classical genetics was gaining momentum. It explained how mutations are inherited and, under the influence of selection or other evolutionary processes, undergo further modifications or remain in equilibrium. Classical genetics provided many fundamental insights into heredity. For a long time, the identity and location of genetic material were unclear. What were the key players— proteins and nucleic acids or their respective constituents, i.e., amino acids and nucleotides? This remained controversial throughout the first half of the 20th century. It was not until Watson and Crick that DNA finally took center stage, ushering in a new era—the era of molecular genetics. In 1953, the two researchers discovered, with substantial preliminary work by others of whom hardly anyone speaks anymore, the magnificent structure of the DNA double helix formed of a combination of four nucleotides. At the end of their short article in the journal *Nature*, Watson and Crick made a terse remark that has become famous: "It has not escaped our notice that the specific pairing we have postulated immediately suggests a possible copying mechanism for the genetic material" (Watson 1968/2005). A more British understatement of such a discovery is not conceivable.

Even before genetics and the fervent search for hereditary material, a new research discipline developed, population genetics, without which Darwin and evolution cannot be understood. In fact, Darwin's theory is discussed on the basis of individuals, their characteristics and their behavior. At least, this is the impression one gets from time to time. However, this is misleading. Traditional evolutionary biology is a population science. It is about populations, for example, the Amur leopards in northeastern China or a population of bacteria in the laboratory. Darwin's thoughts (and Wallace's) should not be understood other than as thoughts on the populations of plants or animals. It is not individuals that adapt but rather the populations of individuals. Variations take place in individuals, who can then inherit them. The adaptation process of such variations, however, takes place at the population level. It takes place over long periods under changing environmental conditions. The principles for this, according to Darwin, are natural selection and the *Survival of the Fittest*. I will explain in detail later, with reference to evolutionary developmental biology, that there is an approach to evolutionary change other than via the path of the adapted population (Chapter 3).

Only slowly after 1900 did a mathematical-methodological awareness develop regarding how heredity and evolutionary processes in nature can be best described at the level of populations rather than at the level of individuals. The mathematical models of population genetics and statistics were and continue to be complicated. Their inventors bear great names. Only their models, with their many differential equations about abstract gene frequencies, created the basis for the great leap to Modern Synthesis that is now followed.

The idea of what exactly a gene is has undergone multiple transformations, and it is now

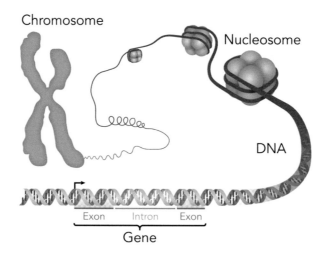

Figure 1.5 Gene – chromosome – DNA. Simplified schematic showing a gene as a segment of the chromosome A eukaryotic gene contains exons (coding sections) interrupted by introns (noncoding). The DNA double strand is wrapped around a set of eight histones (approximately two turns), forming a nucleosome—the basic DNA packaging unit. Finally, it is condensed into several chromosomes, one of which is shown schematically here. The material that makes up the chromosomes, i.e., the DNA and proteins surrounding it, is called chromatin.

more ambiguous than ever. Initially, it was a hypothetical hereditary factor, but it then became a unit of recombination, mutation, or biological function. Later, a gene was understood to be a protein-determining code in the form of a contiguous, then interrupted, segment of DNA (Figure 1.5). This understanding became increasingly unclear when DNA segments were found to be parts of several genes. In no case can a gene or the genome be seen today as the unit solely responsible for transforming heritable things from one generation to the next through reproduction. The view of genes "as causally privileged determinants of a phenotypic appearance" (Nowotny and Testa 2009) triggers more and more discussions. On the contrary, DNA does not transform anything. It is not an active element but only a template. It requires the complex machinery of the cell, particularly the activity of enzymes, to regulate the genes it encodes. Only through interactions among genes, cells, enzymes, and the environment does the multi-step process of evolutionary change take place. These interactions will be the subject of this book.

The question can also be examined the other way around, and it can be said that no protein is produced nongenetically by means of transcription and RNA. According to the definition commonly used today, a gene is a functional DNA sequence. The transmission of genes to offspring is one, but not the only, basis for the inheritance of a phenotypic trait. However, I do not want to get too far ahead of myself here; we will return to this complex subject in Section 3.7, when the gene is presented in a modern, systems biology context.

1.4 SUMMARY

Darwin's theory of natural selection for biological evolution and the idea of *Survival of the Fittest* were epochal. His idea of biological evolution was later accepted as fact. William Bateson emphasized discontinuous variation over the continuous variation espoused by Darwin and thus created the basis for a long-lasting debate pitting the two contrasting views against one another. The search for the carriers of hereditary material, the genes, dominated the first half of the 20th century. August Weismann's early hypothesis that there was no possibility of changing the germ line from the outside determined evolutionary theory for a long time. Where exactly genes are located remained unclear until Watson and Crick elucidated the structure of DNA.

REFERENCES

Bateson W (1894) *Materials for the Study of Variations Treated with Especial Regard to Discontinuity in the Origin of Species*. MacMillan, London.

Bateson W (1897) On progress in the study of variation. Sci Prog 6(5):554–568.

Darwin C (1859) *On the Origin of Species by Means of Natural Selection, or the Preservation of Favoured Races in the Struggle for Life*. John Murray, London. http://test.darwin-online.org.uk/contents.html#origin.

Darwin C (1871) *The Descent of Man, and Selection in Relation to Sex*. John Murray, London.

Leakey R, Johanson D (2011) Human Evolution and Why It Matters: A Conversation with Leakey and Johanson. YouTube. https://www.youtube.com/watch?vpBZ8o-lMAsg.

Nowotny H, Testa G (2009) *Die gläsernen Gene. Die Erfindung des Individuums im molekularen Zeitalter*. Suhrkamp, Berlin.

Onoa R, Shin Kobayashi S, Wagatsuma H, Aisaka K, Kohda T, Kaneko-Ishino T, Ishino F (2001) A retrotransposon-derived gene, peg 10, is a novel imprinted gene located on human chromosome 7q21. Genomics 73:232–237.

Theißen G (2009) Saltational evolution: hopeful monsters are here to stay. Theory Biosci 128:43–51.

Watson JD (1968) *The Double Helix*. Atheneum Press, New York.

Weismann A (1892) *Aufsätze über Vererbung und verwandte biologische Fragen*. Gustav von Fischer, Jena.

TIPS AND RESOURCES FOR FURTHER READING AND CLICKING

Browne J (2006) *Darwin's Origin of Species. A Biography*. Atlantic Books, London.

Dupré J (2003) *Darwin's Legacy. What Evolution Means Today*. Oxford University Press, Oxford, UK. Philosopher John Dupré astutely and critically questions the impact of Darwin's work on human self-image and understanding of human culture.

CHAPTER TWO

The Modern Synthesis: The Standard Model of Evolution

The Modern Synthesis is currently the standard model of evolution. It is assumed that the small variations in heredity (gradualism) are determined through genetic mutations. Furthermore, according to Darwin and Wallace, this model considers natural selection as the main mechanism of evolution. Within a species, those best adapted to their environment survive statistically more often, thereby producing a higher number of reproductive offspring. Their ability to pass on their own genes to the next generation is superior to that of their competitors. Now that we have summarized the abbreviated rendering of how evolution works from the point of view of the Modern Synthesis; let us turn our attention to the hardships with which it came into being and how it has developed up to the present day.

Important technical terms in this chapter (see glossary): Adaptation, chromosome, DNA, *Drosophila*, gene, gene pool, genetic drift, gradualism, macroevolution, microevolution, natural selection, neutral mutation, population genetics, punctualism, punctuated equilibrium, and recombination.

2.1 EMERGENCE AND KEY STATEMENTS

Around 1930, conditions were wholly unsupportive of a unified theory of evolution. There was disagreement on several points—even on whether the characteristics described by Mendel were physical or theoretical entities. These units of inheritance were first designated as genes in 1909. In the 1930s, the great edifice of the Modern Synthesis emerged in the English-speaking world. Today, the Modern Synthesis, also called neo-Darwinism, is the accepted theory of evolution. It combines the concepts of heredity and evolution with the insights from genetic studies and findings of Darwin, Wallace, and Mendel. Evolution occurs in populations, not in individuals. Population genetics expands on this through complicated statistical-mathematical calculations. Neo-Darwinism, originally associated with August Weismann's theory (see Section 1.3), underwent several transformations and has long been equated with the Modern Synthesis, especially in English-speaking countries.

August Weismann's teachings, mentioned previously, merged both the Weismann barrier and sexual recombination into the Modern Synthesis. Important early work on genetics was performed by the American physician and later Nobel Prize laureate (1933) Thomas Hunt Morgan (1866–1945; Figure 2.1). He studied the fruit fly on a level of detail that had not been previously accomplished and has never been eclipsed. *Drosophila melanogaster*, as it is known by its taxonomic name, became famous in the wake of his discoveries, consequently developing into one of the most important model organisms in biology to this day.

Figure 2.1 (a) Thomas Hunt Morgan, (b) Julian Huxley, (c) Sewall Wright, and (d) Ronald A. Fisher. Julian Huxley in front of a portrait of his grandfather Thomas Huxley, Darwin's friend and patron.

2.1.1 Thomas Hunt Morgan and Thousands of Flies

Around 1910, Morgan bred a white-eyed male mutant from otherwise red-eyed flies. The offspring from crosses among white-eyed flies followed the Mendelian rules of inheritance. Morgan was able to determine that the predisposition for eye color was located on the Y chromosome—the male sex chromosome. Innumerable fruit fly crosses constituted Morgan's daily laboratory work after this first success. Probably, thousands of mutants were created in his laboratory. Many (such as those with four wings or those with two legs on their heads) were simply not viable under natural conditions. Doubts began to arise in terms of the relevance of such monstrous results to evolution.

However, Morgan was not deterred by the skeptics. He was able to show the structure of chromosomes by staining the cell nuclei and demonstrated the arrangement of genes on chromosomes. At the time, he was not aware of the exact nature of genes. In 1916, Morgan added upon the observations of Weismann and was the first to discover that sexual reproduction involves the exchange of entire sections of chromosomes. The paternal and maternal (homologous) chromosomes exchange intact segments to generate a new version of the paternal chromosome containing a maternal section and vice versa. This crossing over, as it is called today (Figure 2.2), introduced an enormously expanded view of recombination possibilities in sexually induced cell division—meiosis. Recombination consequently became a second element of chance in the Modern Synthesis, along with mutations. The gene pool of the population remains unchanged, and recombination generates a DNA strand that is no longer completely derived from one parent but comprises parts that have been exchanged between strands of paternal and maternal origin. With 23 pairs of chromosomes in humans, trillions of theoretically possible new combinations exist; thus, the potential for genetic diversity is enormous. In this regard, the occurrence of identical offspring is practically impossible, with the exception of monozygotic multiples (identical multiples). The mechanism of recombination and the genetic diversity it produces is, therefore, a clear merit of sexual over asexual reproduction.

An eminent question in Morgan's time concerned the stability of genes and genetic variation. What if a variation disappeared after a few generations? In such a situation, genetics—as a foundation for evolution—would collapse like a house of cards. In 1908, the independent and timely discovery of two researchers led to the concept of Hardy–Weinberg equilibrium, a simple formula and guaranteed

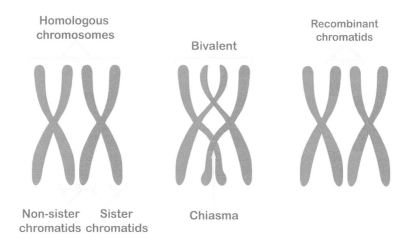

Figure 2.2 Double crossing over. Homologous sexual recombination between two chromosomes. Recombination is an important evolutionary factor and provides remixing of existing allelic material. The separation and rejoining of chromatids by an enzyme must be absolutely precise at the chiasm, so that not a single nucleotide is lost or added. Crossing over is a highly impressive, complicated genetic process.

exam material for every student of evolutionary biology, which mathematically proves that genetic variation in a population can be stable and does not always disappear through mixing. Biology was believed to operate with the exactness of physical laws, and the stability of genes was defended. Genes were soon recognized as fundamentally stable; however, their physical and chemical representation remained unknown. Occasional mutations were viewed as an exception; neither genes nor mutations disappeared by themselves. To ensure that the mutations are perpetuated, external influences—selection and genetic drift—were necessary. Additional information on both of these concepts will be presented shortly.

2.1.2 Statisticians, Zoologists, and Others at the Same Table

Let us turn to the real architects of the Modern Synthesis: The British zoologist Julian Huxley (1887–1975), grandson of the Darwin promoter Thomas Huxley (both Figure 2.1) and brother of the no less famous writer Aldous Huxley, gave the new research direction and thus a whole scientific epoch its stamp. *Evolution – The Modern Synthesis* (Huxley 1942/2010) is the title of his famous book, which has been reprinted four times to date. It is one of the most successful books in the history of biology. Even today, the author's thematic vision is still impressive to the reader, for example, when he addresses the genetic effects on individual development (ontogenesis) in connection with evolution. Admittedly, decades would pass before this particular idea grew into a serious research program. Huxley showed the connection from the external view of Darwin and Wallace, based on natural selection, population genetics and a certain, albeit a very rudimentary, genetic internal view. He also founded the field of evolutionary biology.

The British mathematician and statistician Ronald Aylmer Fisher (1890–1962; Figure 2.1) was a central figure in the Modern Synthesis. His publications began garnering attention in 1918 and culminated in 1930 with the publication of his book *The Genetical Theory of Natural Selection* (Fisher 1930). According to his teachings, Mendelian genetics and natural selection were theoretically compatible. All populations, as Darwin pointed out, exhibit a wide range of variation that is undirected or random—i.e., having nothing to do with a potential adaptive advantage—and genetic mutations are the origin of such adaptive variation in the population. Fisher's genetic theory of natural selection was, in a sense, the first mature synthesis that fused the selection theory of Darwin and Wallace with Mendel's genetic theory of inheritance. Fisher introduced statistical techniques to population genetics. He analyzed traits determined not by one gene but by many gene loci. Genes become abstracted, mathematical frequencies (gene frequencies) in an abstracted population. Mutation rates, selection factors, and gene pools are used, while population evolution is expressed in mathematical formulas.

Fisher was in a heated argument with the American Sewall Wright (1889–1988; Figure 2.1) over which of the following mechanism had a greater influence on species evolution: genetic drift or natural selection. Wright brought genetic drift into play as a random effect and considered it more important than selection in evolutionary processes. According to this doctrine, random extinction of parts of a population can occur. The composition of all the genes in a population (the gene pool) can change statistically with the occurrence of a sudden natural event, such as an earthquake, mountain formation, or flood. Wright lost the debate, i.e., selection carried the day, and Fisher had consequently saved the concept of Darwinism from decline.

2.1.3 Dobzhansky and Mayr: The Practitioners

In parallel with Fisher, J.B.S. Haldane (1892–1964) applied mathematical analysis to real examples of natural selection in the 1920s. Haldane concluded that natural selection can act even faster than Fisher assumed. In 1941 and 1945, the Russian Ivan Schmalhausen (1884–1963) developed the concept of stabilizing selection, i.e., the idea that selection can stabilize the expression of a particular trait. Elsewhere, this is referred to as canalization (see Section 3.1).

The younger but no less creative founders of the Modern Synthesis were Theodosius Dobzhansky and Ernst Mayr (Figure 2.3). Dobzhansky (1900–1975), an American who emigrated from Russia, published the first classic portrayal of the Modern Synthesis—*Genetics and the Origin of Species* (Dobzhansky 1937)—in 1937, before the publication of Huxley's aforementioned book. It was this work that closed the gap between the teachings of naturalists and those of geneticists,

Figure 2.3 Ernst Mayr, 1994, one of the most authoritative evolutionary biologists and naturalists of the 20th century. He was a co-founder and, until his death, a major proponent and passionate advocate of the Modern Synthesis.

which until then had been considered incompatible. The author is considered the forefather of the Modern Synthesis based on this book, and not in the least in relation with the sentence that no evolution book dares omit: "Nothing makes sense in biology unless it is seen in the light of evolution."

Subsequently, the German-American Ernst Mayr (1904–2005) made a major contribution to our current understanding of the conditions under which new species preferentially evolve. This type of speciation occurs through the isolation of small populations. Mayr called the process allopatric speciation (Mayr 1942). A new watercourse or mountain formation may be the cause of such isolation. One of the most frequently cited examples is the Grand Canyon, whose geological formation created a barrier that separated populations of squirrels, resulting in isolation of their evolution, which continues to this day. Gene exchange or unrestricted gene flow was now no longer possible throughout the population. Therefore, on the north and south rim of the canyon, the Kaibab and Abert's squirrels evolved, respectively (Figure 2.4). As the two populations were no longer in contact with each other, they evolved based on different mutations, as described in many biology books.

Recently, however, the Kaibab squirrels were downgraded to a subspecies status. According to molecular genetic studies, the differences between the two populations are not sufficient to consider them different species. Color differences, such as the snow-white, bushy tail of the Kaibab squirrel, alone are not sufficient to distinguish two species. However, the redefinition does not detract from Mayr's significant discovery and the presence of allopatry in squirrels. Allopatric speciation exists in many locations, including on both sides of the Isthmus of Panama, which separated what is now the Gulf of Mexico from the Pacific Ocean 3.5 million years ago. Another example of this is the speciation of chimpanzees north of the Congo River and bonobos on the southern shore. The population of their common ancestor was permanently separated by the river nearly 1.5 million years ago. Furthermore, although these two species are not readily distinguishable externally, they exhibit significant differences in terms of social behavior. Another good example is island formations, which give rise to endemic species that inhabit only one island (Mayr 1942). Some famous examples of endemic species on islands are Darwin's finches on the Galapagos Islands and kiwis in New Zealand.

(a)   (b)

Figure 2.4 Kaibab and Abert's squirrels. The two species of squirrels, Kaibab (a) and Abert's squirrels (b), evolved during the formation of the Grand Canyon by allopatric speciation. The color differences on the dorsal, or back, side of the tail can be seen clearly. However, the separation of two distinct species is not considered complete at present.

"Species are groups of actually or potentially interbreeding natural populations, which are reproductively isolated from other such groups" (Mayr 2001). Therefore, a species is a reproductive community. Darwin does not explain specific speciation as precisely as Mayr: the Darwin-Wallace doctrine emphasizes continuous, gradual change rather than the mechanism through which clearly delimited species are formed. In a way, Ernst Mayr was considered the Darwin of the 20th century—there is no evolution book in which he is not mentioned. He published more than 850 articles and books during his life of more than 100 years. In the Modern Synthesis, Dobzhansky and Mayr embody the counterweight of the naturalists, alongside the dominance of the population geneticists. Before natural selection eventually won a preliminary mathematical victory, Dobzhansky continued to emphasize that there was by no means a consensus from the outset as to whether evolution proceeds continuously or in leaps and bounds. According to him, it was also not clear whether micro- and macroevolution proceed via the same mechanisms, i.e., molecular evolution on a small scale and species-forming evolution.

2.1.4 Role of Dinosaurs and Plants in the Modern Synthesis

Another central figure in the Modern Synthetic theory is the American paleontologist George Gaylord Simpson (1902–1984). Simpson demonstrated that the Modern Synthesis harmonized with paleontology, the study of extinct species. Initially, several paleontologists had strongly rejected the new ideas, principally those related to selection. However, Simpson's work in 1944—*Tempo and Mode in Evolution*—made it clear that the fossil record is consistent with the irregular, branching, undirected patterns advocated by the Modern Synthesis (Simpson 1942). The American botanist George Ledyard Stebbins (1906–2000) also deserves to be mentioned here. He contributed significantly in bridging genetics and botany through his book *Variation and Evolution in Plants* (1950).

The Modern Synthesis consequently emerged as a collaboration between representatives of several disciplines. It is an integrated effort involving the fields of population genetics, zoology, systematics, botany, paleontology, and other research areas in biology. In essence, it is an early interdisciplinary approach to research. Renowned researchers found a common explanation of evolutionary processes, primarily founded on the evolutionary mechanism of natural selection outlined by Darwin and Wallace. This new branch of biology could now compete with the exact science of physics in aspects such as predictability, measurability, and provability. Fisher even dreamed that his theorem might be compared to the second law of thermodynamics.

Historically, the development of the Modern Synthesis has been an extraordinarily complex process. Biologists other than those mentioned here in brief made significant contributions to the present-day theory and practice of evolutionary research. In parallel, new scientific disciplines emerged with a new perspective and through the efforts of their advocates. Unification of these disciplines in the newly formed evolutionary biology was an outcome of the Modern Synthesis. Outside of narrow scientific circles, beginning in the 1970s, the Modern Synthesis became internationally known to a wider audience primarily through the popular science books of the Briton Richard Dawkins. It was also Dawkins who captivated me with his riveting book *Climbing Mount Improbable* (1996) and first piqued my interest in the theory of evolution. However, reading it left me with more unanswered questions than the answers to my prior questions. I have been pursuing these questions ever since.

2.2 VARIATION–SELECTION–ADAPTATION: THE PRACTICE

In the following section, I describe a series of experiments and examples that have been instrumental in underpinning the mechanism of the Modern Synthesis with natural selection at its core. In addition to population geneticists, the study of species and subspecies in nature was an important pillar in establishing the novel form of thinking. Regionally distinct jumping mice were studied, as were subspecies of butterflies with varying caterpillar patterns. In the case of mice, specific breed characteristics appeared to remain constant after hybridization. In other words, such traits were inherited. Prior to this, inheritance had been anything but certain. In the case of caterpillar patterns, it was possible to infer varying degrees of gene action (in modern terms, gene expression) directing the expression of alternate patterns. Local adaptations of species obtained a genetic basis through these experiments; furthermore, it was even possible to map the geographical differences between subspecies on chromosomes. Geography of genes subsequently emerged.

2.2.1 An Ingenious Experiment Confirms the Modern Synthesis

The long stretches of time estimated by Darwin and Wallace for speciation may perhaps be much shorter than expected. For example, the acceleration of bacterial evolution in the laboratory has long been possible, such as through the application of irradiation stress. The high rate of variation and multidrug resistance of bacteria, which can no longer be controlled via antibiotics, has become a major cause for concern worldwide. Even the human immunodeficiency virus (HIV) has adapted to changes in external conditions in just a few years and has gained the ability to survive in a constantly mutating form. We can observe evolution at work, so to speak. The ingenious laboratory experiment of the microbiologist Salvador Luria (1912–1991) and the geneticist Max Delbrück (1906–1981) in 1943 in the USA on bacterial colonies of *Escherichia coli* proved that spontaneous mutations in the heredity of bacteria must first be present before populations react adaptively under the stress of a virus attack, and thus, under altered selection conditions, i.e., become resistant to phages. Therefore, bacterial mutations do not occur as an adaptation however, they are an underlying condition that allows bacteria to adapt. In 1969, Luria and Delbrück were awarded the Nobel Prize for their work.

2.2.2 Guppies with and without an Enemy

Are phenotypic variations in multicellular organisms easily detectable in a short time similar to that in bacteria? One of the most elegant projects demonstrating the effects of selection in higher organisms was John Endler's guppy experiment (Endler 1980; Figure 2.5). He aimed to demonstrate how the point pattern in guppies (*Poecilia reticulata*), a strongly genetically controlled trait, changes when environmental and predatory patterns are modified. In his initial experiment, guppies with different coarse patterns of dots swam in an aquarium with an enemy and a coarse substrate. In another aquarium, the substrate was fine, more evenly colored sand, and an enemy was also present. After about 15 generations, the guppies in both aquariums had already adapted, by natural selection, to their respective substrates, i.e., coarse in the first aquarium and small and spotted in the second. Thus, they evolved over time to become barely distinguishable from their surroundings. The enemy fish swimming above them had a harder time recognizing its prey. In a second experiment, Endler removed the enemy from the tank. The substrate was again coarse-grained or finely structured. Interestingly, the result showed that sexual selection dominated in this experiment, favoring guppy males that,

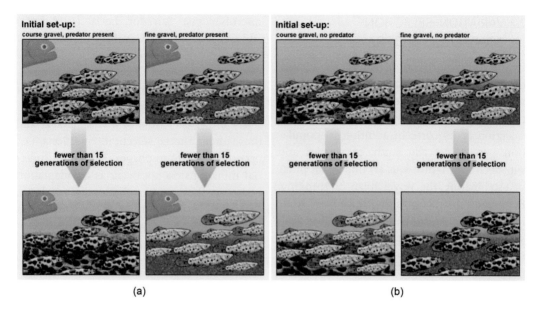

Figure 2.5 (a and b) Evolution live. John Endler's experiment on the adaptation of guppies to their environment ((a) with and (b) without predator) in an aquarium within a few generations. For anyone who has doubts about evolution, the evidence is impressive.

in contrast to before, stood out well from the substrate and were thus more attractive to females. Of course, Endler's experiment was conducted under conditions more restrictive than those in a natural environment; however, the inspiring impression of selective forces remained.

2.2.3 Perfect, Less Perfect, or No Adaptation

An example of an exceptional morphological adaptation is that of the wood frog or ice frog in Alaska (*Lithobates sylvaticus*). It can withstand freezing temperatures in winter. To accomplish this, it is triggered by cold stress and associated with an adrenaline release and produces both increased glucose via the liver and urea within the cells. As a result, the cells do not freeze. Water flows out of the cells into the extracellular space, thereby lowering the freezing point. Approximately one-third of the body's total extracellular fluid can freeze without damaging cell membranes. Heartbeat, blood pressure, and respiration stop completely. Increased urea production ensures survival in the dryer environment. In the spring, the frog thaws out without damage (Larson et al. 2014). The ice frog represents thousands of other prime examples of adaptations in the Darwinian sense. Today, however, the mutation-selection mechanism is no longer considered necessarily sufficient for the emergence of morphological traits. This concept will be revisited in a later section (see Chapters 3–6). In passing, it should be mentioned that frost protection in animals has evolved several times in different, independent ways.

Elsewhere in nature, structures exist for which a selective advantage is not demonstrable and likely nonexistent. A particular example is male nipples. Their evolutionary origin, however, has been elucidated. The development of nipples is triggered in all embryos before sex formation is activated via the Y chromosome and it cannot be stopped. The question remains as to whether nipples in males serve a real purpose. For some evolutionary biologists, the statement *Use it or lose it!* holds true, i.e., structures with no known function are lost in the evolutionary process.

Accordingly, the gene or genes in question inevitably mutate at some point, so that their function eventually disappears. *Use it or lose it* is the thesis of adaptationism in the strictest sense, according to which all characteristics have a fitness function. Doubts regarding this are permitted in the example mentioned.

One such case is that of the female orgasm, and novel hypotheses have attempted to explain this since the time of Aristotle. Nevertheless, to date, its biological function remains unclear. Perhaps it has none, at least not related to reproduction. However, if something persists for a long enough time throughout the process of evolution, evolutionary biologists seek to understand its adaptive advantage. Failure to identify this advantage in the example here has frustrated generations of scientists. The argument that the female orgasm is an "evolutionary accident," or that it only exists as an evolutionary by-product, is rejected today with the reason that its neuro-endocrinological mechanism is much too complex for such a conclusion to be true. Such a complex trait is not thought to arise without multiple steps of stringent selection; however, we will explore examples that contradict this statement in a later section. A novel explanation for the female orgasm was recently provided by a team led by Mihaela Pavlicev, a former colleague of mine at the University of Vienna, and Günter Wagner at Yale (Pavlicev et al. 2019). The team relates the female orgasm to ovulation based on empirical tests in rabbits. Indeed, in rabbits, cats, and camels, orgasm triggers ovulation. Although this is not the case in women, the existence of such a mechanism in other species hints at its selective advantage. Therefore, according to the study, the female orgasm appears to have a long evolutionary history, during which the functional coupling to ovulation, and thus its adaptive value, was lost. Of course, the responsible authors know that their study outcomes are anything but definitive answers to the question of the biological function of the female orgasm.

There are many similarly suspicious examples that call adaptation into question, such as the human chin, which our ancestors of the genus *Homo* did not possess. However, the fitness advantage of the appendix and tonsils, contrary to earlier convictions, cannot be as easily dismissed. Both contain lymphatic cells, and thus, are a part of the immune system.

Let us present a seemingly perfect example of a behavioral adaptation. Coastal wolves in Canada eat salmon (Figure 2.6); indeed, their diet is specialized in prey from the sea. However, they only eat the head, not the rest

Figure 2.6 Dancing with the salmon. A coastal wolf feeds on a salmon, eating only its head. It does not touch the rest. By this amazing adaptation, he avoids ingesting tapeworms that are deadly to him.

of the salmon body, although it would probably taste good to them. The reason is that salmon transmit deadly tapeworms to wolves. Bears also love to eat salmon but, unlike wolves, they consume the entire body. The tapeworms that develop in the bear's body, however, do not survive the bear's winter period of dormancy because, once in the bear's digestive tract, they have no source of nourishment and they consequently die (Darimont et al. 2003). The fact that wolves have evolutionarily learned to eat only the heads of salmon is an example of intraspecific selection and adaptation, although it is not known today what exactly guided this adaptation process.

The fact that macaques on the Japanese island of Kojima have learned to wash sweet potatoes in river water and—according to their taste—in seawater before eating them, is a no less impressive adaptation and one of the earliest described forms of cultural behavior in a non-human species. First observed in 1954 in the female Imo, the behavior became established within 10 years throughout the local population, except for the oldest monkeys. The pattern of imitation described here is culturally inherited (Hirata et al. 2001), which will be addressed in detail in a later section.

An example of non-adaptation is the rock pigeon (*Columba livia*). While drinking unsuspectingly on the banks of the Tarn River in southern France, they are devoured by river catfish. The European catfish (*Silurus glanis*)—the largest pure freshwater fish in Europe—approaches imperceptibly on the gravel bottom just below the water's surface, darting up to capture a bird on dry land. It has been 40 years since the catfish was released there. In that time, the pigeons have not yet learned to be wary of the dangerous predator; even at very close range, they do not notice it in the clear water, and after a wild attack, the surviving birds are soon seen as unconcerned as before. By contrast, the learned, novel hunting behavior of the catfish is an amazing adaptation (Cucherousset et al. 2012).

Notably, perfectly adapted organisms do not exist in nature. Adaptation is always a compromise, because in an organism, it requires the coordination of many biological units and functions, which cannot evolve in isolation. This requires optimization rather than maximum adaptation:optimization (Kutschera 2004). The fact that adaptation is not the best solution that prevails but is the least bad one has been expressed previously. From this point of view, numerous characteristics, for example, of the human body, which we perhaps consider unsatisfactory, but which were not an obstacle to the existence of our species, can be explained. Among these are the weak spine, the narrow birth canal of women, the blind spot on the retina of the eye, and numerous others. Nature is not an engineer; as such, there is no such thing as a design optimized for the overall organism system, neither in us nor in other species. Humans cannot obliterate our evolutionary past, which we share with fish. Nevertheless, you will become familiar with a whole series of intrinsic organismic design mechanisms of evolution in the following sections.

2.3 FURTHER FINDINGS UP TO 1980

The contributions outlined in this section constitute additions to the Modern Synthesis. In addition, some speak in the context of an extension (Kutschera 2004). However, to avoid confusion, I use the term *extensions* in this book only in the context of the extended synthesis in Chapters 3–6. The additions in this section generally do not follow other basic assumptions (random genetic mutation and gradual changes only, among others). The central mechanism of the Modern Synthesis, natural selection, is accepted without question in principle. An exception is sociobiology, which overcomes the reservation that natural selection is always directed only at the individual. Finally, in the context of additions to the Modern Synthesis, adaptation processes are also increasingly critically examined.

2.3.1 Punctualism

The fossil record contains many puzzles. The series of extinct species often shows jumps, creating the impression of abrupt change. Evolutionary theory, however, predicts a continuous progression, according to which changes occur more or less equally and often in small steps. Gaps in the fossil record could, therefore, only indicate that more finds were necessary.

Two researchers vehemently contradicted this assertion: Niles Eldredge and Stephen Jay Gould. In their sensational article *Punctuated equilibria: An alternative to phyletic gradualism* published in 1972, they developed a new view on the evolutionary process. They argued for long-lasting, stable phases in which species changed only slightly or not at all (stasis). These phases, or equilibria, are interrupted by abrupt spurts (punctuations) (Figure 2.7). This finding had a number of "unpleasant" consequences for the Modern Synthesis, which was based on a continuous course of species-level changes. Adaptive processes, which have a "ubiquitous" effect on every mutation at the population level, no matter how insignificant, lose importance in the new theory. Therefore, punctualism is also referred to as non-adaptationism. Today, punctualism is a well-accepted phenomenon, and neo-Darwinists must accept the existence of gaps in the fossil record.

2.3.2 More Criticism of Adaptationism

The paleontologist Stephen Jay Gould (1941–2002; Figure 2.8) emerged as an opponent of strict adaptationism. In 1979, he published a paper with Richard Lewontin dealing with the arches of St. Mark's Basilica in Venice (Gould and Lewontin 1979). Evidently, these architectural features have much to say about evolution. The two researchers took the bricked-up rectangular surfaces between the arches and the wall or ceiling surrounding them—known as spandrels—as an example of how a component of a structure does not necessarily have an architectural function; moreover, it can be painted with beautiful

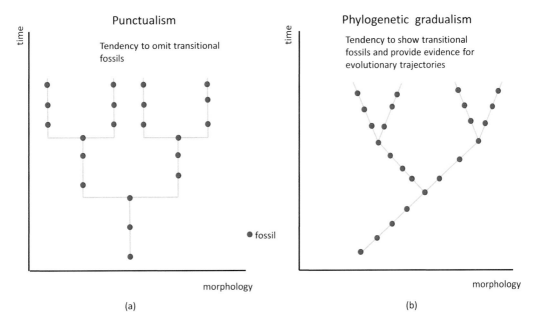

Figure 2.7 Gradualism and punctualism. (a): long periods of dormancy (stasis) interspersed with abrupt, punctuated speciation as described by Eldredge and Gould. (b): classical, neo-Darwinian view with continuous, gradual speciation.

Figure 2.8 Stephen Jay Gould. Gould played a significant role in reconsidering embryonic development as a factor in the evolution of morphological form.

images. The phrase *just so*, used by Gould, was famously taken from *Jungle Book* author Rudyard Kipling's *Just So Stories for Little Children*. In the children's stories, the author tells his daughter, who asks among other things, why the leopard has spots, that it is *just so*.

Applying this idea to evolution, Gould and Lewontin argued that many features of species did not necessarily arise adaptationistically—they do not confer a fitness advantage but exist nonetheless. This was an even greater general attack on Modern Synthesis than the punctuated equilibria proposed earlier. For the two authors, living beings exist in their current form as a result of innumerable earlier adaptations the conditions of which are not necessarily present. Today, there is still a differentiated discussion regarding the scope, limits, and explanatory value of adaptation. The reader will soon discover that modern research in evolutionary developmental biology is veering away from adaptationism and focusing on what occurs in the embryo.

2.3.3 New Worlds: Multilevel Selection and Sociobiology

Since Darwin, the question has been whether selection can be observed at the individual level or whether other levels of selection can exist alongside it, such as that of individual groups. The American David Sloan Wilson led this discussion quite openly and provided brilliant examples, in which the behavior of the group or the interaction between groups has a greater impact on reproductive success than the behavior of the individual. Individual selection is always present; however, selection at other levels (multilevel selection) may predominate. According to Wilson, higher units of the biological hierarchy can be regarded as organisms similar to individuals. This applies to a human society as well as to biological ecosystems. They can equally be vehicles of selection. Levels of selection in this sense include the species, an entire society, a group, kinship, the individual, organs (for example, the immune system), or genes (Wilson 2007; Wilson 2019).

The transition of humans from hunter–gatherers to farmers can be considered an example of selection at a level above the individual. Agriculture, in conjunction with human sedentary behavior, is a group-specific, cooperative activity based on the division of labor, and its selective advantage only comes into play at this group level. Modern humans organize themselves in innumerable small and large groups based on the division of labor. They construct *meaning systems*. They all tell stories of how we try to live together better. In studying them, we learn to the extent to which we are a group-selected species. Group selection also becomes clear in the example of overpopulation, especially in overflowing megacities and mega-societies on earth. Here, it crystallizes the concept that selection pressure is strong on this group level. Only time will tell whether humans can withstand the selection pressure in the long run. According to D.S. Wilson, megacities are transforming our species. Wilson's answers to today's global challenges can be found in his latest book (Wilson 2019), in which he calls for an "evolutionary worldview," embracing genetic as well as cultural inheritance. From this perspective, "culture is evolution" and "culture is biology." The challenge, according to Wilson, is to finally recognize this view. Wilson argues for the unlimited application of evolutionary, Darwinian principles as found in all natural systems with multilevel selection. He asserts that these principles should be applied at all levels of culture and politics. On the basis of the correct theory alone, i.e., the theory of evolution in the form of cultural evolution of complex systems, the human superorganism could, in theory, adapt (for criticism, see Section 7.4).

The discussion related to the extension of the level of selection beyond the individual reached its climax in the life work of another American, Edward Osborne Wilson. The recipient of many honors, E.O. Wilson is the founder of a new scientific discipline: sociobiology. He and the German Bert Hölldobler worked as a brilliant team on superorganisms, i.e., state-forming insects. It was not surprising to them that the evolutionary behavior of individuals is not very clear from the point of view of the Modern Synthesis. By contrast, the superorganism as a whole system, for example, in ant or bee colonies, reveals evolutionary behavior based on natural selection that could not be explained before. If one compares a superorganism with our brain, for example, one could reach the simple conclusion that a neuron cannot tell us anything related to the intelligence of the individual, but the totality of neurons and their cooperation can.

It should be noted that group selection is just as gene-centric as individual selection (Jablonka and Lamb 2014). That is to say that the material provided for selection is, for sociobiologists as well as for representatives of the Modern Synthesis, random genetic mutations. By contrast, however, D.S. Wilson, in the context of multilevel selection theory, strongly emphasizes that cross-generational cultural evolution matters no less than genetic processes (Wilson 2019). Later, we will discuss in detail how increasing group size and cultural evolution drove the evolution of our brains (Chapter 7).

2.3.4 The Neutral Theory of Molecular Evolution

What if most mutations subject to selection turned out to be neutral, rather than detrimental, to organisms? Could this finding threaten the theory of evolution? The neutrality of mutations was analyzed by the Japanese Scientist Motoo Kimura in bacteria in the 1960s. Most genetic changes, he claimed, are neutral to natural selection. Some confusion arose as a result of this claim, which prompted a heated discussion. Selection theory was threatened, but it held fast. The relevant point is that neutral mutations also represent a basis of plasticity, i.e., the ability of the phenotype: to express itself differently under changing environmental conditions. Innovations are brought about more easily if many neutral, hidden mutations are present. This will be a topic of further discussion (Sections 3.1 and 3.3).

2.3.5 Molecular Study of Lineages

Phylogenetics is the study of relationships between life forms and has been greatly impacted by the emergence of molecular technology. DNA sequences can be used effectively to determine lineages more accurately than methods used in the past. Molecular biology, thus has made an enormous contribution to evolution as fact.

2.3.6 Criticisms Remaining of the Modern Synthesis after 1980

According to the Modern Synthesis, large evolutionary changes (macroevolution) arise no differently than small ones (microevolution). Both cases are a matter of many small genetic changes—a bold statement, but from the point of view of the Modern Synthesis, a compelling one, which after all only confirmed the claim of Darwin and Wallace but soon met with criticism. It is immediately obvious and, using the arguments of the Modern Synthesis, well understandable the difference between the brown bear (Ursus arctos) and the polar bear (Ursus maritimus) could have become apparent. The polar bear needs its bright fur for camouflage and for the heat regulation. However, it is not so easy to explain how reptiles and mammals evolved. Reptiles continue to exist, and there is no natural necessity for them to change into mammals. Yet, evidently, mammals evolved from reptiles and amphibians.

To understand the morphological emergence of higher taxonomic lineages and the changes that occur in them, additional criteria must be used other than the distribution of genes in a population. These include climatic and geological change, the internal structure of species, and other factors. A mass extinction, such as that of the dinosaurs, is not accessible to population genetic models. With respect to the gene and its pathway to the phenotype, the Modern Synthesis has been quite brief. According to the Modern Synthesis, one or more genes determine a trait and its expression. In between, however, there was a "blank space" that had to be filled. Criticism of this abbreviated view was consequently pre-programmed.

A concept that remained strongly criticized after 1980 is that of gradualism, i.e., the idea that change occurs primarily in small steps, and gene-centrism, which recognizes genes as the only units of heredity.

2.4 SUMMARY

It is one thing to hypothesize that evolution exists and another to explain its mechanisms. In the Modern Synthesis, the core tenets of Darwin and Wallace were taken up and expanded with new perspectives. The Modern Synthesis aims to explain species change but also major blueprint transformations, such as the transition from fish to amphibians, to lizards and mammals, to humans.

The gene became known as the foundation of all change. Genetic mutations in the inheritance of a few individuals in the population of a species can be neutral, harmful, or beneficial to the survival of the species. They are sorted out at a higher level, the level of the individual. The individual is the "target" of selection; that is to say, selection acts at the level of the individual. If the changes are positive, that is, if they persist in the course of natural selection by heredity and improve the fitness of their carriers, they have a chance of gradually spreading through the population. Further small, heritable changes in subsequent generations can reinforce the original variation if they are in turn positively selected. In the long run, over the course of many generations, this can lead to greater, even complex, variation. Adaptation occurs in the population as a result of individual selection of the best adapted. The population is said to be adapted to a particular environmental condition.

In its earliest form, the Modern Synthesis conveyed a simple causal relationship between genotype and phenotype: genetic mutation changes the phenotype. This concept was refined in later years. However, the following three propositions remain at the core of the Modern Synthesis: first, genetic mutation as the causal explanation for variations in phenotype; second, the statistical-mathematical selection mechanism underlying many small additive steps, and third, the statistical survival of the best-adapted individuals in a population.

By 1980, the Modern Synthesis was undergoing additions. However, these did not shake the basic pillars of the theory, such as random genetic mutation as the sole cause of variation, gradualism, and natural selection as its principal mechanism. The theory of neutral evolution questioned the credibility of positive mutations and adaptation. Eldredge and Gould's theory of punctualism and Gould and Lewontin's critique of adaptationism are directed against overly rigid views of the Modern Synthesis.

REFERENCES

Cucherousset J, Boulêtreau S, Azémar F, Compin A, Guillaume M, Santoul F (2012) Freshwater killer whales": beaching behavior of an alien fish to hunt land birds. PLoS One 7(12):e50840. https://doi.org/10.1371/journal.pone.0050840.

Darimont C, Reimchen TE, Paquet PC (2003) Foraging behavior by gray wolves on salmon streams in coastal British Columbia. Can J Zool 81(2):349–353.

Dawkins R (1996) *Climbing Mount Improbable.* Norton, New York.

Dobzhansky T (1937) *Genetics and the Origin of Species.* Columbia University Press, New York.

Eldredge N, Gould SJ (1972) Punctuated equilibria: an alternative to phyletic gradualism. In: Schopf TJM (eds.) *Models in Palaeobiology.* Freeman, Cooper & Company, San Francisco, 82–115.

Endler JA (1980) Natural selection on color patterns in Poecilia reticulata. Evolution 34:76–91

Fisher RA (1930) *The Genetic Theory of Natural Selection.* Clarendon, Oxford.

Gould SJ, Lewontin R (1979) The spandrels of San Marco and the Panglossian paradigm: a critique of the adaptionist programme. P Roy Soc B Bio 205(1161):581–598.

Hirata S, Watanabe K, Kawai M (2001) "Sweet potato washing" revisited. In: Matsuzawa T (ed.) *Primate Origins of Human Cognition and Behavior.* Springer, Berlin, 487–508.

Huxley J (2010). *Evolution. The Modern Synthesis. The Definitive Edition.* MIT Press, Cambridge, MA. First published in 1942.

Jablonka E, Lamb MJ (2014) *Evolution in Four Dimensions. Genetic, Epigenetic, Behavioral, and Symbolic Variation in the History of Life,* 2nd ed. MIT Press, Cambridge, MA.

Kutschera U (2004) The modern theory of biological evolution: an expanded synthesis. *Naturwissenschaften* 91(6):225–276.

Larson DJ, Middle L, Vu H, Zhang W, Serianni AS, Duman J, Barnes BM (2014) Wood frog adaptations to overwintering in Alaska: new limits to freezing tolerance. J Exp Biol 217(Pt 12):2193–2200. http://jeb.biologists.org/content/early/2014/04/01/jeb.101931. Accessed day: 20. Jan. 2010.

Mayr E (1942) *Systematics and the Origin of Species.* Columbia University Press, New York.

Mayr E (2001) *What Evolution Is.* Basic Books, New York.

Pavlicev M, Zupan AM, Barry A, Walters S, Milano KM, Kliman HJ, Wagner GP (2019) An experimental test of the ovulatory homolog model of female orgasm. *PNAS* 116(41):20267–20273.

Simpson GG (1942) *Tempo and Mode in Evolution.* Columbia University Press, New York. (Dt. (1951) Zeit und Ablaufformen in der Evolution. Wissenschaft- licher Verlag Musterschmidt,, Göttingen)

Wilson DS (2007) *Evolution for Everyone. How Darwin's Theory Can Change the Way We Think about Our Lives.* Delacorte, New York.

Wilson DS (2019) *This View of Life. Completing the Darwinian Revolution.* Pantheon Books, New York.

TIPS AND RESOURCES FOR FURTHER READING AND CLICKING

A recent reference book for the Modern Synthesis: Ernst Mayr (2001) *What Evolution Is.* Basic Books, New York.

An exemplary work on the Modern Synthesis from a somewhat later perspective: Richard Dawkins (1996) *Climbing Mount Improbable.* W. W. Norton & Comp, New York.

CHAPTER THREE

Evo-Devo—The Best of Both Worlds

The interrelationship between embryonic development (ontogenesis) and phylogeny (phylogenesis) was neither obvious nor spontaneously recognizable to representatives of the Modern Synthesis. While an individual process is the focus of the developmental researcher, the focus of the evolutionary researcher is a population one. In comparison with that of evolution, the duration of the developmental process is short. Until 1960, the role of development in an evolutionary context was uncertain. It was the unknown link between the genotype and phenotype. This chapter describes the emergence and core themes of the fledgling research discipline known as evolutionary developmental biology (evo-devo). Evo-devo is one of the four central research domains in the Extended Evolutionary Synthesis (EES): developmental plasticity, inclusive inheritance, and niche construction (Laland et al. 2015).

Important technical terms in this chapter (see glossary): Arrival of the fittest, buffering, canalization, cell signaling, complexity, developmental constraint, developmental gene, evolvability, gene centrism, genetic accommodation, genetic assimilation, heterochrony, Hox genes, inclusive inheritance, innovation, phenotypic plasticity, reductionism, robustness, self-organization, and threshold effect.

This core chapter provides a historical context for the modern-day contribution of evolutionary developmental biology to evolutionary theory. However, the reader may also choose, as a topical introduction, sections of Chapter 4 with evo-devo examples as well as the two interviews with evo-devo researchers, namely Gerd B. Müller and Armin Moczek.

3.1 CONRAD HAL WADDINGTON—EPIGENETICIST AND FOREFATHER OF EVO-DEVO

We owe the term *epigenetics* to the British developmental biologist Conrad Hal Waddington (1905–1975; Figure 3.1), whose theory was a major precursor of evolutionary developmental biology—the central topic of this book. Here Waddington receives the attention he deserves after decades of being ignored and rejected during the heyday of the Modern Synthesis.

3.1.1 Veins in Fly Wings—An Attempt at a Proof

In 1953, Waddington provided empirical evidence in support of his theories of epigenetic inheritance in his paper titled *Genetic Assimilation of an Acquired Character*. Therein, he showed how cross veins, short vessels connecting the main wing veins in flies, disappear. In his novel selection experiment, he exposed fly eggs to short periods of heat shock. He repeated this treatment for several generations, each time selecting the flies that

Figure 3.1 Conrad Hal Waddington. A mastermind of key ideas in evolutionary developmental biology.

lost the largest proportion of their cross veins in response to heat shock exposure. At the end of the series of experiments, some insects lacked cross veins, even in the absence of heat shocks (Figure 3.2). Waddington theorized that epigenetic changes brought about during fruit fly development were subsequently genetically assimilated. That is to say, epigenetic changes in response to an environmental trigger became heritable in subsequent generations. This simple experiment attracted little attention at the time; nevertheless, it was a pivot toward a novel view of inheritance and evolution.

Waddington thus provided some empirical evidence justifying his doubt, expressed as early as 1942 when he wrote: "It is doubtful, however, whether even the most statistically minded geneticists are entirely satisfied that nothing more is involved than the sorting out of random mutations by the natural selective filter" (Waddington 1942).

Fifty years later, Yuichiro Suzuki and Fred Nijhout conducted a similar experiment on tobacco hornworm (Section 4.5). Moreover, Peter and Rosemary Grant described the very-short-term evolution of beak shape in Darwin's finches in association with developmental changes (Section 4.1).

3.1.2 Buffers and Canalization in Development

As early as 1942, Waddington theorized on the relationships in his example of cross veins cited above (Section 3.1). Several established developmental pathways produce a phenotypic trait. According to Waddington, the developmental process is "canalized." Presently, the phenotype is described as

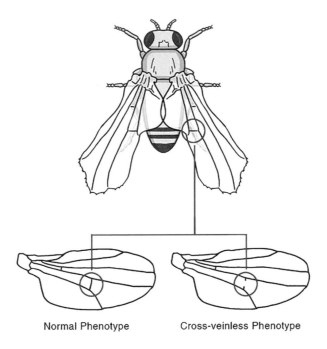

Figure 3.2 *Drosophila* with and without cross veins on its wings. The first experiment to prove that an external influencing factor, namely, brief heat shocks, lasting over several generations, could permanently change the phenotype.

robust to changes in genetic or environmental factors (Flatt 2005). Numerous such buffered alternatives exist because combinations of several genes are involved in the formation of a given phenotypic trait. Waddington explicitly emphasized that selection is involved in the formation of such networks of alternative developmental pathways (Waddington 1942).

By way of explaining the evolutionary origin of a unique feature, Waddington describes the conspicuous naked skin callosities on the chests of ostriches, where feathers are absent. Such callosities protect the bird when it crouches on the hot, rough desert floor. Waddington assumes that, at one time, these callosities did not exist in the ancestral lineage of the ostrich, which raises the question of their origin.

The ancestral bird may have acquired callosities over several generations when the appropriate body parts were stressed during juvenile growth. An environmental factor, such as the introduction of very hot, sandy soil, thus came into play. Before callosities existed, development was canalized to a phenotype without them. Moreover, canalization in wild species is not easily prevented. Several "invisible" mutations (buffers) support the phenotype. In addition, environmental change acts as a stressor and can alter the developmental course of a trait. However, for this to occur, the external influence must be sufficiently strong and last for several generations. Only then can it overcome the buffering, i.e., the hidden mutations, and thus, the canalization to the existing phenotype without callosities. In other words, development is decanalized or re-canalized to the new phenotypic trait. This process requires the long-term presence of the environmental stressor. The previously masked or hidden mutations act on the new phenotype only in conjunction with the stressor. Typically, the new trait does not occur in a single animal but affects several individuals, often the entire population. In the case of climatic heat stress, this can be easily imagined.

3.1.3 Genetically Assimilated

Over several generations, the stimulus for maintaining the new phenotypic trait may become unnecessary or may only be required at a weakened level. As an explanation, Waddington introduces the novel concept already illustrated by fly veins, which revolutionizes previous evolutionary theory: genetic assimilation follows phenotypic variation. Waddington proposes that the new phenotype, in this case callosities, which are produced for numerous generations in the presence of heat radiating from the ground, will eventually be genetically underpinned, in his words, assimilated. In retrospect, the phenotype favored long before by the environmental stressor is selected. This occurs through corresponding mutations, with some of the mutations that were previously buffered, i.e., hidden, already present in large numbers. They are referred to as hidden or masked because they exert no effect on the phenotype without the influence of the environmental stressor. However, in the presence of such a stressor, their effect becomes apparent. Over time, the phenotypic feature is maintained even in the absence of the external stimulus. The ostrich continues to feature callosities, even when it lives and reproduces in a zoo in a temperate climate.

Numerous conceivable variations could have evolved along the path suggested by Waddington. He could just as well have mentioned the callosities on our feet. Similarly, an animal's fear of a new predator, such as a snake, may be initially learned and later genetically stabilized. When we discuss evo-devo, we will be introduced to numerous other examples, for Waddington's view will not be forgotten in the course of the book.

3.1.4 Waddington Was Looking Forward

In a broader context, Waddington's study shows how an organism's genotype can respond to the environment in a coordinated fashion. Development and evolution can interact with environmental influences and respond to them in a directed manner (Waddington 1942).

With the advancement of evolutionary developmental biology, Waddington's views are regaining popularity, and his findings are increasingly recognized. For example, in their book *Evolution in Four Dimensions*, Eva Jablonka and Marion Lamb write about Waddington: "Many years ago, before anything much was known about the intricate ways in which genes are regulated and interact, and long before concepts of genetic networks became fashionable, geneticists realized that the development of any character depends on a web of interactions between genes, their products, and the environment" (Jablonka and Lamb 2014). Gerd Müller cites both Waddington's (genetic) assimilation and the entire field of epigenetics in the Waddingtonian sense as the conceptual roots of evo-devo (Müller 2008).

3.2 EVO-DEVO—HISTORY AND EARLY PRIORITIES

At the beginning of the era of evo-devo, researchers approached the connection between embryology and evolution. The basic idea was that knowledge about development could provide greater insight into mechanisms of evolution, and knowledge about evolution could in turn provide insight into development (Bonner 1982). Development is a powerful aspect of evolution from an evo-devo perspective because it is the pathway by which the genotype is translated into the morphological phenotype. Mutations and genetic variation become the raw material of evolution. The developmental process—morphogenesis—forms variations screened by natural selection, with organisms adapting to their environment over generations.

3.2.1 Embryo Was Already on the Radar in the 19th century

Let us review the historical development of evo-devo. While in the 19th century, the two present-day sub-disciplines of development and evolution were not yet treated separately in biology, Darwin recognized the crucial role of embryonic development in evolution. However, as discussed above, he did not address the emergence of variation, which was considered as a prerequisite by him and did not need to be explored further. The exclusion of the emergence of variation from the theory has clear advantages because it allows characteristics of unknown ontogenetic origin to be treated in the same way as those with known developmental pathways. In this abstract approach, one does not need to know the evolutionary history of a trait (Amundson 2005).

In fact, by the second half of the 19th century, embryology had already advanced beyond Darwin and more than half a century later Modern Synthesis appeared. Wilhelm Roux, a German developmental biologist in Breslau, Innsbruck, and Halle, wrote and reasoned in 1881 that there is no one-to-one relationship and thus no complete determinacy between the heritable material and the developing organism, the phenotype. There couldn't be none (Roux 1881). This does not fit the later picture of the exact genetic blueprint envisioned by the Modern Synthesis.

3.2.2 The Embryo Does Not Fit into the Picture of the Modern Synthesis

The Modern Synthesis did not address embryonic development, although the field of embryology had existed for centuries. The gene-centered, statistical thought space of the Modern Synthesis, contains abstracted population gene frequencies but does not encompass developmental processes. For example, a progeny can deviate from its parent, and something can "go wrong;" nevertheless, a coordinated process can lead to a modified phenotype in an individual organism. The neo-Darwinists ignored embryonic development and did not seek answers to the question of how embryonic development could influence evolution and vice versa.

Nevertheless, during the emergence of the Modern Synthesis, some researchers thought differently from those in the mainstream. At this point, it is useful to mention a few precursors. Waddington's concept of canalization and genetic assimilation has already been discussed (Section 3.1). He referred to canalization as a developmental path analogous to a solidified gutter along which a ball is directed. Just as a developmental path is not easily changed by genetic mutations, the ball cannot easily leave the gutter. The result remains the same despite mutations. Canalization is thus an expression of stability in development.

This reversal of the previous view, in which now externally, epigenetically triggered phenotypic change is followed by genetic fixation, was impossible to accept as a theory apart from the Modern Synthesis. Waddington was dismissed as an eccentric. His empirical foundation may not have been particularly convincing; nevertheless, his ideas have been incorporated into research and the theory of evolutionary developmental biology (evo-devo). He is therefore considered one of the forerunners of this discipline.

Besides Waddington, the British embryologist Gavin Rylands de Beer, together with Huxley, developed modern thoughts for his time (1930) in *Embryos and Evolution*.

In 1953, as already described, Watson and Crick elucidated the structure of DNA (see Section 1.3). Genes are composed of DNA, and their double-stranded structure is perfectly suited to a copying mechanism of inheritance. It was not clear, however, how gene expression could be switched on and off, in other words, how spatiotemporal gene expression took place. The two Frenchmen Francois Jacob and Jacques Monod, both later Nobel Prize winners, took up this topic, and in 1961, their efforts were successful. In the bacterium *Escherichia coli*, they discovered the intricate mechanism of genetic regulation, laying the foundation for understanding the process in multicellular organisms.

The first of these was the fruit fly, in which gene regulatory events that take place during the development of larval segmentation were described. Over time, this discovery has provided constantly increasing insight into entire chains of gene regulation typical of embryonic development, so-called gene regulation networks. We will address the role they play in evo-devo and whether they can be used to explain the structure and form of organisms.

3.2.3 Developmental Genes—The Embryo Is Rediscovered

Developmental biology experienced a rapid surge in popularity beginning in the 1970s with the discovery of developmental genes, including the well-known group of Hox genes in vertebrates. In *Drosophila*, genes responsible for the spatial structure and shape of the fly larva had already been on the radar since the 1930s. These genes, which control the formation of characteristic body structures, were called homeobox genes.

Developmental genes are expressed exclusively during embryogenesis and may never be needed again. As it turns out, developmental genes are a molecular construction toolkit common to most living things. Their existence alone is convincing and significant evidence of evolution. The Basel molecular and developmental geneticist Walter Gehring discovered that one such gene, *Pax6*, initiates eye development in all vertebrates. In addition, it controls eye development in the fruit fly and other insects. Several other steps follow the first step of *Pax6 expression* in a gene expression cascade. The *Pax6* double gene knockout mouse is born without eyes (Figure 3.3). In a heterozygous embryo, where the gene is mutated on only one parental chromosome, small eyes develop. Thus, the *Pax6* gene became known as the "eye gene."

Unexpectedly and impressively, *Pax6* exemplifies the relationships among animal taxa; however, it is only one of several developmentally important genes. We must examine

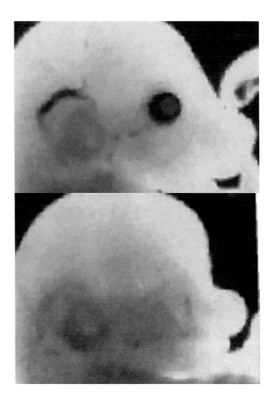

Figure 3.3 Eyeless mouse. A double knockout of the *Pax6* gene in mice prevents eye development. If only one *Pax6 gene* is mutated, the embryo has smaller eyes; therefore, this allele is also called *Small eye* or *Sey*. Eye malformations based on *Pax6 mutations* also occur in humans.

the Hox genes in further detail. In vertebrates, 13 different Hox genes are always present as four coherent complexes (A through D) in all vertebrate cells. Thus, for example, Hox genes A9, B9, C9, and D9 are located on different chromosomes of the same cell. Strikingly, the arrangement of these genes in the genome corresponds to the order in which they are expressed to form the structural elements of the body.

When a Hox gene is in a series with others downstream in a chromosome, its gene products are expressed toward the caudal end of the body (Figure 3.4). Hox genes in vertebrates help determine how the vertebrae are shaped or at which positions the ribs appear. Hox genes function in tandem for complex structural tasks, i.e., several of them produce

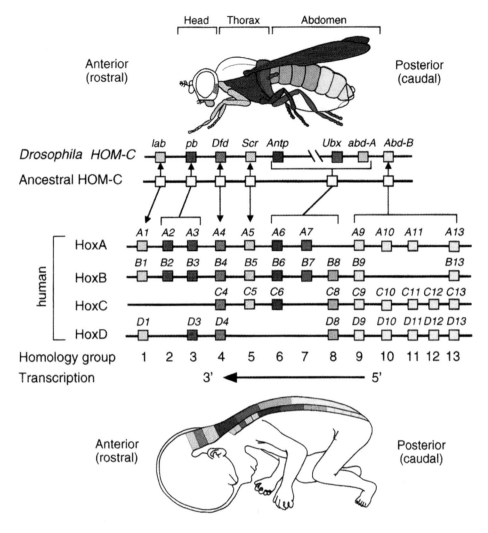

Figure 3.4 Hundreds of millions of years of kinship. Humans (below) and flies have similar developmental patterns despite their distant common ancestry. The Hox genes in the image are an essential part of a successive segmentation cascade of gene expression responsible for the local arrangement of the head, thorax, and abdomen as well as antennas, wings, and legs. Their exact position in embryonic segments has been determined. The picture shows the homologous genes of *Drosophila* and humans. While the fly possesses a single cluster of homeotic genes, mice and humans have four Hox gene clusters. They are labeled A to D.

one complex phenotype (Nüsslein-Volhard 2004).

Sonic Hedgehog, a well-known key developmental gene, played a critical role in the research lives of the 1995 Nobel laureates, German biologist Christiane Nüsslein-Volhard and two other researchers. In the nascent organism, developmental gene expression can show wide spatial and temporal variability. Thus, development is more efficient than it would be if a new gene were needed for each function; after all, genes are limited in number. Sonic Hedgehog (abbreviated Shh for the gene or SHH for the corresponding protein) was named after Sonic the Hedgehog in the American animated series *Adventures of Sonic the Hedgehog* because the gene produces a hedgehog-shaped mutant in the fly larva. Shh is required for hand development in vertebrates; lung, tooth, and face formation; and neurogenesis, i.e., in the brain. Moreover, the evolutionary history of *Shh* dates back hundreds of millions of years and is highly conserved between fruit flies and humans. In *Drosophila*, it is jointly responsible for larval segmentation.

Sonic Hedgehog, *Pax6*, and the Hox genes are all striking proof of the evolutionary relationships among all vertebrates, in particular, those between humans and flies and other insects—even worms. This recent revelation of the stringent interrelationship of animal families is one of the most impressive achievements of modern biology—indeed of biology in general. In 2011, I was visiting the Max Planck Institute for Developmental Biology in Tübingen, Germany. I had been invited by Hans Meinhardtt, whom I admired and who has unfortunately since passed away. The developmental biologist introduced me to his research on pattern formation with Turing mechanisms. I was hoping to exchange a few words with the Nobel laureate Christiane Nüsslein-Volhard, whose office was diagonally opposite his, but unfortunately, she was not there.

Nüsslein-Volhard and her colleagues and fellow Nobel laureates Eric Wischaus and Edward B. Lewis are genetic evo-devo biologists par excellence, although they are not likely to describe themselves as such. Certainly though, research on how the control of individual evolution of living things has evolved in phylogeny is also a central evo-devo topic (see Section 3.4).

As Walter Gehring discovered, homology between species occurs not only in individual genes but also in groups of them. In 1984, he achieved notoriety (McGinnis et al. 1984) with his discovery of the homeobox (Figure 3.4). The homeobox provided a "toolbox" of genetic elements that regulated developmentally important genes in a fixed pattern during embryogenesis. The full complement of Hox genes is highly conserved across the animal kingdom and follows the same developmental principles from arthropods to vertebrates. It is not an exaggeration to state that kinship in development spans geological periods. The same developmental principles operating in evolutionarily distant tribes with the same genes and similar homeoboxes are solid evidence of their evolutionary relatedness. The surprising revelation of these developmental similarities represents a pinnacle that rarely occurs in research—an early, striking breakthrough for evo-devo.

The subset of homeobox genes known as Hox genes are control genes (also called master genes or master control genes). These are genes that coordinately manage the expression of other, functionally related genes during development. Hox genes control axis formation, segmentation, and appendage development in particular. Clearly, mutations in Hox genes, especially those in the homeobox, are extremely critical and often lethal for the embryo. The reason for this is that basic structure formation takes place with the help of these master genes in the early phase of embryonic development. An error at this stage can easily disrupt the structure and function of the whole organism. This is another way of framing the high degree of conservation of these genes.

It is clear at first glance that the genetic toolbox has been preserved over hundreds of millions of years of evolution and that this high degree of conservation indicates its indispensability to every organism. Nevertheless, evolution occurs. How then, against this seemingly immobile background, can variation arise at all, whether continuous or discontinuous, adaptive or not? How can organismic diversity, as we know it, arise from a rather rigid-looking genetic construction kit and its developmental pathways? These questions stimulated further inquiry in the minds of researchers. The diversity of life forms in nature is almost unbelievable, which raises the question of how the contradictory principles of conservation and change can coexist. The early evo-devo researchers faced a completely different situation from those in the initial phase of the synthetic theory. The representatives of the Modern Synthesis had no answers to these elementary questions of evolution.

In the initial phase of evo-devo, other obvious questions came to mind: What would happen if a developmental gene was naturally expressed earlier or later, and what would be the consequence if the gene remained active a few minutes longer and thus reached higher levels of expression than in the normal case? This phenomenon, referred to as heterochrony, received considerable attention. What is the effect of such temporal developmental changes on the embryo? Other questions followed: How can the developing organism generate complex, discontinuous variation, consisting of various cell and tissue types with diverse functions? Can a structure such as an extra finger or toe arise embryonically, in a single generation? What is the range in phenotypic shape (developmental plasticity) of complex variation, for example, the beaks of Darwin's finches (Section 4.1)? When and why does this range exist? Why is there often stability instead of plasticity?

Several of these questions can be placed under the heading of "evolvability." The concept of evolvability, or evolutionary capacity, is intended to capture the extent to which organisms can generate viable phenotypic variation (Kirschner and Gerhart 2005). This, of course, implies the opposite consideration of whether evolvability can be constrained in principle by developmental mechanisms. Such constraints—of which there are different types—are called developmental constraints. The questions posed up to this point already constituted several problem areas or challenges. They have been the subject of hundreds of dissertations and publications over the past three decades. It should also be mentioned that other questions arose as evo-devo evolved and became a pillar of the EES, which I will discuss below and in Chapter 6.

One thing is certain: the Modern Synthesis did not provide answers to all these questions.

If one wishes to condense the new evolutionary biological goals into a single picture, then the Modern Synthesis is concerned with the *Survival of the Fittest*, while evo-devo is concerned with the *Making of the Fittest*, i.e., with how the best adapted came into being in the first place. Some also speak of the *Arrival of the Fittest*, which means the same thing. Both represent the evolutionary developmental path up to this point.

3.2.4 Ontogenesis and Phylogenesis—How a New Research Discipline Emerged

In his landmark book entitled *Ontogeny and Phylogeny* (1977), the influential American author Stephen Jay Gould wrote about numerous topics that would underpin the connection between development and evolution. Using the concept of heterochrony, which received in-depth treatment in this book, Gould reasoned that changes in the timing of development can be a source of major evolutionary transitions. The massive work is considered a catalyst for the emergence of evolutionary developmental biology. In 1979, Gould, along with Richard Lewontin, followed up with a memorable paper, in which they rejected the idea that every phenotypic trait must be adapted (Gould and Lewontin 1979). Mistakenly, they argued, traits were considered individually in the Modern Synthesis, which did not present an accurate picture of evolutionary events. In fact, according to the two authors, because of multiple developmental constraints, different developmental processes could only vary together or be shifted in blocks; therefore, they could also evolve only in blocks. The disintegration of the rest of the embryo is thereby avoided; there is no chaos.

A clear demonstration of this exists today, for example, in the formation of additional fingers or toes (polydactyly). A finger constitutes such a "block." It consists of several different cell types in the form of bones, muscles, nerves, blood vessels, etc. Not only can it be regenerated and inherited with a single genetic point mutation in one generation; but in numerous cases, it is also one hundred percent functional (Lange et al. 2014). We will examine more such impressive examples in detail.

Returning to the historical progression, in the 1980s, researchers were undergoing changes in perspective, and the discussion turned increasingly toward the fact that heterochrony in development is one of the most effective ways to trigger major phenotypic changes with little genetic change effort. For some phenotype changes, for example, a change in eye color or hemoglobin, the developmental pathway was clear. However, the aforementioned shifts in the timing of developmental events due to heterochrony were more interesting variations in phenotype, which could cause a major structural change in the embryo.

In addition to heterochrony, other evolutionarily effective developmental mechanisms have been discovered. The location of gene expression can also shift, a phenomenon known as heterotopy. The term is seldom used, but in this book, you will find at least two impressive examples of it. In this context, one also speaks of ectopic gene expression. For example, Armin Moczek's lab succeeded in generating new eye tissue in beetles (Zattara et al. 2017, see Section 3.9). In Section 4.8, the reader will learn how the expression of *Sonic Hedgehog* at a site of the hand bud, where it is normally never expressed, is responsible for the formation of one or more additional fingers.

The fact that a gene product, i.e., a protein, is produced in different amounts—via a mechanism called heterometry—can give rise to phenotypic variations. We will examine this in more detail in Section 4.1 in the evolution of beak shape in finches. In summary, we have several degrees of freedom in evolution, such as time, place, quantity, and others, that can facilitate evolution. (I thank Armin Moczek for this reference).

The new research discipline of evo-devo emerged around these themes. The term was introduced in 1996. Development refers to embryonic development, in several cases, early embryonic development, including the phases in which the basic structures of the organism, for example, the preforms (anlagen) of the extremities or the organs such as the nervous system, heart, and so on, are formed. Evo-devo is a prime example of modern interdisciplinary research. It investigates how control mechanisms in development can influence biological evolution at the genetic and epigenetic levels.

A coordinated, intensified debate on the relationship between evolution and development was initiated at an international conference in Berlin in 1981, the Dahlem Conference *Evolution and Development*, chaired by John Tyler Bonner. Bonner, a developmental and evolutionary biologist, died in February 2019 at the age of 99 as I was writing this chapter for the German edition. At that time, he had brought together 50 biologists in Berlin from several subfields to present what was then known about development and evolution. The interdisciplinary aspect was explicitly at the heart of the conference. Although the door to a discussion on the ultimate role of genes was not opened, role of genes in the various, complex phenotypic shifts during development was questioned. According to Bonner, the key to this is still completely unknown. Beyond genes, the enormous superstructure produced by genes was a seminal discovery. In this superstructure, some events seem to occur that cannot be predicted by genetic information alone (Bonner 1982). This pointed to the role of development as a complex, interrelated system. This idea was the tenor of the conference.

The realization that development must be part of the explanation of the emergence of the phenotype in numerous, interdependent steps is, of course, not a sufficient justification for a new theory of evolution—not if one remains attached to the idea that this path is completely given or readable in the genome and that the genetic blueprint determines the phenotype. However, at the 1981 Dahlem Conference, the conventional reductionist and deterministic model of thinking, including the notion of genetic blueprint, was challenged. In addition, serious doubts about the unlimited possibilities of natural selection were formulated (Bonner 1982): Development was discussed as a hierarchy of levels of complexity, in which each level maintains open lines of communication with all the others. The novel concept that developmental constraints must be seen within these hierarchical levels of the organism emerged. Moreover, due to the complexity of the various levels, the capacity of selection to change them is limited. Moreover, the selection is always limited by the events of the preceding developmental stage. Finally, the more complex the organism, the less direct the effect of phenotypic selection on the genome.

The outdated picture of evolution determined solely by natural selection was thus far from being consigned to the archives, but the 1981 conference, together with the immediate preliminary work of Gould (1977) and Gould and Lewontin (1979), marked the beginning of a new research discipline, even though it did not yet bear the name evo-devo. Increasingly and consistently, this program furnished answers to questions, such as "How can the organism itself generate variation? Are there intrinsic mechanisms of variation generation in development, namely, without selection and without adaptation?" These were and still are some of the fundamentally novel questions that arose against the background of the Modern Synthesis, which itself did not provide a good access point for these new ideas. Evo-devo was thus to become the foundation of a comprehensive new way of thinking about evolution.

3.3 GENE REGULATORY NETWORKS

After the turn of the millennium, evo-devo underwent a two-way evolution. Building on the progressive technique of comparative genome sequencing, an early group of geneticists focused on developmental genetics and gene regulation. Sean B. Carroll (2006), USA, Andreas Wagner (2014), Switzerland, and Paul Layer, Germany, to name a few, view gene regulation processes with changing combinations of gene switches as the primary factors in organismal development and change. Indeed, trans- and cis-acting regulatory elements are the more easily mutated genes themselves.

3.3.1 Genetic Toolkit

Sean B. Carroll is an award-winning evo-devo researcher in the United States who has been popularized in the media. His very successful book *Endless Forms Most Beautiful* (Carroll 2006) introduced the genetic version of evo-devo to a wider audience outside the specialist community. There are several YouTube presentations in which Carroll demonstrates how the genome uses a genetic "toolkit." The important tools in this toolkit are mainly the Hox genes, responsible for body form and structure. As mentioned earlier, Hox genes are conserved over long geological periods and have remained very similar across animal phyla. How then, is evolutionary change possible?

Evolution uses so-called gene switches to generate phenotypic shape and variation out of highly conserved genes. For example, the *Bmp* gene family encodes a set of signal proteins called bone morphogenetic proteins, of which approximately 20 have been identified to date. The gene product BMP-5 is involved in skeletal construction, more specifically that of the ribs, outer and inner ear, fingertips, and nasal cavity. Developmental genes are surrounded by numerous switches through which their spatial and temporal expressions are regulated. In the case of BMP-5, some gene switches direct their expression in the ribs and others in the construction of the inner ear. The switches themselves are short DNA segments to which proteins can bind via a specific "lock and key" mechanism in each case. Altogether, there are a few hundred such locks. The key protein function is to activate (switch on) or repress (switch off) gene transcription. Every step in development, and thus every new switch combination, occurs in a controlled sequence. There are millions, perhaps billions, of steps in the entire development process. In this way, according to the ideas of gene regulation researchers, with the involvement of numerous developmental genes (e.g., the Hox genes), their switches, and the corresponding lock proteins, the three-dimensional body is formed.

A single gene is often regulated by 10 or more differently structured switches, which can occur in an immense number of possible combinations. The combinations of switches are more highly mutable than the genes themselves. For a three-way combination of 500 switches present in an embryo, there are $500 \times 500 \times 500$, or 12.5 million possible combinations. If four switches are combined as instructions for the expression of a specific gene at a very specific point in development, there are already more than 6 billion possible combinations. This gives some idea of the immeasurable design schemes that are possible with the help of gene switches (Carroll 2006). Carroll uses numerous insect wing shapes and patterns to demonstrate the seemingly endless repertoire that can be produced with the same arsenal of genetic tools. In summary, Carroll states that evolution is mainly about how "old genes learn new tricks" (Carroll 2007; Figure 3.5).

According to Carroll, several of the secrets of evolution stem from changes in genetic switches. The smallest changes in the sequence and combination of switches can result in an enormous effect and diversity in the design of embryos. This makes it much easier to understand why humans get by with so few genes (approximately 23,000, compared to the nematode with about 19,000). The gene switch combinations are, for Carroll, the music makers. According to this hypothesis, it is far more efficient to expand the coding capacity of the genome via combinatorial regulation than by maintaining hundreds of thousands of genes. Thus, the evolution of organismic forms depends not so much on which genes an organism has, but on how they are used. The antiquated view that a new set of genes is needed for every morphological variation is considered outdated.

However, because so incredibly many switch combinations are possible, there can also be numerous combinations that produce the same or approximately the same phenotypic output, i.e., proteins with the same function. This is the canalization or robustness toward the same output that we are already familiar with. The consequences of multiple identical phenotypic outputs are made clearer by Andreas Wagner.

Without question, macroevolutionary change can also be explained by the principle of gene regulatory networks, according to Carroll's theory. Changes in form and function can be explained in detail in living beings such as arthropods, as can the emergence of new forms, which in the end are not fundamentally new but are created and rebuilt from existing ones, such as claws from the forelimbs of crayfish or wings from the forelimbs of birds. For example, at the beginning of vertebrate evolution, there were numerous duplications and shifts of a single Hox gene cluster. Changes in the DNA sequences of the gene switches that regulated the Hox genes were always involved. For Carroll, new structures in evolution are modifications of existing ones. However, the essence of change or innovation, no matter how extensive or how new, is the duplication of existing genes and new combinations of gene switches.

A strict Darwinist, Carroll repeatedly uses the phrase *use it or lose it*. A gene or element that is not used is, according to him, inevitably lost. It mutates, and thus, gradually but surely

Figure 3.5 Wing shapes and pigmentation patterns of higher bipeds. After 70 million years of evolution, a seemingly infinite spectrum of color and structural patterns has been revealed.

loses its function. Selection occurs on the gene level, with the incessant oversight of natural selection. This is Carroll's adaptationist signature. Regarding the evolutionary development of increasingly complex structural forms, we learned earlier that genes often function in interrelated blocks, and developmental processes can often only be shifted in blocks. This view integrates developmental modules; however, Carroll spends little or no time addressing this point and provides no answers as to the mechanisms evo-devo can use to rapidly provide complex variation. He does not concern himself with pattern formation processes and self-organization on the cellular level, for example, in the context of limb formation—he even rejects them. The environment figures only marginally in his work but plays a crucial role in more recent treatments of the topic (Section 3.8).

The complexity of the developmental system in Carroll's view is a modest one and much less systemic than that of the subsequent, more epigenetically oriented group of researchers presented in the next section. Carroll's discoveries are consistent with Darwinian selection; he does not advocate a theoretical shift.

3.3.2 The Arrival of the Fittest

Like Carroll, the Austro-American biologist, Andreas Wagner, focuses on gene regulation networks. Wagner does not use the term evo-devo but instead focuses on the robustness and innovation in biological systems. Wagner shows how the two interact—a supposed contradiction, since robustness is a rather conservative attribute. Robustness implies the maintenance of the organismal form in its current state. Innovation, on the other hand, is a progressive or explorative concept. Herein lies the contradiction Wagner seeks to resolve: How can both be possible simultaneously?

We learned about robustness when Waddington introduced the concept of canalization, the ability of a biological system to withstand perturbations such as DNA mutations and environmental change. Language, for example, is robust. When I omt sme ltrs hre, the reader may bristle for a moment but understands my meaning nonetheless. By comparison, computer programs are much less robust. A single mistake in a line of code, and the program performs differently or crashes altogether. In biology, for example, gene redundancy and duplication are sources of robustness to mutations. Approximately half of the genes in the human genome are redundant. Natural selection can maintain this redundancy and the resulting robustness (Wagner 1999, 2000).

In addition to duplicated genes, proteins, gene regulation, and metabolism are particularly robust (Figure 3.6). Wagner uses the example of lysozyme, which acts as a bactericide in saliva and tears. In laboratory experiments, the polypeptide sequence of lysozyme was slightly modified to generate 2,000 protein variants. Curiously, no less than 1,600, or 80% of all artificially modified protein variants, retained their bactericidal ability. Wagner refers to such modified protein variants as "neighbors" if they occur in the organism and share functional redundancy. Neighbor proteins occur frequently, according to the study. "The more such neighbors with the same phenotype, the more robust the organism is" (Wagner 2014). Robustness exists above all in metabolism, a complex, biochemical system where alternative biochemical metabolic pathways exist should a gene or protein fail altogether. Metabolic pathway redundancy is essential to life. Importantly, natural selection maintains this diversity, thus preserving robustness. Wagner depicts robustness as a state of disorder in which several pathways can perform the same function. In the absence of pathway redundancy, the organism would be programmed like a computer and enormously susceptible to errors.

Why then, is robustness alone, not sufficient to drive evolution? Here Wagner arrives at the progressive, explorative aspect of innovation. The connection between the two becomes clearer immediately: If humankind invents something new, and it does not work as intended, the unsuccessful invention can be abandoned for a new one. Nature, however, cannot afford this luxury. It must preserve what works, the old and the new. Without the stable preservation of their functions, organisms die. Simultaneously, they must be prepared to cope with new situations. I will use the example of Apollo 13. It is well known that a life-threatening fire occurred during this flight to the moon. Aborting the mission and turning back was out of the question, so the flight crew found a way on the flight to cope with both the fire disaster and the scarcity of oxygen. They did so *on the fly*, in the truest sense of the word: the crew mastered the problem brilliantly with the resources available on board.

A similar situation exists in nature. At any time, an organism can be confronted with a new situation for which novel characteristics are required. It resorts to the neutral genetic mutations identified by Motoo Kimura (Section 2.3). These mutations, long considered useless, thus become the basis for later evolutionary innovations (Wagner 2008).

Suppose environmental conditions have changed, necessitating species adaptation. This is where neutral mutations come into play. Although neutral under the old conditions, they can quickly contribute to novel and sometimes advantageous phenotypes under the altered conditions. Neighboring proteins already present can be used for this purpose. Selection ensures the preservation of neutral mutations as the basis of new form

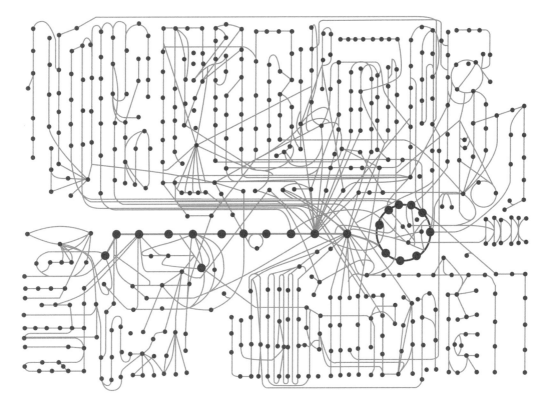

Figure 3.6 Robustness. Eukaryotic metabolic network. Dots indicate metabolic intermediates (metabolites), and lines indicate conversions by enzymes. Many metabolites can be produced via multiple pathways; hence, the organism is robust to the loss of certain metabolic enzymes.

and function. Where neutral mutations are absent, and robustness is lacking, nothing new emerges (Wagner 2005, 2011, 2014).

In addition, with only minimal genetic mutations, different protein neighbors can be generated next to functionally identical ones when innovations and adaptations are required. Long protein chains often only require a difference of a single amino acid to exhibit a new function. The same applies to variation in metabolism and gene regulation. As an example, Wagner cites the synthetic biochemical pesticide pentachlorophenol. In an impressive innovation, the bacterium *Sphingobium chlorophenolicum* can use its own enzymes to convert the pesticide into a metabolically useful molecule. Wagner likens the conversion of a highly toxic substance into food to turning a weapon into a bar of chocolate. Another common mechanism in bacteria is to feed on the very antibiotics that have been circulated to destroy them (Wagner 2014).

Wagner's point is that innovations are often easy to realize. There is a hyperspace of large networks of genotypes capable of encoding the same phenotypes. Populations of organisms can explore this hyperspace, and astronomical numbers of possible solutions are available there, many of which are much closer than imagined. However, Wagner does not mention that constraints on the development of body structures block the arbitrary conquest of this morphospace, and developmental biases that we will address shortly (Section 3.4) can impose significant constraints on innovation, given enough time (Minelli 2015; Newman 2018). He thus reopens the door to a way of thinking about old adaptation.

In 2011, based on decades of laboratory work with his teams and these mechanisms, Wagner proposed a theory of innovation in which the ability of living systems to innovate is a consequence of their robustness. The ability to innovate in nature, as we have seen, arises from the organism's engagement with a constantly changing environment (Wagner 2011).

To summarize, Wagner brought forward interesting and valuable explanations at the level of gene regulatory networks and protein folding. With his modern understanding of the necessary connection between robustness and innovation, he is reminiscent of Waddington. However, Waddington more clearly established the link between environmental perturbation factors and the pathway to genetic assimilation. The connection thematized by Wagner is no longer disputed in science today. "Robustness and plasticity are complementary and intertwined and must be considered together. They should not be seen as being in opposition to each other" (Bateson and Gluckman 2012).

Wagner does not require higher levels of biological organization above proteins, and like Carroll, he does not address them. In the eyes of others, the *Arrival of the Fittest* in evo-devo is much more than Wagner's version of it.

3.4 SYSTEMIC-INTERDISCIPLINARY VIEW—A RESEARCH DISCIPLINE ACQUIRES ORDER

It is a fact that modern, practical evo-devo research is dominated by developmental genetic topics around gene regulation and gene networks. This may create the distorted image that evo-devo is restricted to this field alone, but such a conclusion would be made in haste. Nevertheless, the second group of evo-devo scholars envisages more far-reaching goals. They are taking a systemic, interdisciplinary approach and aiming to paint a larger evo-devo picture. Men and women such as Mary Jane West-Eberhard (2003), Denis Noble (2006), Marc Kirschner (Kirschner and Gerhart 2005), Armin Moczek et al. (2019), and Gerd B. Müller (2007) view the entire developmental apparatus as a system that interacts in complex ways at the genetic and epigenetic levels of organization (DNA, proteins, cells, cell communication, cell aggregates, the organism, and the environment). According to this group, the entire system of embryonic development is itself evolved and is in a complex, systemic relationship with the environment. The beliefs of this group thus represent a new broader perspective.

To explain phenotypic formation and change, one must consider not only higher levels of organization than genes and gene regulation but also new principles—those of pattern formation, self-organization, developmental bias, and so on. These principles allow predictions to be made about evolutionary variation. Before addressing these topics in detail, we will describe the evo-devo research program pioneered by the above-mentioned group, which extends far beyond the knowledge of gene regulation. Chapter 4 presents some examples, and then Chapter 6 deals in detail with the EES, which builds on the concepts described here.

According to Müller (2008), evo-devo research is divided into three thematic blocks (Figure 3.7), some of which contain individual questions that the scientific community is only beginning to address.

The first block contains three evo-devo questions (1–3), directed from evolution to development:

1. The first topic deals with how the developmental program in recent species could have arisen during evolution. Development itself is a system that evolved over hundreds of millions of years. The process, still far from being understood in terms of its interaction with the outside world, which we see and analyze today in higher developed

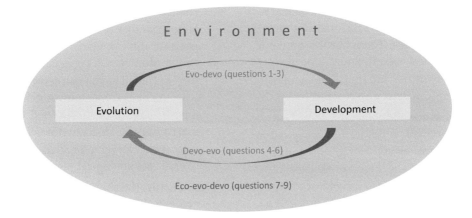

Figure 3.7 Questions at the interface of evolution and development. Evo-devo queries mechanisms of action of evolution on development (1–3); of development on evolution (devo-evo, 4–6); and of the interaction between development, evolution, and the environment (eco-evo-devo, 7–9). The theory of evolution is thus methodologically extended and complex.

species in the embryonic phase, was not always present as this routine, finely adjusted interplay. There must have been a long evolutionary process leading up to this, which continues to this day. Starting with the first multi-cellular animals (Metazoa), selective fixation and genetic routinization did not emerge until much later in those robust forms of development and in the reliable Mendelian forms of inheritance that we see in modern-day organisms.

2. *Evolution of the developmental repertoire* refers, on the one hand, to the evolution of genetic tools. As we know by now, Sean B. Carroll (2007) speaks of the genetic toolkit. The question is how this toolkit could emerge and evolve or how, for example, genetic redundancy, new gene functions, or modularity could emerge at the genome level. On the other hand, the evolution of the developmental repertoire, which emerged over millions of years, also includes a complex variety of epi-genetic processes. These processes of developmental interactions were simpler hundreds of millions of years ago. Today, they include sophisticated, well-rehearsed mechanisms that precisely regulate, for example, cell–cell interactions.

3. How does evolution affect specific developmental processes? One such process is heterochrony, which involves not only a temporal shift of developmental processes but also changes in developmental control. Evolutionary modifications in the segmentation and regional differentiation of larger sections of anatomical structure, for example, are accompanied by shifts in domains of Hox gene expression (Müller 2007).

The second block of evo-devo questions (4–6) concerns the influence of development on evolution, i.e., the inverse of those posed in the first block. These are the novel questions of evo-devo. They reveal the causal interactions between development and evolution for the first time and challenge the prevailing theory of evolution (Figure 3.7). These include the following:

4. How does development influence phenotypic variation? In evo-devo, we often speak of the developmental constraints already mentioned. Such constraints, in several cases, prevent

the occurrence of large evolutionary variations and are thus hurdles during development. These hurdles can be of a physical, morphological, or phylogenetic nature, and they lead to robustness (canalization) or stability.

5. What does development contribute to phenotypic innovation? This question concerns evolutionarily new formations, such as the inner skeleton of vertebrates, mammary gland of mammals, bird feather, insect wing, or turtle's shell. Natural selection cannot produce anything new; it can only act upon the existing. That which is selected out, the anatomical solution to problems, is contributed by the developmental system. In other words, if selection on its own cannot reshape organic forms, there must be some other way in which organismic innovation arises. The answer can only lie in development. More on the exciting topic of innovation in a moment.

6. How does development affect the organization of the phenotype? The question of the organization of body plans in development is not directed at the origin or variation of individual body traits, as proponents of the Modern Synthesis assumed, but at how the organism can be produced as an *integrated system*. Morphological structures and organ systems vary and evolve as integrated units, with numerous changes occurring in a coordinated fashion so that complex variation is possible without rendering the organism nonfunctional. Supernumerary fingers (polydactyly) can serve as a convenient example here. A new finger does not appear without blood vessels or muscles. It is physiologically fully intact. Development ensures this overall morphological performance.

Finally, the third block, namely the eco-evo-devo set of questions (7–9), concerns the causal relationships of development and evolution with the environment. This type of question was also newly introduced by evo-devo, since the Modern Synthesis cannot deal with such mechanisms of action (Figure 3.7).

7. How does the environment interact with developmental processes?
8. How do environmental changes affect phenotypic evolution?
9. How does evolutionary development affect the environment?

A theme central to questions 8 and 9 is phenotypic plasticity. Plasticity means that a genotype or, better stated, development, can produce different, possibly strongly deviating phenotypes under different environmental conditions. Examples of this will follow (Section 3.8), but first, other evolutionary theorists who advocate a systemic evo-devo will have their say.

In Conversation with Gerd B. Müller

Professor Müller, you studied medicine as a young man. Why did a trained physician with the best career prospects at the time dive into a new world and complete a second dissertation in zoology?
During my medical studies, I became more interested in the causes of the biological structures and processes we were to learn than in the therapy of their disorders. I was initially fascinated by the possibilities of experimental embryology and then became increasingly involved with evolutionary theoretical issues, moving further and further

away from applied medicine. Eventually, this turned into a parallel degree in zoology. However, I still think that a sound knowledge of biology and evolution is also a basic prerequisite for understanding pathological phenomena in human medicine.

The Nobel laureate Konrad Lorenz worked at the University of Vienna in your early days there. I believe he read in the same lecture hall as you. You received your doctorate under Rupert Riedl. Your indispensable compulsory reading at that time was Chance and Necessity, **a book by another Nobel laureate, the Frenchman Jacques Monod. Looking back, how did you experience this time, which was so strongly marked by the synthetic theory?**

Lorenz returned permanently to Austria in 1973, the year the Nobel Prize was awarded, and at the invitation of Rupert Riedl gave guest lectures at the University of Vienna on his evolutionary approach to behavioral research. These lectures eventually led to a long-running seminar at his family residence in Altenberg. Attending these lectures and seminars laid the foundation for my later involvement with evolution. At the beginning of his first lecture, Lorenz recommended three books, which he held up emphatically in the lecture hall: The Logic of Scientific Discovery by Karl Popper, Chance and Necessity by Jacques Monod, and–amazingly–Nicolai Hartmann's The Teleological Thinking. I was amazed. My medical studies had never mentioned any textbooks other than subject-specific ones, and certainly not philosophical texts. After the lecture, I purchased the three books at the university bookstore around the corner, and it became clear to me for the first time how closely natural science and philosophy are connected. In the Altenberg Circle, all discussions were characterized by this mutual relationship.

Riedl was the driving force behind these events. His lectures finally brought me into biology, which I finally completed with him with a dissertation on evolutionary experiments in developmental biology. Riedl was an evolutionary rebel who saw himself surrounded by dogmatic representatives of the synthetic theory, i.e., the evolutionary theoretical position that wanted to attribute all evolutionary change exclusively to population-genetic mechanisms within the framework of the basic Darwinian model. Riedl countered this with his systemic theory, which was based on reciprocal interactions between different causal levels and referred primarily to morphological evolution, which was largely left aside by the standard theory. Although Riedl's idiosyncratic conceptualization often meant that he was not listened to very much, there was an enormous amount to learn for us students from this confrontation of different intellectual positions. From today's perspective, it can be said that Riedl was right on many points.

You have witnessed the emergence of the evo-devo research program from its inception in the 1980s, and you have been able to shape its development. Embryonic evolution staked its claim: it had and still has something to say about evolution. What fascinated you personally so much about evolutionary developmental biology?

Reflections on the role of embryonic development in changing the species have a long history that goes back to pre-evolutionary times. However, in the population-genetically influenced synthetic theory, the principles of individual evolution had not been reflected at all. Personally, the insight that phenotypic evolution can only occur via a change in developmental mechanisms fundamentally motivated me to enter basic research and contribute to the elucidation of this relationship. The methods of experimental and molecular developmental biology emerging at the same time, as well as new, three-dimensional, computer-assisted imaging techniques, also made it possible for the first time to address questions in evolutionary developmental biology empirically and quantitatively. The great potential of this new approach to evolutionary biology was clearly evident in the early 1980s. Here, a new departure was possible, a chance to redevelop and shape an entirely new field of research.

What gave you the impetus to believe that evo-devo is not just another empirical research discipline but contains "ammunition" for a fundamental extension of evolutionary theory?
I would see it somewhat differently: the limitations of the standard theory were known and widely discussed, and evo-devo was clearly one of those fields from which a theoretical renewal could emerge. It was therefore necessary to develop appropriate questions, methodological approaches, and experimental procedures that could demonstrate the influence of developmental biological conditions on the evolutionary process. This has since succeeded in many instances and has provided the data for those concepts which, when brought together with results from other fields of evolutionary biology, have provided the impetus for a new theoretical synthesis. It is important to note here that many of the crucial evo-devo concepts were formulated before the molecular genetic turn in developmental biology and not because of it, as is often claimed today.

You met all the great evolutionary theorists in the postwar period. Which was your most impressive and lasting encounter?
In fact, it was a wonderful experience to meet Ernst Mayr, John Maynard Smith, Richard Lewontin, Edward Wilson, Stephen J. Gould, and many other theorists in person. Of this generation, Lorenz and Riedl certainly had the most lasting impact on me. Interestingly, however, colleagues of the same age as me were even more determining for me, those who had not taken the detour via medicine and were therefore already active in developmental and evolutionary biology research before me. These include Pere Alberch, Günter Wagner, Eörs Szathmáry, Eva Jablonka, Stuart Newman, and a few others. Newman was a crucial encounter because he brought a fundamentally different view of cellular and developmental biological processes to the evolutionary theoretical discourse. Our meeting resulted in many joint articles, symposia, books, and a friendship that continues to this day.

Since you were appointed to the University of Vienna in 2003 and founded the Department of Theoretical Biology there in 2005, you have accelerated your efforts toward an EES. In 2008, you organized the important Altenberg-16 Symposium with 16 world-leading evolutionary biologists and theorists from different disciplines. I was allowed to accompany you on this exciting journey for a while. Tell me: How did this initial meeting develop into an expansion of theory, and what was the outcome of this important meeting, which was certainly not characterized by homogeneous ideas?
In 2007, I met Massimo Pigliucci, an evolutionary theorist of the time, who was then teaching at Stony Brook University, at a meeting in Exeter, and we discovered that we

both had an article in print whose title included the term "Extended Synthesis" in the sense of a "new evolutionary theoretical synthesis." This coincidence seemed to us symptomatic of the situation in evolutionary theory at the time, and we decided to take this as an opportunity to think more broadly about innovative evolutionary theoretical concepts and their relevance to the overall theory. I was able to set up a workshop at the Konrad Lorenz Institute for Evolutionary and Cognitive Research in Altenberg near Vienna on the topic Pigliucci and I had planned, with the goal of coming out with a book on the current state of evolutionary theory in the Darwin Year 2009. We narrowly missed this goal; the book did not appear until 2010, but it did much by bringing together the first major assembly of concepts such as developmental bias, epigenetic inheritance, niche construction, etc. and placing them in a common theoretical framework. Of course, not all participants were in complete agreement, and some even later participated in counter-articles to the EES, but the book signaled to the outside world a new departure that was, by and large, very well received.

Is the EES an extension of the standard theory, or is it a reconstruction?
We are interested in a comprehensive synthesis of the known evolutionary factors and theoretical concepts in evolutionary biology, which consists of many more components today than it did half a century ago. Of course, large parts of the standard theory are included here, such as the rules of population genetics, but the inclusion of new components, such as evo-devo or niche construction, also creates a new theoretical structure that is based on a different logic and leads to different predictions and explanations than the classical theory. Some elements of the standard theory are also not adopted or are given a different function, such as natural selection, and therefore, EES cannot be interpreted as a mere peripheral extension of the standard theory.

Some of your colleagues see the efforts of the EES representatives as fighting windmills. What do you say to the accusation that considerations of the EES are basically already implicit in the Modern Synthesis?
The representatives of orthodoxy will always insist that an existing explanatory system already includes all phenomena not explicitly considered. In the present case, it is usually argued that some of the new components of EES have been considered before or have been applied here and there. But these components have never before been placed in a concrete theoretical context. The novelty of EES is not that all its elements were unknown before but that these concepts are synthesized into a new theoretical structure. The excitement about this, in some respects, is unfounded. Scientific theories are always changing, sometimes in small steps and sometimes in somewhat more radical paradigm shifts. One day, EES or some differently named extended theory will be the standard theory and will in turn be reworked again.

Another question that has remained open since we met: Synthetic evolutionary theory is a theory about populations. This approach has been the reason for its success. Evo-devo also aims to provide evolutionary insights, but its theoretical approach is based on events in individuals. How does that fit together?
Populations are also made up of individuals. Evolutionary changes at the level of populations are caused by changes in the individuals. We must therefore understand the

regularities in the construction and function of individuals in order to be able to explain the changes in populations. These rules determine what is "feasible" in the evolutionary change of species in the first place. This is precisely the essential contribution of evo-devo.

What are the greatest hurdles that need to be overcome for the EES to gain the desired broad acceptance? How long does such a process take?
The greatest hurdles are dogmatic insistences. These are based not only on cherished theoretical convictions in the heads of some protagonists of classical theory but also on the priorities of the current scientific establishment. At stake are interpretive sovereignty, power, influence, positions, research funds, etc. When newly emerging currents begin to tie up resources, there is understandable resistance to the decline of traditionally successful fields. However, I do not have the impression at all that the acceptance of EES is so low. Especially from our younger colleagues, we receive mostly positive reactions, and large parts of the evolutionary theoretical discourse deal very seriously with the arguments of the EES. The problem of persistence—if it is one—will solve itself.

Eight years ago, I asked you whether practical science is ready for methodological rethinking. How is research today adapting to the fact that complex theoretical models are required and that interdisciplinary, networked, and systems thinking is needed? Are these efforts in your subject, i.e., in evolutionary theory, increasingly flowing into research and publications today?
We often hear from applied research that no matter what theories may exist, we move on. Such attitudes exist even in evolutionary biology and are increasingly supported by the trends toward quantification and digitization in all areas of research. In some places, the impression has arisen that theory-free research is also value-free and thus more objective. After all, one could always look later to see if the data had any relationship to the theories, in the sense of "data without theory meets theory without data," as the philosopher of science Werner Callebaut once said. I think this is a fundamental misunderstanding of the role of theory in the natural sciences because it is the theories that determine what we can find, whether we are aware of it or not.

Unfortunately, however, even the large public universities, whose role would be to provide education, knowledge, critical thinking, and interpretations of our world (and thus theories), have been completely sucked into data production and have surrendered—at least in the biological fields—to a largely economized model of science. Research is to be done where large amounts of money can be procured, which in turn brings reputation and further economic advantages to the universities. Theoretically oriented directions that do not follow this logic are increasingly pushed into the background at universities. For this reason, other independent institutions are emerging that actually take over the original functions of the universities, such as the Santa Fe Institute in the USA or the Konrad Lorenz Institute in Klosterneuburg near Vienna, in which I am also active. Such institutions offer the opportunity to think in an interdisciplinary and networked way and to work on complex issues. They promote discourse on theory development and philosophical questions in the natural sciences and preserve free science.

Professor Müller, thank you for the interview!

3.5 FACILITATED VARIATION—THE PERSPECTIVE OF CELLS

Marc Wallace Kirschner is a cell and systems biologist at Harvard Medical School. He, along with John C. Gerhart at the University of California at Berkeley, developed the *theory of facilitated variation*. This is a systems theory that put evolutionary development on a new footing. The authors thus belong to the evo-devo group of systemically thinking scientists. The theory was presented in 2005 in the book *The Plausibility of Life. Resolving Darwin's Dilemma*, co-authored by the authors Kirschner and Gerhart. The following remarks refer to this book.

For a brief review, the synthetic theory argues at the level of the genome, with gene frequencies abstracted from the organism, and examines mathematically–statistically the occurrence of mutations in these genes, which are assumed to be random, on the population level (Section 2.1). This is a strictly reductionist method because the gene frequency approach alone, together with natural selection, serves to explain evolution. Kirschner and Gerhart aim to overcome this. For this purpose, the theory of evolution is better thought of as consisting of the following three equal pillars or partial theories:

- a theory of phenotypic variation
- a theory of inheritance
- a theory of selection

Let us take a closer look at how the authors integrate these three subtheories. According to Kirschner and Gerhart, the theories of inheritance and selection have traditionally been described in detail. Therefore, they focus on the first component, the novel theory of phenotypic variation, which has been missing up to this point. It combines the existing doctrine of inheritance and natural selection with the phenotype. Different processes act in cells to realize phenotypic variation. The cellular point of view is also a novel one. I will therefore examine these processes individually in more detail. In the next chapter, the theory will then be illustrated using the example of different beak sizes and shapes in Darwin's finches (Section 4.1). The multiple functions of the HSP90 protein also fit into this picture (Section 3.8).

Cellular events can be viewed as

- preserved core processes in cells
- exploratory processes in cells
- weak regulatory couplings between cells
- compartmentation, the formation of modules or blocks of cells.

Let us begin with the long-term preservation of what Kirschner and Gerhart call the preserved core processes in cells. Some of the terms we are already familiar with, especially the Hox genes, reappear here.

3.5.1 Preserved for Hundreds of Millions of Years

Humans share 15% of their genome with the bacterium *E. coli*, 30% with the fruit fly, and 70% with the frog, Kirschner says. In the cells of recent species are processes that have remained unchanged for as long as these cells have existed. This does not mean that all chemical processes in recent cells are as they were in the beginning; however, there are a few hundred fundamental processes, core processes that are so elementary that changing them would spell death for the cell. By core processes, we mean the biochemical processes known to occur in different cells, several of which are identical. We are dealing with a larger but limited set of core cellular behaviors. The term "core processes" does not necessarily refer only to the cell nucleus and thus to the DNA but should be understood beyond that as "main processes," which can also be, for example, intercellular signaling pathways.

However, this limited set of conserved core processes may very well change in the way the individual processes (let us call them "Lego bricks" in advance) interact.

However, the individual processes do not change the process. Cell processes can therefore be recombined or used differently. Only very rarely, however, do truly new processes appear. Rather, we see novel combinations of the established core processes; these allow the evolution of new phenotypes. Kirschner and Gerhart call this flexible possible combination of cell processes "adaptive cell behavior" (Figure 3.8).

The authors give examples of preserved core processes in cells. These include the cell cytoskeleton, i.e., its internal structure, metabolic reactions (Sections 3.3 and 3.3.2), and signal transduction, i.e., the numerous signal transmission chains within the cells, as well as the mechanisms of gene expression. The fact that the cell has highly conserved processes on the one hand but is simultaneously adaptive is only made possible by three prerequisites, all of which are based on the following stable core processes:

- protein synthesis based on an identical genetic code for all living things.

- an identical, permeable function of the cell membrane in all cells of all living organisms; This enables cells to communicate and cooperate with each other and with their environment.

- identical functions of the Hox genes, the gene family responsible for important aspects of the body plan. Embryos are built from compartments (subplans), in vertebrates, for example, for the head, spine, and tail.

The stringent conservation of these core processes means high constraints, i.e., barriers to unwanted change. The core processes are also compared with Lego bricks. Their exact dimensions are their constraints. They only fit together with other Lego bricks. Kirschner says that "Constraint deconstrains variation." Lego bricks can be combined in several ways with other Lego bricks, precisely with their constructional restriction. This is their decisive advantage.

Thus, the conservation of master genes to protect against unfavorable mutation must

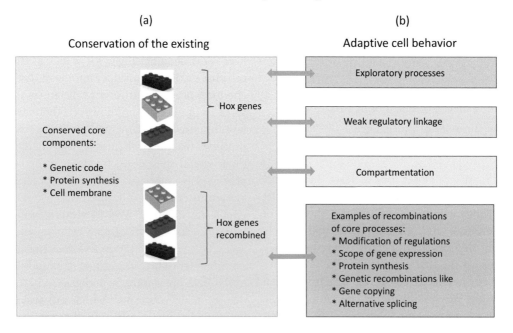

Figure 3.8 Facilitated variation. Evolution requires both constancy and variability. Kirschner and Gerhart therefore distinguish two types of cell processes—those that maintain the status quo (a) and those that allow change (b). Organisms have evolved over long periods of time in such a way that both sides of a species are in a certain balance between preservation and change.

be ensured, among other things, to maintain the continuous flow of life. This is demonstrated in evolution for extremely long periods of time. If today we find the same genes in *Drosophila* as in humans, for example, *Sonic Hedgehog*, they must have existed before our predecessors split off from those of the fruit fly. Earlier, we even learned about an active protection against unfavorable mutation with the heat shock protein *HSP90*. In the absence of stress, it can buffer unfavorable mutations and thus maintain stability. Simultaneously, it can allow variation in the presence of heat shocks (Section 3.9). In the author's view, conserved cell processes are primarily maintained by natural selection. Unfavorable changes in the core processes of cells would be consistently eliminated by selection because they do not allow the cell to survive. For this reason, they are termed "core processes."

3.5.2 And Yet Change Is Possible

How does this conserved landscape contribute to variation? When Kirschner and Gerhart speak of recombination of the core processes of cells, they primarily mean modification of regulation, for example, with respect to time and place, the circumstances or extent of gene expression, RNA availability, or protein synthesis. However, genetic rearrangement, such as the formation of gene and gene segment copies, is also involved in the recombination of core processes. In addition, alternative splicing plays a key role in the recombination. The term alternative splicing indicates the fact that the transcription of a gene—i.e., transcription of the genetic information—into RNA (more precisely, messenger RNA or mRNA) and translation of the RNA into a three-dimensional protein does not necessarily result in an identical protein every time. Rather, the protein structure may vary. For instance, a gene with coding segments a, b, c, d, and e could be read as a-b-c, a-b-d-e, b-c-d, or other combination. Alternative splicing thus increases the number of components available (think of Lego building blocks) to create newly assembled proteins from existing coding genes. It is thought that 95% of human genes with multiple exons are alternatively spliced (Pan et al. 2008). Thus, 500 different mature mRNA alternatives can be produced from a single gene transcript, even one from a human gene.

However, there are other mechanisms that generate the phenotypic differences between humans and chimpanzees, even though the overall difference in their two genomes is only approximately 1%–2%. The reason lies in what Kirschner and Gerhart call the regulatory diversity or combination diversity of the core processes of the cells: Genes are expressed in different locations, at different times, under different circumstances, and in innumerable different combinations. Such changes in gene expression can easily occur during development, and the effects on the phenotype can be significant, as in the chimpanzee and the human. The stable core processes allow for some forms of expression or idiosyncrasies that are essential to enable precisely the facilitated phenotypic variation that we discuss here. These are, first, the exploratory behavior of cells, second, their weak regulatory couplings, and third, compartmentation in the embryo. I discuss these three peculiarities of variable cell responses in more detail below to illustrate how facilitated variation is made possible. Now that we have discussed the conservation of cell processes, the following three sections will deal with change, i.e., with the question of what is facilitated or what makes the facilitation of variation possible.

3.5.3 The Exploratory Behavior of Cells

Behind every phenotypic change, for example, in the skeleton, are other necessary changes that must occur in an orderly, sometimes simultaneous fashion to maintain the system. For example, in the evolution of the vertebrate extremity, flexible adaptations of tendons, muscles, nerves, and blood vessels are necessary in addition to bone remodeling. Kirschner and Gerhart speak here of exploratory behavior, when cells, depending on their extracellular environment, exhibit a broad spectrum of alternative reactions. If one examines the finely branched blood vessel system, it quickly becomes clear that there is no deterministic specification in the DNA for supplying each individual cell in the body with sufficient oxygen or nourishment, certainly not with different amounts of either one. This is because the required oxygen supply varies from one tissue to another. This is regulated by exploratory behavior. The advantage of this is that in our capillary system, which has a total length of 100,000 km, the local needs of all cells in the body can be supplied in different ways. Capillaries can form at any time because cells react accordingly to a lack of oxygen. By contrast, a system with a genetic default regulated in detail would have to be infinitely complex.

The readers may be familiar with the hypothesis that the papillary ridges on the underside of the fingers, those that produce the fingerprint, are not genetically determined. Have you ever wondered how they are formed? Perhaps you suspect epigenetic involvement in this case, and it could well be called that. Equally good, but unfortunately rarely mentioned, examples are the blood capillaries (Figure 3.9) or the neural networks, which are even more differentiated. Their courses are all nongenetically determined, as are the details in the beautiful coat pattern of my Maine Coon cat. If you have ever doubted the notion of a genetic blueprint or genetic program, you may find your belief in it reinforced. The sense in which the term "genetic blueprint" was introduced decades ago may not be accurate anymore. The genome needs permanent feedback from the cells. Even the smallest developmental step in the embryo is determined by interactions of the genome with cells and of cells with their environment.

In this way, the exploratory process generates an unlimited number of result states (for example, an infinite number of papillary ridge patterns). This property of cells, namely their ability to form branched structures, is based on heritable changes in regulation. Thus, the plasticity of the possible formation of the structures (blood vessels, nerves, etc.) is what is important to produce a complex organism, and it is what the cells have mastered for billions of years.

The authors illustrate this concept clearly again with another example. The evolution of the vertebrate extremities forms very different shapes and sizes, from the bat wing to the elephant foot and many other forms for swimming, grasping, climbing, etc. During the remodeling process, it must always be determined which cells will differentiate into specific cells of bones, tendons, muscles, nerves, or blood vessels. It is

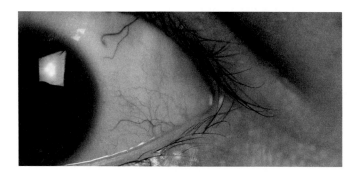

Figure 3.9 Exploratory behavior. The courses of small blood vessels (capillaries) are not genetically determined in detail. They follow explorative behavior.

almost inconceivable that random mutations and individually timed rounds of selection can accomplish this. There must be a more effective way that does not involve minute, selective testing. This is where the principle of exploratory cell behavior comes into play. Supportive cell functions are made available wherever they are needed, even in the face of major skeletal changes. We are familiar with the case of polydactyly, supernumerary fingers and/or toes in vertebrates, including humans. Here, a new unit, a finger or toe, arises in a single generation. In the place where it is formed, nothing existed previously, not even a single cell. But when the additional finger is developed and the baby is born, it has all the physiologically necessary cell types and tissue forms. No lengthy mutation and selection process is required; the morphological adaptation of the components occurs spontaneously. One tool for this is the ancestral preparedness of the cells for what is to come and their exploratory behavior. The new finger is functional in every respect.

Exploratory processes must also be imagined in the growth and formation of brain cells. Axons with their long nerve fibers search for and locate each other via biochemical exploratory behavior. On YouTube, you can find a video showing how axons move toward each other through the extracellular spaces, finally locating each other and forming new synapses (see tips for further reading). This is a daily occurrence. If you have read this chapter carefully, then read it again, and hold some of it in your mind the next day, then you would have experienced such a process. The wiring and thus the physics in your brain would have changed.

Structures in an organism are thus variably formed with exploratory processes. Genetic determination analogous to a program with an exact scale is unnecessary. This "program" would be inefficient because it would be much more complex. It would therefore take far too long to make adjustments. This is what Kirschner and Gerhart call "Darwin's dilemma." The Viennese zoologist Wolfgang Wieser, who died in 2017, had already written in 1998 without any doubt: "The genome does not provide the blueprint for a living being, but only a map with an average scale" (Wieser 1998), but even that would still be a kind of blueprint. However, from the point of view of systems thinkers, the genome is not a blueprint at all. We will approach this radical idea in a moment (Section 3.7).

3.5.4 Weak Regulatory Couplings– Cells in Loose Conversation

Other cellular mechanisms are needed to initiate phenotypic variation. How can cells communicate with each other so that core processes can be recombined for evolution? What kind of cell signaling substances must be present so that new combinations of promising core processes can occur?

This is where that which Kirschner and Gerhart call "weak regulatory coupling" comes into play. It is weak in the sense that the biochemical specifics of a cell signal have only a weak relationship to the specific output on the other side. The receiver can be the same cell or a different cell. What exactly occurs in the target cell is determined by its own regulatory processes and not solely by the signal substance sent. Only at the target site is the response maximally prepared and ready for retrieval. However beware that the expression "weak" also implies the absence of "strength," i.e., there are "no strong couplings." What Kirschner and Gerhart mean by "weak" is that the types of connections can easily be changed evolutionarily to fulfill different functions. Take as an example the electrical outlets in a home. The system is highly adapted. All sockets are identical in construction, and the current is the same. The electrical current (information) that flows from the socket into your laptop does not need to know how this terminal device works. You can also plug a power cord into your dishwasher. It is the same current flowing, and it makes all your appliances work. Thus, in terms of facilitated variation, that would be weak information for a prepared process at the end of the information flow. The information "120 V alternating current" is even suitable for devices that have not been invented yet!

The reason why regulatory couplings are weak is that they are an indirect, undemanding, low-information type of regulatory connection that can easily be reversed or converted to suit other purposes. Thus, evolutionary changes do not require alterations in a highly integrative complex process—only in the intensity or site of action of simple signals.

To take an example from biology, if an organism reacts to the intake of sugar by increasing insulin secretion, then a weak, indirect coupling is present. There are even multiple weak couplings between sugar and insulin. The different reactions of the organism cannot be performed directly by the molecules of sugar and insulin. Even the registration of the blood glucose level itself consists of multilevel, weak couplings, as does the triggering of trembling or sweating in a diabetic when the blood glucose level is too low or even the secretion of glucagon, an antagonist of insulin, which occurs in extreme cases as a survival mechanism of last resort. Glucagon acts quickly to stabilize blood sugar again if, despite the aforementioned physiological signals, glucose is not supplied externally. These are only a few of the countless self-control mechanisms operating in the body—in this case, a metabolic process based on a complex chain of weak links. Only the stable core processes in the cells make it possible that such loose couplings have formed.

This has also been vividly formulated in the following way: Cells and genes do not have a cause-and-effect relationship to each other like the gas pedal and the car engine. The cells interpret the DNA in a kind of "consensus process," but they do not obey it. Cells and individual tissue areas are autonomous. They behave as a whole that interprets stimuli, but not like a machine that is unilaterally dependent on DNA commands. In such a scenario, evolution is made possible by the fact that the cell does not take direction from genes but interprets their instructions loosely. In this way, there is no need for a meticulous "plan" for all the details of a body feature.

The tendency of biological units to unite and move toward the construction of cooperative systems based on the division of labor can be observed along all lineages of evolution. Here, cooperation stands in contrast to egoistic behavior, as presupposed by the Modern Synthesis. Wolfgang Wieser also drew attention to this as early as 1998.

3.5.5 Compartmentation–The Modular Solution

During development, specific cells are formed for individual tissue types (skin, muscles, nerves, etc.). Initially, the cells are not specialized. Therefore, how does the specialization come about? Kirschner and Gerhart speak of compartments. By compartment, they mean a region of the embryo in which one or a few specific genes of the cells are expressed, and certain signaling proteins are produced at a certain stage of development. However, a compartment can also be defined at the cellular level with specific intra- and intercellular processes. The ability to activate differently conserved core processes at different locations in an organism and create these reaction spaces in the first place is called compartmentation. Which compartments do we know?

An insect embryo, for example, forms approximately 200 compartments in the middle phase of development. Science can determine compartment maps with the spatial arrangements of the compartments of an animal, the quasi framework for the arrangement and construction of complex anatomical structures. Each animal phylum has its typical map, which is highly evolutionarily conserved, much more than the detailed anatomy and physiology that emerges or builds upon a compartment.

The neural crest region at the edge of the central nervous system is a good example of a concrete compartment. The neural crest cells migrate throughout the body during development and proliferate or differentiate. Thus, bones, cartilage, nervous tissue, or even parts of the heart develop from originally undifferentiated cells of the same type. What exactly happens depends on cell signals and other factors. Just how spectacularly diverse the differentiation options are can be seen from the fact that the neural crest region in animals could give rise to such completely different head outgrowths as antlers, horns, or trunks. Kirschner and Gerhart also cite the enlargement of the human skull in the context of brain enlargement as an example of flexible, yet regionally determined, cell differentiation.

3.5.6 A Theory with New Informational Content

In summary, Kirschner and Gerhart argue that the generation of complex phenotypic change is facilitated by existing processes in cells that have been conserved for millions of years. This reduces the amount of genetic change required to generate phenotypic novelty. These same core processes are repurposed in ever-changing combinations. However, it may also be true that proteins are generated in places where they had not previously been produced, with genes that were previously active in other domains. On the one hand, the stable core processes severely limit the amount of variation; on the other hand, they create the needed space for phenotypic variation. Facilitated variation has evolved adaptively, that is, natural selection has promoted it (Figure 3.8).

In other words, facilitated variation means that phenotypic variation must be possible in development based on the construction of an organism itself. Evolution does not work effectively in the manner proposed by Darwin and the Modern Synthesis. According to Darwin and neo-Darwinian theory, every aspect of an organism is subject to possible mutation. This view must be updated to account for the fact that in the course of evolution, some things change and others do not. Changes have occurred in spurts over 3 billion years, interspersed with periods of conservation (stasis, Section 2.3). Basic, fixed processes are passed on to descendants in variable combinations. Stabilization and diversification are closely associated.

Carroll also speaks of conservation of Hox genes, and Wagner emphasizes the interplay between the preservation of existing processes and structures and possible innovation. Only Kirschner and Gerhart, however, transfer the connections to the cell, to groups of cells that send out and receive signals, and thus, to the interaction and cooperation between cells. Such signaling pathways allow the cell to act and react in its specific environment. Cell assemblies are the phenotype. We will learn about a computer simulation based on cell–cell interactions in Section 4.8. Using the concrete case of evolutionary limb development, we can thus understand which phenotypic patterns can be induced by cell–cell interactions.

The theory of facilitated variation extended the statements of the Modern Synthesis. Evolution is not a mutation–selection process occurring with equal frequency everywhere in the organism. Most importantly, the new theory is broader, richer, and more open. A theory can be judged either by its informational content or by its explanatory value. The theory of facilitated variation contains novel information about stable cellular processes and interactions among the different levels of organization of the organism.

The theory presented makes it clear that the origin of phenotypic variation is less dependent on very lengthy gradual Darwinian trial-and-error processes that may take far too much evolutionary time to be truly effective. Rather, Kirschner and Gerhart show us that organisms themselves play a major role in determining the nature and extent of variation. The organism is prepared for change. It has a "response-ready response system" by which it responds to mutations with highly organized processes. In other words, the organism has intrinsic properties to generate variation. Better yet, it has the capacity to generate ordered, intrinsically coordinated variation. In this way, the capacity for adaptation is greater and more flexible than in the conventional mutation–selection scheme. "Variation is facilitated largely because so much novelty is available in what is already possessed by the organism" (Kirschner and Gerhart 2005).

3.6 INHERITANCE IS MUCH MORE THAN MENDEL AND GENES: INCLUSIVE INHERITANCE

Eva Jablonka, professor emeritus of genetics (and an epigeneticist) in Tel Aviv, and her London colleague Marion Lamb have researched inheritance in depth. They also address taboo subjects without hesitation. For the first time in the history of the theory of evolution, epigenetic inheritance has been explained thoroughly and systematically from several perspectives.

Jablonka and Lamb target the statement generally propagated in the Modern Synthesis that genes are the true and only units of inheritance, and acquired characteristics cannot be inherited as described by Lamarck. On this point, all proponents of the Modern Synthesis agreed. However, Jablonka and Lamb expand on this narrow view by arguing that multiple routes of information flow to subsequent generations exist, all of which are relevant to evolution. The novel components of this view are genetic inheritance of nonrandom mutations and epigenetic inheritance (Figure 3.10). In addition, there is cultural inheritance in the form of learned behavioral inheritance and, in humans, symbolic inheritance through signs, writing, the Internet, etc. (Jablonka and Lamb 2014). All these forms are highly relevant to evolution and are grouped under the term "inclusive inheritance" (Danchin et al. 2011). Inclusive inheritance is one of the four central research areas in the EES, along with evo-devo, developmental plasticity, and niche construction (Laland et al. 2015). Of course, inheritance can also be treated as a subfield of evolutionary development, as is done in this chapter. Let us examine in detail the different forms of inheritance.

(a) (b)

Figure 3.10 (a and b) Epigenetic inheritance. The field of epigenetic inheritance has experienced a surge of interest in the last decades. In this context, parental care is an exciting topic. The interactions between genetic and epigenetic mechanisms contribute to parental care. Brood care is not limited to mammals, for example, in herds of elephants, where nonparent relatives also care for the young. The male strawberry poison frog (*Oophaga pumilio*) carries his tadpoles on his back, one at a time, to watering holes in the funnels of bromeliads. The mother provides the young with an unfertilized egg, which she lays in a watering hole of the plant and which serves as food.

3.6.1 Genetic Mutation Does Not Always Occur by Chance

Ambiguities already exist in genetics. On the one hand, identical genes can lead to different phenotypes, which we have come to know as plasticity. On the other hand, combinations of different genes or different levels of gene activity can produce the same robust phenotype. Genes cannot be considered in embryonic development without reference to their environment. In fact, there is a constant interaction between genes, their products, proteins, and their environment. The environment consists of not only the cell itself and neighboring cells but also of more distant cell tissues, organs, and the external world. Recall August Weismann's barrier between the soma and germ line. In the long term, the idea should not take shape in the form he imagined, namely that once information has left the germ line, i.e., has been incorporated into somatic cells, it cannot re-enter the germ line. Interactions between the somatic cells and the germ cells are a hot topic in research today. The idea is by no means new, but it is still not a part of the neo-Darwinian canon. After dealing with Waddington, it became less possible to ignore such interactions in the long run.

Against this background of broader interaction, we must examine "chance variation" more closely. One historical passage illustrates, particularly well, how positions on this have changed. I reproduce a passage written by the French Nobel laureate Jacques Monod in 1971. His book *Chance and Necessity*, for a long time, required reading for biology students. He says: "It necessarily follows that chance alone is at the source of every innovation, and of all creation in the biosphere. Pure chance, absolutely free but blind, at the very root of the stupendous edifice of evolution: this central concept of modern biology is no longer one among many other possible or even conceivable hypotheses. It is today the sole conceivable hypothesis, the only one that squares with observed and tested fact" (Monod 1971). The doctrine of the Modern Synthesis cannot be expressed more aptly.

The use of the term "chance" or "random" in the theory of evolution is ambiguous. The above picture of evolution as the result of random processes shall be revised. Monod is to be understood as saying that chance is integral to the synthetic theory of evolution. However, this can carry different meanings. First, it can mean that mutations are copying errors that occur during inheritance and cannot be eliminated despite elaborate, effective DNA repair procedures. Second, mutations are random in the sense that it is not possible to predict which gene will be mutated and at which location. Third, random mutation can be undirected with respect to its advantages or disadvantages for an organism. According to classical theory, the effect of a mutation on the fitness of a population (positive/negative/neutral) only becomes clear during the selection process.

A fourth and contrasting statement of the Modern Synthesis is that genetic change is random with respect to its (physiological) function (Noble et al. 2014). This view is now considered outdated. By means of various developmental mechanisms that we have come to know and that interact constructively at different levels, random mutation may very well produce functional alterations. The EES (on which more will be discussed in chap. 6) would even go so far as to say that development contributes to the functional integration of positive effects of mutations on the phenotype. This brings back into focus the genotype-phenotype relationship and particularly the physiology of the phenotype (Noble et al. 2014).

A critical examination of the synthetic theory reveals that the last two ideas of chance lead us on a tangent. We learn that genetic mutation and phenotypic variation can also have a bias. Jablonka and Lamb (2014) emphasize, "No longer can we think about mutation solely in terms of random failures in DNA maintenance and repair." However, fundamentally, mutations are not always biased either. The truth lies somewhere in the middle, the authors say. They cite several cases that give a better sense of the complex issue and shed new light on mutations.

It has long been known that bacteria show an increased mutation rate under stress. To be sure, each mutation can be viewed as random. The overall response of the genome, however, is a biased, adaptive process. Nobel laureate Barbara McClintock observed a similar phenomenon in plants, primarily in corn, beginning in 1948. Here, contiguous sections of DNA could be identified, so-called "jumping genes," better referred to as transposable elements or transposons, which are copied "in one piece" and moved to a different location in the DNA. McClintock interpreted this as adaptive behavior. For a long time, the scientist was not taken seriously, to put it mildly. Today, it is debated whether the increased mutation rate is a side effect, a so-called byproduct of the stress situation to which organisms are exposed. However, the fact that this results in an ordered copying of contiguous DNA segments makes one wonder. The fact is that 44% of the human genome consists of transposons or transposable elements, of which only a small proportion is active today. However, it is precisely this small proportion that is responsible for the genetic diversity of human populations (Mills et al. 2007).

The next case, local hypermutation, is more difficult to criticize. It occurs at a site in the genome where mutations are reasonably likely to be advantageous; these sites in the genome are called *mutational hotspot*. The genes in these regions code for proteins involved in important and distinct cellular functions. It appears that certain DNA regions have been virtually selected for mutation. Such striking local hypermutations have been observed in pathogenic bacteria, as well as snails and venomous snakes.

The third case deals with induced local mutations. Induced refers to the fact that they originated in the environment. Surprisingly, these mutations are observed at locations in the genome where they can help the organism cope with a change in the environment. Such mutations are known, for example, in the bacterium E. *coli*.

All three forms of mutation mentioned here have an adaptive and thus nonrandom potential, that is, they can arise during a selection–adaptation process for entire populations and promote fitness. Research on this topic is ongoing. Like Kirschner and Gerhart, Jablonka and Lamb also conclude that instructive processes for generating variation exist. The authors summarize: "It would be very strange indeed to believe that everything in the living world is a product of evolution except one thing—the process of generating [genetic] variation" (Jablonka and Lamb 2014).

An early proponent of nonrandom genetic variation is the American biologist, James A. Shapiro, the discoverer of transposable elements in bacteria. Some of the concepts mentioned in this section are based on his work. He introduced his own field of research, which he called *natural genetic engineering*. In his book *Evolution: A View from the 21st Century* (2022), he calls for a set of interactive and information-based evolutionary principles. Among other things, he states that living cells and organisms are cognitive, sensitive entities that act and interact for specific purposes to ensure survival, growth, and propagation. They possess corresponding sensory perception, communication, information processing, and decision-making abilities. Cells are built to cooperate, he says. They can rapidly change their inherited traits through *natural genetic engineering* and epigenetic processes, as well as through cell fusion. Evolutionary innovation, he says, emerges through new cellular and multicellular structures as a result of functions of cellular self-modification and cell fusion (Shapiro 2022). This leads very clearly toward a new view of the genome not only as read-only memory but also as read–write memory, toward a new notion of evolutionarily more flexible adaptive capacities and thus the role of the cell in evolution.

3.6.2 Epigenetic Inheritance–A Widely Discussed Topic in Scientific Literature and Media

Epigenetic inheritance has received more attention than evo-devo in public media. Epigenetic processes regulate gene expression. In simple terms, nongenetic modifications such as DNA methylation and chromatin remodeling (see Figure 1.4) determine whether the transcription enzymes can access particular genes. Such modifications are called epigenetic markers, and alterations in these due to external influences are termed epimutations.

From an evolutionary point of view, epigenetic inheritance is particularly interesting. In this context, the Lamarckian concept of inheritance of acquired characteristics comes into view. Nutritional studies conducted in Sweden found that if grandfathers were undernourished during their slow growth period before puberty, their grandchildren were four times less likely than the average to develop diabetes. The opposite is also true: grandchildren whose grandfathers were well nourished during that same period of life were significantly more likely to die from strokes and other vascular diseases. Different nutritional situations affect structural proteins that organize DNA into chromatin. This is a case of epimutation. Emma Whitelaw, an Australian molecular biologist and leader in epimutation research, epigenetically bred twin mice with different appearances (Figure 3.11). In 1999, she successfully demonstrated epigenetic inheritance by the next generation in agouti mice (Morgan et al. 1999), and in 2018, it was reported that abnormalities of mice traumatized in early childhood are detectable even after three generations (van Steenwyk et al. 2018).

In 2014, the discovery that paternal stress can be passed on to offspring via small non-coding RNA fragments known as microRNAs caused a sensation beyond scientific circles (Gapp et al. 2014). Sustained stress alters the microRNAs produced in the epididymis. One sperm alone consists of hundreds of different types of such microRNAs. These RNAs are transported together with the sperm to the egg cell, wherein they alter the RNA in the cytoplasm, disrupting the cellular processes of the oocyte. Such newly emerging life is thus epigenetically shaped by influences that may predate its conception. This mechanism is responsible for the fact that paternal trauma can be transmitted epigenetically to the following generation and beyond. Epigenetic inheritance can mean that children feel the traumas of their parents or grandparents without sharing any experiences with their ancestors.

In other contexts, however, it may also be the case that parental epigenetics provide for stress resilience, i.e., resistance, and that the following two generations are less susceptible to stress. Let us turn our attention to the epigenetic inheritance of positive experiences: In the past, people might have laughed at a mother singing to or talking to her child during pregnancy, convinced she was doing something beneficial. In the meantime, biological evidence has been found that the child is epigenetically imprinted—in a positive sense—by the experience.

Another example of epigenetic inheritance from among the many ways of responding to environmental stress (Ellis et al. 2017) is that stressed mother rats in the laboratory spend less time licking their young than nonstressed ones. In haste, one might judge these mothers to be neglectful parents; however, this would be incorrect. The young females benefit from love deprivation by gaining greater social dominance over those that have been licked more; they are more attractive to males and have a greater chance of reproducing. Males, in turn, become more involved in juvenile

Figure 3.11 Agouti mice. The discovery in 1999 that fur color differences and tail shapes in mice can be inherited epigenetically, without DNA-level changes, caused a sensation.

play fights when they have been licked less. As adults, they are more combative. Moreover, females that are licked less also lick their own young less. One would expect natural selection to be predictive. This is an example of adaptive stress resilience.

What is the evolutionary significance of this? Even with epigenetic inheritance, we are members of the same species; nevertheless, it is the small but significant steps and changes within the species that can only be investigated empirically. We speak here of intra-species evolution. It is difficult enough to achieve scientifically verifiable, stable results with humans, in which way a characteristic or a behavior is inherited over several generations—the generation gap is considerably large. Decades-long experiments are impractical and financially unfeasible. Consequently, it is the clear evidence of epigenetic inheritance mechanisms in the case here that motivates us to recognize and discuss possible evolutionary changes.

Today, several mechanisms of epigenetic inheritance are known, and they influence evolutionary development. It is assumed that epigenetic inheritance can also be adaptive under certain conditions, which is of course particularly interesting for evolutionary biologists. However, there is some debate as to how many generations of epigenetic inheritance can span. Take, for example, a study that examined the effects of two insecticides commonly used in agriculture on rat fertility. The chemicals injected into pregnant females during the early development of their embryos impaired the embryos' gonadal development. As a result, the male offspring developed oversized testes and less capable sperm. Amazingly, however, the males passed on the developmental defect to their own offspring via the germ line. The DNA was not changed, but the methylation of the DNA in the sperm of the offspring was changed. The epigenetic changes extended into the fourth generation (Skinner et al. 2010). This research result, which was published in *Science* magazine in 2005, created significant impression.

One could raise the objection that after four generations, the environmental influence described here has been overcome. However, this would be shortsighted, because to learn something about environmental influences on inheritance, we must consider that in the case of agrochemical substances, kerosene, commercially available plastics, or nutrient deficiencies, we are often dealing with permanent stresses in large parts of the population. In such cases, however, inheritance by means of alteration of epigenetic markers that do not follow the classical genetic rules of inheritance may well be evolutionarily relevant. Epimutation is now associated with an increased risk of obesity, changes in personality structure, and changes in social behavior in mice. Consequently, Michael K. Skinner speaks of the classical view of evolution as a rather inert product of random mutations. According to Skinner, this view must be expanded to include epigenetics with the possibility of faster evolutionary reactions.

The topic of epigenetic inheritance is a subject of ongoing research (Lind and Spagopoulou 2018). Multiple lines of evidence are provided that what is passed on to the offspring via the egg and sperm is more than just genes, as has been preached to us for nearly 100 years. Epigenetic markers and microRNA are the interfaces of the genome with the environment during organismal development. One of the most exciting topics in evolutionary research will be whether epigenetic pathways allow faster adaptations to environmental changes than random genetic mutations. Genetics and epigenetics may not yet see eye-to-eye. Nevertheless, epigenetics and epigenetic inheritance are becoming very influential in medicine and evolution.

3.6.3 Epigenetic Inheritance by Learning from the Ancestors

It may be questioned, take, for example, the silk bowerbird: how does the male of the species create his elaborate, beautiful bower (Figure 3.12)? Is the skill passed down through many generations? As it turns out, this is partially true. In fact, male bowerbirds watch other more experienced birds, experiment for years, and constantly improve their skills until they can build beautiful arbors (it is not a nest!) and exhibit extraordinary and complex courtship to attract females, which are very demanding. He then leaves many small gifts for her in front of the arbor, all in the same color. Similarly, the subsequent generation must learn the arbor-building technique, which is not handed down in detail from its predecessors. This form of inheritance of a learned behavior is predominantly nongenetic.

The media has repeatedly reported on amazingly coordinated, seemingly anticipatory

Figure 3.12 Great silk bowerbird with love arbor. The intricacies of arbor construction must be relearned by male birds in each generation; initially, they often experience rejection by females. In front of the arbor lie gifts for the wooed. The ability to create such a work of art is epigenetically inherited.

behaviors in animals. For example, a BBC documentary shows a perfectly synchronized pack of dolphins driving their prey fish from shallow water to the muddy shore, pouncing after them, and picking them up on land before crawling back into the water. Orcas work with perseverance as a team to tip seals off small icebergs by creating high waves. Such group behavior is passed on by older animals to younger ones. Finally, it should be mentioned here that tits, finches, nightingales, and other young songbirds learn to sing from their parents and other birds by imitation until they master their own plastic song with many complicated stanzas and numerous regional dialects. Their vocal apparatus, the syrinx, vastly surpasses ours. Evolution has enabled tits, with their small bodies and even smaller lungs, to sing continuously without suffocating. Of note, Aristotle introduced the term dialect (*dialektos*) while discussing variations in the songs of birds.

Reflecting on whether Lamarck has any lasting relevance in biology, the conclusion at the end of this chapter is that the nongenetic inheritance of learned behaviors, and, furthermore, the inheritance of symbols such as writing and language, are highly Lamarckian when viewed in this light.

I will discuss in Chapter 5 how the learned behavior of organisms, passed down through generations, changes their environment and, via feedback, helps determine their own evolution.

3.6.4 Decoupling of Evolution from Biology

Jablonka and Lamb extensively deal with humankind in the context of evolution. They view the ability of *Homo sapiens* to deal with symbols as unique.

Richard Dawkins. introduced the term *meme* in analogy to the gene. A meme is behavioral information, a content of consciousness, which can be exchanged between individuals (Dawkins 1976). A meme can be a new fashion, a style of clothing, or a new game. Memes can spread through a population very quickly, orally, in writing, or electronically, but they can also disappear just as quickly.

The nature, scope, and complexity of how we humans acquire, organize, and communicate information is unique. Our symbol-based culture is constantly changing. We hardly think consciously about the meanings of signs, pictures, signals, beeps, and cell phone sounds. These symbols are hereditary.

This view is not particularly Darwinian, but it is relevant to the evolution of humankind. The brain has opened up the possibility for humans to change their own evolution through adaptive cultural action. Moreover, humans are decoupling from biological evolution, as Konrad Lorenz recognized early on. As humans, we have taken our evolution into our own hands with medicine and genetic engineering. Soon—whether we want to or not—humans will overcome the hurdles of rebuilding their own genome with *genome editing*. However, mistakes are almost inevitable. We are already intervening in epigenetic processes where the genome fails, for example, in cancer therapy. With increasing medical knowledge, harmful mutations can be eliminated, diseases brought under control, or their outbreaks prevented. The selective forces of nature have long been pushed back by the technology available to humankind. Whether one calls humankind's intervention in evolution a natural process or not (see the interview with Eva Jablonka) remains a matter of definition and opinion. I had the opportunity to discuss this point with Eva Jablonka in person. In the process, the reasons for our differences in thinking became apparent; nevertheless, both points of view are legitimate. In this book, I make a strict distinction between natural processes and human interventions (which, of course, have evolutionary foundations). This distinction allows us new fields of consideration when it comes to our own evolutionary future (chap. 8).

3.7 THE MUSIC OF LIFE

Nothing is static. Systems biology is marked by change. The model of a gene regulation network can look very complicated, but the totality of biological complexity is greater. Such networks are often designed like computer circuit diagrams. However, computer schematics are deterministic; biology is not.

Today, we have reached a point where 500 years of thinking in linear models contrasts with nearly 50 years of only tentatively beginning cross-disciplinary thinking in complex models. However, the future belongs to the understanding of complexity. Complex models will find their way into all applied disciplines. They will use random processes. But beyond that, they will also deal with unpredictable, uncertain environments that cannot even be predicted by probabilistic modeling (Mitchell 2009).

Organismic systems biology is a young discipline that aims to understand complexity and causal interactions. One scientist who advocates for its necessity and fights for a new world view in biology is Denis Noble, an Oxford emeritus cardiovascular physiologist. From him, one can learn how infinitely complex a single biological function, such as the heartbeat, can be. After reading his slim book *The Music of Life* (Noble 2006), even a biologist is amazed at the intricacies of organisms and biology.

3.7.1 Questionable Determinacy

I am a type 1 diabetic and have been for more than 50 years. I am equipped with the most advanced systems available: a targeted, closed-loop system involving an insulin pump connected to my body. A sensor under my skin continuously measures the glucose level in my blood at 1-minute intervals. The pump and the sensor are connected via Bluetooth. Using my smartphone, I can monitor all my values using the most sophisticated platform. The smartphone in turn reports information back to the pump, for example, whether corrections to the insulin supply are necessary. If so, the smartphone independently doses corrections in amount and duration. The system makes automatic insulin adjustments every 5–10 minutes. If there is even a suspicion that my blood glucose level is above or below the predefined tolerance range, an alarm is sounded and—if I want it to—an automatic message is sent to my adult children wherever they happen to be in the world. In an emergency, they can call the emergency service, which will be at my apartment with a key in 15 minutes. This is an example of modern technology controlling a biological function. For as long as I can remember, I have been using my own intelligence to accurately predict my insulin requirements. But I have given up on trying to do so perfectly and live with very good compromises. Exact determinacy between the variables insulin amount, physical or mental exertion, health condition, carbohydrate intake, insulin action time, injection site and depth on the body, time of day, ambient temperature, etc., and the blood glucose level, to name only a subset of factors, does not exist. There are far too many interactions with the environment. I wanted to find out the exact correlation, but the project always failed. Under similar conditions, my blood glucose measured an hour later would be high one time and low another, but not where it should be. A molecular biologist would certainly explain to me at the molecular level what exactly is going on in the body during insulin delivery. What is clear up to this point is that lifelong concentration and discipline are necessary, but not sufficient, to achieve truly useful compromises when one has diabetes. Perhaps the generations of pancreatic systems that are just now emerging with artificial intelligence will be a great improvement. However, this can be viewed critically.

3.7.2 The Genetic Program Is an Illusion

Noble makes it unmistakably clear: A genetic program does not exist. To explain biology with a genetic blueprint is a mistake. The genome is not the "book of life" from which physiological functions can be read. This was how it had been interpreted even after the decoding of human DNA at the beginning of the new millennium. The old reductionist explanatory model of synthetic evolutionary biology is a one-way causal chain of genes => proteins => cell signals => subcellular mechanisms => cells tissues => organs => environment. We certainly need such monocausal one-way streets to navigate our daily lives. Science today, however, takes on a different appearance. Noble rejects this approach, as well as those that replace the genome with proteins (the proteome) or other entities. Such approaches only lead to a shift of the same thought pattern to another level. Molecular genetics tells us little about life, and life is not a "protein soup" either. We must learn to think in reverse and stop imagining that gene expression is comparable to the reading and playing of a CD. The challenge in our century, according to Noble, is to understand that the genome is "read" by the phenotype, not vice versa. However, even that does not quite hit the mark, because science has made great strides in underlaying protein sequences with coded DNA sequences. "But sometimes we seem to have forgotten that the original question of genetics was not what makes a protein, but rather 'what makes a dog a dog, a man a man.' It is the phenotype that stands in need of explanation. It is not just a soup of proteins" (Noble 2006).

With this drastic, naturally simplified picture, Noble strikes at the core of what has already been addressed in the preceding chapters. Today, evo-devo research in a systems context approaches evolutionary development in the way that Noble demonstrates. Kirschner and Gerhart agree with this (Section 3.5), as do Jablonka and Lamb (Section 3.6), Gerd B. Müller (Section 3.9), and Armin Moczek (Section 6.5).

There is no biological function that arises from the coding of a single gene, and each gene in turn is involved in numerous biological functions. Physiological functions are not found at the gene level. Genes, Noble argues, are blind to what they do. Proteins, cells, tissues, and organs are similarly blind. None of these levels determine function by itself. For a time, it seemed obvious to shift the focus from genes to proteins. But even that is insufficient, because biological functions also require molecules that are not encoded by genes, such as water or lipids (fats). "A lot of what (the) products (of genes), the proteins, do is not dependent on instructions from the genes. It is dependent on the poorly understood chemistry of self-assembling, complex systems." We will address self-organization in a later section.

3.7.3 We Inherit Cellular Machinery from Our Mother

Noble also voices what cannot be said often enough: We inherit more than just our parents' genes. We inherit a complete, fertilized egg cell from our mother (zygote). This cell contains the machinery necessary for DNA to function. "The cell is an evolved structure which, far from being assembled through instructions contained in the DNA, is a product of several billion years of evolution" (Dupré 2010). This machinery is one hundred percent prepared to enter the nucleus and begin its work, the transcription of parental DNA into RNA and the translation of RNA into proteins. These proteins are, therefore, encoded by the maternal genome. The elements provided by nature, primarily water, are usually not mentioned at all in this context.

"Without genes, we would be nothing. But it is equally true to say that with only genes we would also be nothing" (Noble 2006). According to Denis Noble and John Dupré, understanding inheritance as the inheritance of genes is an outdated dogma. "Today, more than 50 years after the discovery of the double helix, it seems that the identification of the genetic basis of inheritance has separated this necessary component of the process from the rest, equally necessary" (Nowotny and Testa 2009). What is lost and rendered invisible in the process, they argue, is the context in which genes function. In the same tenor, they say elsewhere that the order of genes and that of the products they realize, i.e., the order of the genotype and phenotype, have been separated during the development of molecular developmental biology (Müller-Wille and Rheinberger 2009).

As a consequence of Noble's assertion, the picture of a one-way street of causality leading from the genome to the organism would have to be replaced by one that emphasizes the interactions of the levels (Figure 3.13). This picture also has arrows in the opposite direction, from the organism to cells and from cells to genes. In the new image in Noble's mind, however, the genome is not replaced by another level of organization, such as the cellular level specified by the work of Kirschner and Gerhart.

3.7.4 Against Reductionism?

First, Noble wants to make the reader aware that reductionism in the form of gene centrism in a system model is not discarded only to be replaced by another model. At this point, he points out a certain slippery slope: a systems thinker who engages in rigorous systems-level analysis need not discard the meaningfulness of successful reductionist models. Rather, he uses the explanatory power of such models to integrate them into his own. By contrast, reductionists, Noble argues, usually claim intellectual hegemony. Models are not models of something; rather, they are models for something. If models are misrepresented in the first way, they are assumed to have a "partial or complete structural identity of a model and what is being modeled." However, in reality, models are simplifications (Honnefelder and Propping 2001). This is how even highly interacting systems biology models should be understood: they can and will always only sketch the complexity of the true living world in a limited way.

We now know what Noble does not accept, namely a bottom-up approach. But what view does he espouse? What is the positive core of his system idea? Genes and proteins do not "know" what they do for higher levels

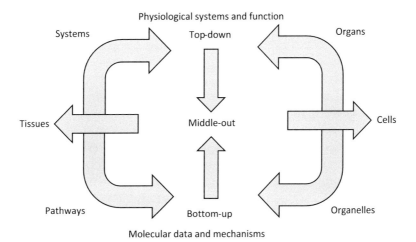

Figure 3.13 The middle-out pathway. According to this proposal by Nobel laureate Sydney Brenner, all cause–effect directions for understanding biological functions are equal and coexist.

of function. The individual parts of a system perform their tasks without knowledge of the overall system. Thus, the cell in the early bud of the hand in vertebrate development does not know, of course, that it is to become part of a hand. I will return to this later when I present an evo-devo computer model of the hand.

Is a top–down approach more appropriate here? With such models (which were no less reductionist than the bottom-up approach), physiology was able to achieve great success, for example, with the explanation of oxygen transport in the blood. One could identify red blood cells, hemoglobin, and many more components. Noble makes it clear with explanations that once one has reached the molecular level in thought, one may believe for a moment that one has finally achieved the goal of understanding biological function. However, ultimately this belief gives way to the realization that one must follow the upward path again. It is equally unsatisfying to assign an overall explanation to one causal chain or the other, Noble concludes. A biological system functions differently. Analyzing the biochemical process of how insulin regulates blood glucose does not help me understand exactly what to do and when to do it in order to manage my blood glucose well.

3.7.5 Biological Function from a System Perspective

Noble picks up on an idea from developmental biologist Sydney Brenner, who died in April 2019 (just as I was working on this chapter) and was awarded the Nobel Prize together with colleagues in 2002 for the discovery of apoptosis, or programmed cell death. Brenner, preferentially classified as a strict reductionist, proposed the middle-out pathway (Figure 3.13). According to this hypothesis, biological functions take place at every level. Each level—be it cells, proteins, or genes—can be used as a starting point for a causal process or model. In an interaction model, there are no preferred alternatives. Consequently, there are both bottom-up and top–down directions of explanation, as well as causal chains in any other three-dimensional direction, and all paths are equal.

The Viennese zoologist Rupert Riedl, always thinking in a systems context, anticipated this to some extent when he wrote, in *Order in Living Organism—A Systems Analysis of Evolution* (1978), "Interdependence proves [...] to be a form of order that permeates the whole organism. It extends from the control of the adaptive possibilities of the individual traits to those of the organismic strains and from the regulation of the individual dependencies to the image of the harmony of the whole individual."

Every biological function—the heartbeat, breathing, insulin secretion, etc., operates on this basis in a healthy person, and this is also how evolutionary development works. All this is explained with many metaphors from above, from below, from levels, reductions, and machines. Without metaphors, the project is not feasible, says Denis Noble. I have thus inevitably chosen the same way as his to explain it.

In his most recent book, *Dance to the Tune of Life* (Noble 2016), the then 80-year-old Noble once again outlined a determined and courageous overall critique of 20th-century gene-centric biology. In clear language that few of his colleagues are capable of, he sums up his teachings on life: (1). Genes are passive templates. The genome is not a program. The error of the gene-centric approach is to view genes as causative and controlling. (2). The DNA sequences are used by the organism, not vice versa. The organism tells the genome which proteins to make. Patterns of gene expression are determined by the organism, not the genome. (3). The environment does not exert a passive influence on variation, but a direct one. (4). Organisms help direct the evolutionary process. Evolution (variation) is thus neither random nor "blind," as the Modern Synthesis conveys. Accepting this, according to Noble, is the greatest hurdle to

In Conversation with Denis Noble

Denis, please briefly describe what systems biology is.

Systems biology is an *approach* to biology. We don't just study biological components in isolation. That kind of reductive analysis is important, but it does not address the question how those components relate to other components. Those relationships are processes, not things. Systems biology is therefore concerned with processes. The heartbeat, for example, is a process. Since organization at higher levels in organisms necessarily constrains the processes that can happen at lower levels, systems biology is ideally multilevel. Thus, the pacemaker model I developed first in 1960 connects processes occurring at both cellular and molecular levels.

As a young student at the age of 22—you did not have your PhD—in 1960, you published an article in *Nature* regarding your understanding of how the heartbeat develops. What fascinated you about the heartbeat and what did you learn by studying this specific topic?"

I was also fascinated since the heartbeat is so important to life. The idea that I might get the experimental information necessary to formulate the first biologically-based mathematical model was very attractive. But I encountered a major problem. The guardians of the huge computer (a Ferranti Mercury) did not think I knew enough mathematics to do it. So, I spent a few months attending the maths courses for Engineers at UCL. That enabled me to convince people that I knew enough maths to program the computer, but it also taught me something I did not know. Differential equation models require initial and boundary conditions before it is possible to obtain solutions to the equations.

I succeeded in developing the model to explain heart rhythm, and that led to the two publications in *Nature*. But it also set me thinking.

That makes me curious to ask further. Scientists want to explain their object of investigation as simply as possible. Instead, you describe biology and evolution as "complicated." Why?

Exactly so. Scientists want simplicity where they can find it. So did I. I wanted the simplest possible equations to explain heart rhythm. But where did those initial and boundary conditions come from? Were they just arbitrary constants in the equations? The answer is that they can't be arbitrary. Most of the possible constants would not work. The specific constants come, of course, from the biological organization at higher levels. The cell, in all its unimaginable complexity, is where the constants are determined. So, I ended up not explaining heart rhythm simply from molecular level knowledge of how channel proteins work, but rather how those proteins are also constrained to do what they do by the higher level organization. I did not know it at the time, but that fact eventually became the fundamental reason why I doubted the way in which molecular biology, was answering the question "what is life?" When Francis Crick formulated the Central Dogma (of molecular biology, the author) he said that the "Secret of life" came from DNA sequences. I knew for an experimentally-determined fact that this could not be true.

Twenty-five years later, in 1985, when much more complex mathematical models of the heart became possible, we also found that the complexity involved is in fact necessary to produce a robust pacemaker mechanism. By robust, I mean that it is not sensitive to *particular* genes if they are knocked out. Modern genome-wide association research shows why that is true. For each biological function, there can be as many as hundreds of genes that are associated with that function.

Does that mean that the reductionist method of dissecting material objects (e.g., phenotypes) into their smallest constituent parts (e.g., genes) is outdated?
No. I think reductionist science serves an important function. It is brilliant at working out how biological components, proteins, networks, cells, tissues organs, etc. work *in isolation*. By definition, the alternative *integrative* approach is necessary to understand how biological components work *together* in networks.

The Modern Synthesis originally pursued a population genetics approach that relied on abstracted, mathematized genes and simple causality, with the genes serving as the only information carriers. Today, we say that organisms help direct the evolutionary process and that the environment plays an active role in it. Why cannot the two theoretical approaches, which differ remarkably, coexist?
They can coexist. But that co-existence depends on the fact that they refer to completely different definitions of a gene. Population genetics developed before it was known that DNA is the genetic database containing template codes for making proteins. The pioneers of population genetics were working with a *functional* definition of a gene, not a molecular biological definition.

Wilhelm Johannsen introduced the functional definition of a gene in 1909. It was a functional definition since it was defined in terms of the functioning phenotype itself. Essentially, this was also Mendel's definition, although Mendel did not use the word 'gene'. Peas can be wrinkled or smooth, green or yellow, etc. They both thought that something in the organism determined which character was displayed by the organism. Johannsen specifically called it *ein etwas*, meaning *anything* that determined which characteristic is displayed. It wouldn't actually matter whether that was DNA, RNA or some property of a network. In some cases, that can be DNA.

Cystic fibrosis is a good example of a disease state that depends on a single gene. But in most cases, it requires many DNA sequences and much more than DNA itself to specify which characteristic is displayed. All the epigenetic processes can be involved too. So, we have:

a. functional definition: A gene is DNA plus epigenetic factors plus any environmental factors involved
b. molecular biological definition. A gene is DNA coding for specific proteins.

The two are clearly different. If you try to make them the same, we end up in a great muddle. This is what Richard Dawkins. does. When pressed on the existence of contributions to inheritance outside DNA itself, he says "well, that's OK. If it really contributes to inheritance then it can be welcomed as an 'honorary' gene". That makes nonsense of his distinction between DNA as the replicator and the rest of the cell as just the vehicle.

Where and when did the erroneous concept of the genome as a "program" emerge?
The problem began with Schrödinger's *What is Life?* in 1943, with his idea that the genome can be compared to a crystal. A crystal grows (replicates) accurately in a determinate manner. DNA replication is not like that (see Noble 2017, the author). The replication

process produces millions of errors which require a complex mismatch error correction process. The comparison should not be with a determinate program. DNA is just a database, and like all databases, it needs careful maintenance.

What do our offspring actually inherit from us? Do I hear Lamarckian undertones?
Yes. There are now whole books on transgenerational epigenetic inheritance. Physiologists like me are also familiar with all the evidence for paternal and maternal effects. Now that we know that exosomes correspond to Darwin's 'gemmules' I think the wheel has come full circle back to Darwin's acceptance of the inheritance of acquired characteristics. That is why I say that Darwin was not a neo-Darwinist. Anyway, he died (1882) just before Weismann formulated his Barrier idea (in a lecture in 1883).

You emphasize the ability of biological systems to self-organize. Surprisingly, the EES program does not include self-organization capability. Why? How important is Alan Turing's discovery of self-organization ability in evolution?
I think self-organization and agency are the key factors that mean that we need a break with the spirit of neo-Darwinism. The break, as I see it, involves accepting two non-neo-Darwinist assumptions: (a) that organisms can choose, they are not determinate machines; (b) inheritance of acquired characteristics. Darwin accepted both of these. Agency features in his sexual selection theories, while his theory of gemmules explains how he imagined Lamarckism to work (see Noble 2019).

In *The Music of Life*, you say: "The original question of genetics was not what makes a protein but rather 'what makes a dog a dog, a man a man'. It is the phenotype that stands in need of explanation." This sentence became the motto of my dissertation. Could you please elaborate in more detail what you mean by this?
The phenotype is not just the vehicle for transmission of DNA. It is both the target and the means of evolution. It is the target because, for multicellular organisms, it is the organism itself that survives or dies. DNA just follows what happens to the organism. It is the means because the organism is in control both of expression of DNA and of DNA changes when organisms are under stress. Life is no more determined by sequences in DNA than is my text determined by the QWERTY keyboard on which I am writing this text. It is the text that has meaning, not the tapping of keys. It is the phenotype that has meaning, not the DNA sequences. The organism gives meaning to the DNA.

When and how did you realize that the conventional explanation of evolution is inadequate?
My first doubts go back to 1960 and the use of circular causality to explain heart rhythm. That introduced me to multilevel causation. Everything is not caused by the molecular level alone. Then in 1976 when I organized the first debate on *The Selfish Gene* (Dawkins 1976, the author). Dawkins was asked by the philosopher Anthony Kenny whether it would be possible to understand Shakespeare if all one knew about the English language was its alphabet. Richard replied: "I am not a philosopher. I am a scientist. I am only interested in truth." He either did not understand the question or did not wish to answer it. He lost me at that point.

Then in 1985 when I first realized that the robustness of the heart pacemaker is a function of its complexity which gives it the ability to buffer genetic changes. That is a property of networks not of genes. Then to 2009 when I chaired the debate between Lynn Margulis and Richard Dawkins.. Richard was the cleverer debater. But to my mind his reactions to Lynn's great discoveries on symbiogenesis (the fusion of two or several different organisms into a single, new organism, the author) seemed, once again, to miss the point.

Do you go so far as to say that the standard model of evolution is wrong?
Yes, because it insists that it is right! That is a cryptic remark. So, let me explain. Neo-Darwinism is a simplification of what Darwin thought. That simplification was presented as necessary truth. But outside of mathematics and logic, there can be no necessary truths. All scientific hypotheses are just approximations to the truth. Weismann, like the later neo-Darwinists, insisted that the Weismann Barrier prevents the inheritance of acquired characteristics. We now know that the 'barrier' is no longer a barrier. They also insisted that random variations in the genetic material are the sole cause of change.

By contrast, I think that organisms themselves harness and use randomness. Organisms and their interacting populations have evolved mechanisms by which they can harness blind stochasticity and so generate rapid functional responses to environmental challenges. They can achieve this by re-organizing their genomes and/or their regulatory networks. Evolution does therefore have partial direction. The direction is by organisms themselves. Harnessing stochasticity is the way in which organisms become free agents. A free choice is both unpredictable in prospect and rational in retrospect (see chap. 4.8, the author).

Lamarck and Darwin both produced theories to explain how acquired characteristics could be inherited. They realized that something would need to transmit information from the soma to the germ cells. Their theories were very similar (see Noble 2019, the author). The MS actually excluded two very important ideas that Darwin espoused: inheritance of acquired characteristics and the agency of. organisms, as exemplified by sexual selection. In my view, the Modern Synthesis made a major mistake in excluding them.

Could one expect a consistent theory to emerge in a field as complex as evolution? Will there not always be new perspectives to describe complex realities?
I doubt it. I think we will find that Nature has exploited whatever works best at each and every stage. Evolution was not the same mechanism before DNA evolved. It changed again when it did. It changed yet again when eukaryotic cells formed through symbiogenesis. It changed again when multicellular organisms developed. It is worth remembering that the Weismann Barrier, one of the cornerstones of the Modern Synthesis, was not even relevant before the last stage. The Modern Synthesis, even if true, would only have been relevant to perhaps 20% of evolutionary time. So, I think we will find we have a patchwork of mechanisms, all of which interact.

Denis, thank you for this very interesting interview.

overcome in the current understanding of evolution. A summary of Noble's thoughts on the topic can be found in Noble (2015).

The idea of organismic systems biology represented here by Denis Noble relies on integration to the same extent as do the teachings of Müller or those of Kirschner and Gerhart. With this approach, modern systems models are created, and attempts are made to understand biological organisms in their entirety. This branch of biological sciences aims to obtain an integrated picture of all regulatory processes across all levels, from the genome to the proteome and organelles to the behavior and biomechanics of the whole organism. However, in contrast to the aforementioned organismic approach, numerous universities today understand systems biology as a purely molecular discipline. In this case, the focus is on complete DNA sequencing and thus on the totality of all gene regulations in a cell. For example, cellular metabolism is investigated, and it is asked which genes contribute to this and in what form. The corresponding gene regulation processes are then simulated in complex computer models.

In 2011, while writing my dissertation on the topic of evo-devo mechanisms in polydactyly, I asked my supervisor how consistently complex views and systems thinking were reflected in master's theses or dissertations. His answer did not surprise me. There is a lot of talk about it, he said, but when it comes to dealing with concrete topics in scientific papers, unfortunately, it still often comes down to classical deterministic explanations. I was therefore pleased that a cellular, nondeterministic, complex model with 20 million simulated, interacting cell reactions about the embryonic development of the vertebrate limb found its way into my dissertation and was published in an Oxford journal edited by Denis Noble (Lange et al. 2018).

At the end of this chapter, I would like to pose the question of what life really is. Clever minds like the Nobel Prize winner Erwin Schrödinger and others have attempted to explain this. But it is true that what really constitutes life is not, indeed cannot be, revealed to humankind. Denis Noble was aware of this. His book title *The Music of Life* is a beautiful metaphor that conveys a noble aspiration. Noble knows the extent of his explanation. Humankind is incapable of expressing with the brain and language what life actually is. I do not know what makes me feel alive and enjoy spring, and maybe that is a good thing. Let us remain humble.

3.8 PHENOTYPE PLASTICITY AND GENETIC ASSIMILATION: *GENES ARE FOLLOWERS*

As promised, I will now return to eco-evo-devo and take a closer look at the revised role of the environment in comparison with that in the Modern Synthesis. These are points 7 and 8 in the third block of Müller's list of questions on evo-devo as a discipline (Section 3.4). The environment takes on a new role from the perspective of evo-devo. One may say that the environment "makes the music" in evo-devo. Thus evo-devo becomes ecological evolutionary developmental biology, eco-evo-devo. The environment has always been thematized in the theory of evolution, say the representatives of the Modern Synthesis. It is true; but there is a fundamental difference between ascribing to the environment a passive selection role and an active, shaping role that directly changes the organism and its descendants in evolutionary development.

I would like to name a few scientists who fundamentally contributed to revising the role of the environment in evolution. A pioneer of the debate was Conrad Hal Waddington, who early on awakened understanding of the connection between the environment and genetic downstream fixation with the term genetic assimilation (Section 3.3.1 "Genetic Toolkit"). Then, there is Mary Jane West-Eberhard's (Figure 3.14) contribution with her book *Developmental Plasticity and Evolution* (2003). She approached the complex topic from different

Figure 3.14 (a) Mary Jane West-Eberhard, (b) Scott F. Gilbert.

angles and created an 800-page basic work. In 2009, the textbook *Ecological Developmental Biology. Integrating Epigenetics, Medicine and Evolution* by the two authors Scott F. Gilbert (Figure 3.14) and David Epel was published. Important work has also been conducted by Fred Nijhout, who teaches at Duke University in North Carolina. He deals with complex traits whose variation is caused by numerous genes and environmental factors and whose inheritance does not follow Mendelian rules, as stated on his website.

Armin Moczek, a professor at Indiana University Bloomington, is the leading expert in horned beetles (*Onthophagus*). He and his team lead the world in cumulative evolutionary knowledge about these animals. Moczek knows why these beetles have horns and how they came to be, why the horns are large and why, of all things, the males with the largest horns have the smallest penises. Horns, whether attached in pairs to the head or centrally to the thorax, are evolutionary innovations. The paths of their respective emergence are highly variable (Moczek 2008).

Incidentally, Moczek's team always studies the entire life cycle of a species. Evolution does not transfer one adult form into another. In between lies the developmental path of an entire life, and this is naturally seen as integrated into the environment. Anyone who has heard Moczek deliver a live lecture on the horned beetle. will never forget it.

From 2010 onwards, the concerns of the ecology side, specifically the eco-evo-devo side, in terms of additions, reinterpretations, or corrections to the Modern Synthesis can be gleaned from the American scientific literature. I adhere closely to two reviews in this section, one on phenotypic plasticity by Armin Moczek et al. (2011), the other focusing on genetic assimilation by Ian M. Ehrenreich and David W. Pfennig (2016).

3.8.1 The Environment in a New Role

The environment has always played a major role in the theory of evolution: the reader will agree that changing environmental conditions always drive evolutionary processes, according to the Modern Synthesis. Strictly speaking, according to the Modern Synthesis, the first requirement is that suitable mutations are already present in at least some individuals; indeed, they must be present so that

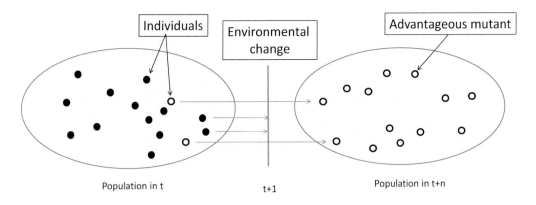

Figure 3.15 Adaptation according to the synthetic evolutionary theory. Mutations (white) are present in the population at time t in a few individuals. A change in environmental conditions at t+1 leads to the selection of these mutants, which have a reproductive advantage over their predecessors. The population at t+n consists mainly of mutant individuals. It is thus adapted. There is no active influence of the environment on the process.

new environmental conditions can affect a population and achieve adaptation. Suitable mutants are the foundation for new adaptation. This was impressively demonstrated by the Luria-Delbrück experiment with E. coli (Figure 3.15). However, here, the environment itself is not a factor that actively influences evolution. The Modern Synthesis overlooks the interactions between the internal and external worlds. Therefore, we are introduced to a novel line of reasoning. For evo-devo, evolution is not realistic without the active connection between organisms and their environment. The environment changes organisms, and the Modern Synthesis is a theory of genes. What is missing, West-Eberhard argues, is a theory of phenotype and phenotypic change. The phenotype, she continues, is a creation of both the genotype and the environment. It is thus the mediator of all genetic and environmental influences on development and evolution. Let us now examine these one by one.

3.8.2 Developmental Plasticity

Phenotypic plasticity is the ability of an individual organism to change its phenotype in direct response to environmental stimuli or inputs. Plasticity, in this view, is an extensive functional response of organisms that helps them cope with an unpredictable environment (Laubichler 2010). We are familiar with a variety of examples of animals and plants.

The environmental factor is undisputed among evolutionary biologists. However, the key role of phenotypic plasticity in evolution is discussed. If the focus is on development, it is better to speak of developmental plasticity. Developmental plasticity is one of the four central research areas in the EES, along with evo-devo, inclusive inheritance, and niche construction (Laland et al. 2015). I treat the topic here under the evo-devo heading of Chapter 3 because plasticity requires consideration of development.

The variability in characteristics can be low or high. Growth differences, for example, can range from inconspicuous to drastic. The skinks (*Scincidae*), a species-rich lizard family in Southeast Asia, are a good example. They vary in size and, their extremities range from short to absent. Another typical example is color variations, such as those found in the African noble butterfly *Precis octavia* (Figure 3.16). It is orange in summer and blue in winter. In the bluehead wrasse (*Thalassoma bifasciatum*), a fish species, females

(a) (b)

Figure 3.16 Phenotypic plasticity. The African butterfly *Precis octavia* demonstrates seasonal plasticity with a summer form (a) and winter form (b). Both are produced by the same genome.

change into males when the latter are absent. The tiger salamander (*Ambystoma tigrinum*) only leaves its larval stage for metamorphosis when its humid environment is no longer suitable for inheritance.

Developmental plasticity destroys the image of genetic determinism, i.e., the idea that the genome precisely determines the phenotype. This gene-centric view is typical of the Modern Synthesis. However, the higher the reaction norm, i.e., the possibility that the phenotype reacts to environmental influences, the less the phenotype is determined by its genotype.

Different types of environmental factors can cause plasticity, such as temperature, nutrition, or even parasites. Temperature often affects the phenotype via enzymes, as almost all enzyme activity is temperature dependent. Food contains chemical signals that can affect the phenotype. A change in light intensity can stimulate plants to form different leaf shapes. All of this has been known for a long time, but the question is why plasticity can promote evolution.

Let us begin with one extreme: Suppose a species completely lacks plasticity, and then environmental influences act on it. Such influences are always present in the form of fluctuation. However, fluctuations destabilize an organism if it has not evolved to be able to absorb them. If its phenotype is evolutionarily prepared for such influences by selection, it reduces the mismatch with its environment. It is plastic; the better the adaptation, the greater is the fitness of the species. Normally, it is not a fixed trait that is the best solution to environmental demands, but rather a more flexible endowment that allows an organism to perceive signals from the outside and respond accordingly. Plasticity will therefore evolve if the cost of doing so is not too high and does not cancel out the fitness advantage. Natural selection will ensure that plasticity emerges. Natural selection favors genotypes with plasticity as the superior response to a changing environment. Therefore, plasticity is found in some form in just about every natural species (Ehrenreich and Pfennig 2016).

3.8.3 Concrete Plasticity—A Goat on Two Legs and Tunnel-Digging Beetles

The discussion up to this point has been very theoretical. Let us now ask the specific question: What does a domestic goat have to do with phenotypic plasticity? Repeatedly, striking phenotypes have been described, including the fascinating bipedal goat. This animal had front-leg paralysis, which forced it to learn to hobble upright on its hind legs. The environmental factor was an infectious disease or an accident. The peculiarity was not so much the goat's bipedal posture, but the adaptation of the pelvic floor and muscles to it. Not only a larger muscle but also a new tendon developed at a point that connected the extended thigh muscle to the pelvis. Simultaneously, the shape of the ischium of the pelvis and the bones of the hind legs changed. Such an extensive morphological change could not be expected, certainly not within 1 year, after which the animal died. A high capacity for phenotypic integration made the involved components functional for the new task without any genetic variation (West-Eberhard 2003; Gilbert and Epel 2009). West-Eberhard introduced the term phenotypic accommodation for this: phenotypic plasticity enables organisms to evolve functional phenotypes. They manage to do so even in the face of major perturbations. Kirschner and Gerhart described this similarly with facilitated variation (Section 3.5).

Beyond the goat, empirical examples of populations with plastic behavior are necessary if one wants to draw conclusions about what plasticity can mean for evolution. The following is an example from the wondrous world of beetles. Have you ever observed a dung beetle at work? To be honest, neither have I. This is probably because beetles are rarely seen anymore. Yet there were so many a few decades ago—the loss is disturbing! Even more fascinating are the stories that exist about them, such as the following.

Female bull-headed dung beetles (*Onthophagus taurus*) dig holes of various depths where they deposit the brood ball of cow or sheep dung, which contains an egg. The beetle larva feeds on the dung in the ball, which is a perfect nest. At first glance, this does not seem significant; however, fascinating new evolutionary connections have been discovered here. A team led by Armin Moczek first found that the depth of the tunnels, which ranges from 10 to 40 cm, affects the size of the larvae. The hotter it is at the surface, the deeper the female dig tunnels. Deeper in the earth, where it is cooler, the beetle larvae grow larger. Larger larvae also grow into larger adults and lay more eggs than smaller ones. The unexpected effect is that the smaller beetles, i.e., those for which tunneling was not deep enough and who were born closer to the surface, invest less in their own children regardless of the temperature; they dig less deeply themselves and have smaller children. All this occurs without the presence of heat as a stress factor for the second generation of children. Thus, only a one-time temperature stress is required to affect development over several generations—an epigenetic mode of inheritance with significant effects. At the same time, this is a nice example of a devo-evo effect (Section 3.4), i.e., evidence that the earlier development of individual organisms can influence subsequent evolution.

However, if temperature stress is always present, as in Western Australia, where it is often very hot, and the population density of dung beetles is also high, then other evolutionary mechanisms come into play. Although there is high competition for food resources, the females dig less deeply despite the heat; nevertheless, they have large offspring. we conclude that the evolution of plasticity, in this case, the tunnel depth, can be very rapid (Macagno et al. 2018). We will return to dung beetles later (Section 5.2) because they have much more to offer in evolutionary terms.

3.8.4 Genetic Accommodation

During evolution, plasticity may be further enhanced or reduced or may disappear altogether. Evolutionary biologists have long suspected that such amplification or attenuation of plasticity is possible in the presence of environmental perturbation and is an important prerequisite of innovation. This is quite possible, for example, in the context of an environmentally induced trait that is quantitatively genetically regulated. Quantitative gene regulation occurs when a trait's expression depends solely on the intensity of expression of one or more genes. On this basis, natural selection can easily influence the expression of the original environmentally induced trait in a bidirectional manner, i.e., by strengthening or weakening it. Today, this process is called genetic accommodation" (West-Eberhard 2003).

When phenotypic accommodation, as described above for the bipedal goat, is followed by genetic accommodation, plastic responses in development can be induced by environmental changes and subsequently influence significant genetic changes evolutionarily. Thus, the trait becomes adaptive because genetic control is assumed for its expression, which typically occurs in several stages.

If in the extreme case, plasticity disappears completely, and the genome has complete control over the expression of the trait in the direction of a fixed phenotype, we speak of genetic assimilation as a special form of accommodation (Minelli 2015). The trait is then genetically fixed and robust to the environmental factor. Genetic assimilation is thus equivalent to acquired robustness for a specific case. The environmental stimulus, which initially has sole control over a trait and must be present permanently over several generations, is no longer required once genetic assimilation has occurred (cf. the example of tobacco hornworm, Section 4.5). Gilbert and Epel (2009) emphasize that the mechanism of genetic accommodation or assimilation can occur frequently because several traits are influenced by both the environment and genetics.

This mechanism was first described in detail by West-Eberhard (2003). She formulated the provocative hypothesis that *Genes are followers*. In other words, it is not mandatory that genetic change precedes phenotypic change. This may be cautiously understood as an alternative to the Modern Synthesis. The cautious West-Eberhard quotes: "The idea that genes can directly code for complex structures has been one of the most remarkably persistent misconceptions in modern biology" (Nijhout et al. 1986).

It stands to reason that a genetically accommodated or assimilated trait can represent a novel phenotype in the interplay of the environment and gene expression. If the process evolves from a persistent environmental stimulus, and quantitative genetic control can be adopted, there is an opportunity for evolutionary innovation.

3.8.5 Environmental Influences Or Mutations—Which Is More Likely?

There is evidence that the initiator of phenotypic change can originate with the environment and not with genetic mutation, although the latter possibility cannot be excluded. When the environment is the initiator, first, an environmental factor—for example, a forced change in the diet—affects several or all individuals of a population simultaneously. Second, the influencing factor may persist over several generations. These conditions override possible random mutations. Third, phenotypic plasticity includes cryptic genetic variation.

A cryptic mutation can be viewed as an evolutionary contingency or buffer. In the presence of an environmental stimulus, the mutation can suddenly function as a switch mechanism. Complex genetic networks contain numerous cryptic mutations. Waddington spoke of cryptic mutations and referred to them as "buffering mutations." Furthermore, the comparable concept of alternative gene regulatory mechanisms has been postulated by proponents of gene regulatory networks (Section 3.3). All three reasons mentioned above suggest that environmental factors may trigger phenotypic variation rather than a genetic mutation process, which may be lengthy.

3.8.6 Heat Shock Proteins Buffer Mutations in Drosophila, Corals, and Darwin's Finches

A laboratory experiment was conducted to illustrate the following chain of events: buffering of alternative gene regulatory pathways => canalization => genetic assimilation. In fruit flies, the protein HSP90 fulfills several roles. In normal, unperturbed development, this protein functions as a chaperone, assisting in the folding of other proteins. HSP90 also aids in DNA repair, thus performing an essential task in development. However, HSP90 is also a *heat shock protein*. As such, in addition to its folding function, it plays a protective or buffering role: under heat stress, it is produced in increased quantities, ensuring that genetic mutations (as another possible consequence of stress) do not lead to phenotypic deviation. The protein slightly corrects the mutated protein structures of these genes. Thus, genes are not expressed in their mutated forms. In other words, they remain buffered. HSP90 thus acts as a canalizing factor.

However, in the laboratory, inhibiting HSP90 expression in *Drosophila* unmasks a series of genetic mutations. HSP90 can thus act as a capacitor for evolutionary change, allowing genetic changes that were previously accumulated and buffered to take effect. Again, researchers generated mutants that exhibited phenotypic changes, even when the protein HSP90 was present in sufficient quantities. They had apparently been genetically assimilated. In this way, epigenetic canalization, buffering, and genetic assimilation were visualized for the first time in 1998 in a way that can occur not only in the laboratory but also in nature. HSP90 appears to confirm Waddington's concept of developmental canalization and genetic assimilation (Gilbert and Epel 2009).

Corals must withstand temperature fluctuations. Various coral species proportionally increase HSP expression to survive increasing water temperatures. Thus, the production of HSPs is environmentally dependent. Some corals even produce HSPs at temperatures to which they have never been exposed in nature. In such situations, HSPs are buffered (Gates and Edmunds 1999). However, in general, evolution is not well equipped for extreme cases that represent a sudden, novel situation.

Could Darwin's finches (discussed in Section 4.1) represent a similar case? Could mutations that allow them to adapt their beaks in a short time be buffered, and could persistent nutritional stress expose such hidden variations? Plenty of material remains for future research. In humans, the way HSP90 suppresses the potential consequences of mutations is unknown. In a sample of 1,500 mutations, those responsible for cancer predisposition were less pathogenic when HSP90 was dominant (forming a protective buffer or masking the mutations) and vice versa: they were more pathogenic in the presence of lower HSP90 expression. The correlation of HSP90 with disease progression suggests a plausible mechanism for the variable presence of HSP90 and the environmental sensitivity of genetic diseases (Karras et al. 2017). HSP90 is also known to play a similar role in other diseases. Thus, there is increasing evidence that this key protein acts as an important buffer to control mutations in eukaryotes than previously thought.

3.8.7 Molecular Mechanisms

The molecular mode of action of phenotypic plasticity constitutes three basic steps. First, an individual sensory system receives and translates information from the environment. Second, the received signal is translated into a molecular response at the biochemical level, leading to cell responses. Third, target cells, organs, or tissues undergo a change in phenotype, accompanied by changes in gene expression. In particular, transcription factors can undergo mutations that alter their DNA-binding properties or the genes they bind to and activate. Such newly activated genes can trigger the formation of hormones that eventually bring about a phenotypic change. This molecular process has already been described by Carroll as the genetic toolkit, but Carroll and Andreas Wagner did not bring the environment into play to the same extent (Section 3.3). Such events take on complex forms when several genes, environmental factors, and threshold effects are involved that trigger the hormone response mentioned above (Ehrenreich and Pfennig 2016).

Plasticity can be realized by means of an existing or new gene regulatory network or by intermediate forms. Moczek cites horned beetles. as an example of an existing network that can express alternative phenotypes. They can produce males with different horns using the same developmental mechanism, namely programmed cell death. The alternative is environmentally regulated gene expression, which can produce plastic phenotypes. According to Moczek et al. (2011), this is a frequent occurrence. As an aside, an adult bull-headed dung beetle has immense strength; it can support a thousand times its own body weight. The reason for its strength appears to be the elaborate mating battles that occur among males, in which *Scolopendromorpha* (centipedes) hook horns, and a male can throw his opponent out of the nest. Moczek and colleagues also found that the larger the horn, the smaller the penis (Moczek and Parzer 2008). All combinations can lead to advantages and disadvantages, depending on the environment and regional population sizes. In this context, new species can emerge in a few decades.

Corresponding molecular mechanisms are proposed not only for phenotypic plasticity but also for genetic assimilation. To clarify, this requires no new genes; rather, a novel gene regulatory pathway can emerge that renders gene expression robust to the environmental stimulus. Quantitative effects on the expression of one or more genes are likely to be involved.

Several years ago, when I first read about phenotypic plasticity, genetic assimilation, and the concept of *genes as followers*, I was slightly confused. Despite the numerous examples given by West-Eberhard and others, the mechanisms were presented in a purely theoretical way. However, the significance of genetic assimilation in natural evolution has been questioned for a long time, and the key question became: how often might it occur in nature? Moczek et al. (2015) reported that evo-devo research was arriving at a deeper appreciation of the interactions between the developing organism, the environment, and ecological conditions. "Developmental plasticity [...] was once considered a special case observable in a subset of taxa, but is now recognized as the norm, and ecological conditions are recognized as being able to influence developmental outcomes at all levels of biological organization. Interactions between developing organisms and ecological circumstances therefore have the power to shape patterns of selectable variation available in a given population."

Among the numerous recent examples are bacteria and aphids, which are predisposed

In Conversation with Armin Moczek

Professor Moczek, we were writing in English by e-mail when you said that you are German and grew up in north of Munich. I did not know this, and it gives me the opportunity to meet you in person today. How did you become a professor of evolutionary biology in the USA?

In short, it boils down to a combination of a fascination for Biology, a lot of hard work, and even more dumb luck. I grew up in Hasenbergl in a public housing project ("Sozialwohnung") as the only child of a working class family. My dad was an electrician and my mom started as an office assistant. Neither had the equivalent of an abitur, in fact my dad barely finished 9th grade, a victim of the postwar circumstances of his childhood. So I wasn't exactly prepositioned for an academic career. But I was always good in school, I loved biology, and after finishing high school and civil service it was difficult for me to envision to start a regular job, so instead I moved to Würzburg and studied biology. I caught a big break after the end of my second year when a research group needed an assistant able to climb with ropes and harnesses (which I was comfortable with) to help study arboreal arthropod communities in Sabah, Borneo (which I had no clue about, but was about to learn). So, I became trained in tropical ecology and spend 6 fantastic months on two separate trips in Borneo, growing convinced that I would become a tropical biologist. My second big break came when I won a DAAD scholarship to attend Duke University in Durham, North Carolina in my final year as a masters student, where I met my future wife Laura, as well as my future PhD advisor Fred Nijhout,. Through him I became introduced for the first time to the notion that one could learn about why evolution unfolds the way it does by understanding how organisms develop and differentiate during ontogeny. This introduction to what we now call evolutionary developmental biology forever altered my view of living systems and how to approach and study them, and after a brief return to Germany to finish my masters degree I was back at Duke within months. From that point on it was a more conventional career at least as far as the US academic system is concerned. I received my PhD in 2002, postdocked for 2.5 years at the University of Arizona with Lisa Nagy and Diana Wheeler, and then started my faculty position at Indiana University in 2004 where I remain today.

You and your team focus on horned beetles.. How did you encounter these unusual animals, and why are they so exciting?
Let me say upfront that we do not just work on horned beetles., but also on a variety of other insects, such as membracid treehoppers and their helmets, and fireflies and their light-producing organs. But it is true that we do the majority of our work on beetleshorns and horned beetles. I selected this group of animals as much as they selected me. I first started to work on a particular species, the bull-headed dung beetle *Onthophagus taurus*, because I needed a study system for a class project in an Animal Behavior class while I was an exchange student at Duke. But within weeks I realized that these are really interesting organisms, and just as important many of the things that made them interesting were experimentally accessible. At first this was the ability of males to develop into alternative morphs depending on larval feeding conditions, which opened up opportunities to study the developmental basis of plasticity and its evolution across populations and species. Later we added the study of horns as an evolutionarily novel morphological trait to learn more about the origins of novel complex traits, and now novelty may emerge from within the confines of ancestral variation, an overarching research program that continues to this day. More recently our foci on innovation and novelty have gotten us interested in the role of nongenetic inheritance and developmental symbioses, and again the beetles stood out as a useful study system with which to investigate the role of maternally inherited gut microbiota in the evolution of novel feeding modes and local adaptation to novel conditions. Similarly, they have proven excellent organisms with which to experimentally assess the significance of niche construction. So in summary, while I can't deny a non-rational love for these organisms that has little to do with science, they also very much constitute powerful study organisms with which to advance key issues at the forming edge of evolutionary biology.

Evolutionary innovation, one of Darwin's specializations, Is among the most exciting topics that your team researches. Why is that?

The origin of novel traits is among the most intriguing and enduring problems in evolutionary biology. It is intriguing because it lies at the heart of what motivates much of evolutionary biology: to understand the origins of exquisite adaptations and the evolutionary transitions and ecological radiations that they enabled. It is enduring because it embodies a fundamental paradox. On the one hand, Darwin's theory of evolution is based on descent with modification wherein everything new, ultimately, must come from the old. On the other hand, biologists are captivated by complex novel traits precisely because they lack obvious homology to preexisting traits. How, then, does the first eye, insect wing, feather, placenta, butterfly wing pattern, etc. arise from within the confines of ancestral variation? But there is another major reason why the nature of evolutionary innovation is so captivating: it is the inability of conventional, population-genetic focused evolutionary biology to offer a satisfying answer, despite a century of great advances elsewhere. In fact, population genetics is so unable to confront the explanatory challenges posed by evolutionary innovation that it has stopped asking the question, a void that is now increasingly filled by evolutionary developmental biology and eco- evo-dev.

One could joke that your school has published more on horned beetles than any other animal by other researchers. Do the many findings on *Onthophagus* apply to other genera?

Our work focuses on the origins of novel complex traits, on the role of developmental plasticity in facilitating and biasing innovation, and the contributions of niche construction, developmental symbioses, and nongenetic inheritance to that process. These are all issues of fundamental significance to all forms of life, and as such the results emerging for our work on *Onthophagus* beetles are broadly applicable and relevant to most other groups of organisms. What makes *Onthophagus* stand out is the experimental accessibility and manipulability of many of the phenomena we are interested in, even though in some instances the focal traits we study may perhaps not be the most exceptional incarnation of a particular phenomenon. For example, while horns are an evolutionary novelty, they pale in complexity to the light-producing photic organs of fireflies we have also studied. But the horn's relative simplicity is a strength in disguise because it made it easier for us to reconstruct their developmental and evolutionary origins through comparative work. Similarly, *Onthophagus* may not be the world's most extreme niche constructors, but what matters is that we can experimentally manipulate this phenomenon in a standardized, replicable, and comparative manner. The same goes for host microbiome interactions – our beetles stand out as one of the relatively few groups where we can remove, modify, or replace microbial partners across populations and species. In so doing we can go well beyond simple correlations, which is where most other work, including on human-microbe relations, is usually stuck.

You are one of the most cited evo-devo researchers in the world. Evo-devo is both empirical and theoretical. What is the "explosive" of evo-devo for the theory of evolution?

Before evo-devo existed, evolutionary biologists recognized the extraordinary organismal diversity that surrounds us and concluded that the same must apply to the diversity of developmental genetic processes that underpins phenotypes. What could a fly possibly have in common with a mollusk or mammal? This perspective had a number of consequences. For example, it suggested that the origin of true innovation in evolution, whatever that may be, requires the evolution of new genes, pathways, developmental processes, and that therefore novelty in evolution should be definable as the absence of homology and homonomy (serial homology). It also meant that if one wanted to study for instance human heart development and disease, there was nothing to be gained by studying say fruit flies. It's perhaps important to recognize that none of this was based on data, only on intuition and logic, it just seemed to make sense.

Evo-devo upended all this. Clearly, organismal phenotypes are incredibly diverse, but how they are made during development is highly conserved. The same genes, gene networks, pathways, morpho-genetic processes, tissue types, etc. help instruct the making of similar and different traits in different organisms. Through that fundamental realization, evo-devo turned organisms and their component parts into LEGO creations, where diversity and novelty emerge less by adding new parts and instead by re-using and re-combining the same set of parts over and over again in different ways. Many implications emerged, for instance, it called into question the dichotomy between homology and novelty: it used to be black and white but evo-devo added shades of gray and layers. Homology could now exist on the levels of genes and pathways but not location, or on the level of cell type but not organ. Suddenly novelty emerged no longer in the absence of homology, but through it, guided much less by new genes and pathways and instead by the rules of assembly. Identifying what those rules are and how they bias the genesis of organismal diversity in development and evolution is one of the current frontiers of evo-devo. It also meant that the developmental biology of animals, including ourselves, shared deep affinity across phyla. If one was interested in studying human heart formation and disease, all of a sudden, a lot could be learned from flies.

You are a member of the Extended Evolutionary Synthesis (EES) project team, the subject of my book. Our colleagues not only welcome the concepts and goals of the EES but also criticize them. Is this criticism justified?

I can understand where some of the criticism is coming from, and a subset of it is justified. For example, some of the work leading up to the coalescence of the EES framework too easily dismissed the value of population-genetic thinking in our understanding of the ease or difficulty with which adaptive mutations may spread through populations, or the role of developmental plasticity in fostering the accumulation of mutational variation. But a lot of criticism I find unjustified and more grounded in ignorance (few population geneticists have a solid understanding of development, how traits come into being in ontogeny, and what a genes "is" or "does") and in turf protection (admitting that "the other side" has a valid point feels like a weakness). But developments in fields such as epigenetics, host microbiome biology, evo-devo, or cognitive sciences have already fully embraced EES perspectives and generated mountains of data in their support, so to me it's only a matter of time until more traditional evolutionary biology will do similarly. What I personally care about the most is that the current generation of masters and PhD students considers EES positions and then critically puts them to the test, and that is already happening across the globe.

Unlike Modern Synthesis, the EES uses new basic assumptions to arrive at different predictions. Thus, environmental factors increasingly influence development and heredity in contrast to synthesis. Such considerations go far beyond mere "additions." The EES cautiously describes the extended theory on its website and does not want a revolution. Is the synthesis nevertheless "out" from your point of view?
No. In my view nothing the Modern Synthesis posits is wrong, in fact all of it is highly relevant toward understanding why and how evolution unfolds the way it does. But it is incomplete, and therefore referring to the EES as an extension rather than a revolution strikes me as correct. Mutation, selection, drift, and migration all matter tremendously as evolutionary mechanisms capable of changing the allelic composition within a population. But the EES rightfully asks if this is the most informative definition of evolution, or whether it should be broadened toward a change in the heritable variation within a population, to make room for nongenetic inheritance and niche construction, for instance. Or to include the significance of developmental bias and plasticity in shaping the phenotypic variation available for selection to act upon. This makes the EES an extension that is non-trivial, but still not a revolution. That said EES thinking can facilitate revolutionary insights and progress: for example, our search for the genetic basis of many diseases has come up surprisingly empty, frustrating the development of treatment options. The realization however that many disease phenotypes may emerge and differentially spread in populations in response to incorrectly constructed internal and external environments, for instance as envisioned by the hygiene hypothesis, has facilitated novel treatment approaches that are helping a growing patient pool. These conceptual medical developments have occurred independently of our work on the EES, but they are fully compatible with it, yet incongruent with simple Modern Synthesis thinking.

How do you envision the future of evolutionary theory? Will the new evo-devo key concepts of constructive development and reciprocal causality be accepted?
Yes, I think so, but probably at different speeds in different sub-disciplines. Where evolutionary biologists integrate development into their work, and not just developmental genetics but actual development complete with cellular biology and morphogenetic movements, etc., a mindset that makes room for or even emphasizes constructive development is increasingly already in place. The same goes for any evolutionary biologist who studies host-microbial interactions: reciprocal causality is everywhere. To ignore it is to step backward. But these are empirical observations and the resulting theories have so far been largely general and verbal. Quantitative counterparts remain to be developed. Therefore, other segments of evolutionary biology will likely take a bit longer before they open themselves up to EES positions. In some cases, this may require less convincing of current skeptics and more retiring and replacing of past leaders of the field.

The theory of evolution as it is taught in schools and universities today is rather simple and clear. This is no longer necessarily true for the EES which includes reciprocal cause–effect relationships that can be difficult to understand
I have two responses to this. First, I would say there simply is no other choice. Simple may be easy but if it's wrong or incomplete then it needs to go or be revised. But perhaps more importantly, I would not underestimate the public's ability to comprehend

complex relationships like those posited by the EES. When I talk to lay people about EES perspectives a very common response I get is that it makes perfect sense that inheritance is not restricted to genes, and that when environmental modifications impact the success of offspring this too can influence future evolutionary trajectories. This is usually followed by some surprise that these perspectives have yet to be integrated, and that they are subject to so much debate. It would not surprise me if similar attitudes emerge when EES positions are taught in schools and universities.

How should evolution be taught in schools and universities in the future?
In many ways, evo-devo and its increasing transformation into eco-evo-devo are the future of evolutionary biology, because they promise to advance the key questions that evolutionary biology has failed to answer, and do so at a time when such answers are increasingly desperately needed, as organismal diversity confronts an increasingly changing planet, and human health increasingly depends on the environmental conditions we are able to restore. So, from basic to applied perspectives, eco-evo-devo ought to be what current students need to be exposed to and trained in.

Professor Moczek, thank you for this interview!
(I interviewed Armin Moczek on July 18, 2019 in Munich.)

to environmentally sensitive gene expression and mutation accumulation. In water fleas, cave fish, tadpoles, snakes, and other animals, environmentally induced phenotypes are reported to be stabilized by subsequent selection of genetically variable factors. At the same time, organisms help to shape the ecological niche in which they develop. In this way, they direct the selective environment during their own lifespans and for subsequent generations (Moczek et al. 2015; on niche construction, see Chapter 5). Finally, in the next section, I would like to address question 5 from Gerd B. Müller's list of points on the evo-devo discipline (Section 3.4) concerning evolutionary innovations.

3.9 INNOVATIONS IN EVOLUTION

3.9.1 What Do We Mean by "New" in Evolution?

Evolutionary innovations have left impressions on the evolution of all earthly life forms. Not all the decisive changes that have occurred during evolution since the beginning of life fit easily into the picture of continuous change; some were radical innovations. These include, for example, the linking of individual replicators to chromosomes or the transition from an RNA- to a DNA-based genome and the genetic code, as well as the evolutionary pathways from prokaryotes to eukaryotes, from solitary individuals to colonies with non-reproductive castes, and from primate societies to the appearance of human societies with language. Evolutionary theorists John Maynard Smith and Eörs Szathmáry devoted their famous book to the topic of system transitions (Maynard Smith and Szathmáry 1995). This section discusses relatively modest innovations, the emergence of which is no less impactful. Here, the primary focus is on the concrete phenomenon of adding a new element, such as the shells of turtles or the feathers of birds, to an existing organismic form.

Evolutionary innovations provide biologists with food for thought. As stated by Armin Moczek (2008), innovations are the focus of numerous biological disciplines, and it is remarkable how little we know about the processes of their emergence. The discussion and teaching of evolutionary innovations pose the same challenges faced by other sciences. Indeed, if one investigates the lowest level of organization, the genes, to explain the causes of the emergence of innovations, one may not discover the principle of phenotypic novelty or the elements that reductionists use to explain how new superordinates are assembled from them. Therefore, other perspectives are sometimes required; perhaps a novel trait characteristic of a species, such as the horn of a horned beetle., lies at a certain level of organization in development. For example, some parts of a gene regulatory network may be homologous while others may not. Thus, the following question needs to be answered: What was already present in the developmental process, and what was not? As Moczek explains, homology is like an onion, with multiple layers of skin that must be removed to determine its core (Moczek 2008).

At this point, a recent article in *Science* magazine from Moczek's team appropriately fits into the discussion (Hu et al. 2019): The article made the cover of the world-renowned magazine in November 2019 (Figure 3.17), an indication that exploring innovation in biology is currently considered one of the major issues in all the fields of science. The evo-devo team targeted 3 of the 2,400 known *Onthophagus* species. The authors describe how the formation of the horn in *Onthophagus* requires a series of original tissues and genes that are also essential for the development of the insect wing. This is referred to as serial

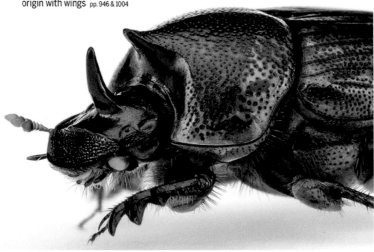

Figure 3.17 Beetle horn in the spotlight. The cover of *Science* magazine reports on the innovation of the *Onthophagus* horn and the newly discovered deep dependence on its homology to the insect wing.

homology. Some of the same early tissues responsible for the construction of the lateral body appendages, namely the wings, in one of the beetle species (*O. sagittarius*) became a new feature along the midline of the body during evolution. Moreover, the horns of *Onthophagus* are considered a classic example of drastically different innovations of the anterior body in different insects. According to the result from Moczek's laboratory, the horns on the anterior thoracic segment (prothorax) are homologous to the wings. The next question arises: Did the evolution of the horn predate that of the wing or vice versa? Thus, it is difficult to determine which structures are truly new in evolution and which ones are not (in other words, partly homologous). The horn phenotype is indeed new, but its developmental genetics and tissue forms are not. We will address this problem in the following sections of this chapter.

Up to this point, we have not received a conclusive answer to the question of how the phenotypic shape of the horn develops. It has been experimentally proven by the authors that tissue arises in the form of two ectopic wings, i.e., from two components in the middle. Therefore, from this fact and from the tissue analyses, evolutionarily, tissue from both sides of the body—namely, from the regions of the insect wings, which are evolutionarily much older—must have been pushed together to the middle and further remodeled. Why then did a completely different external shape emerge from the wing, when important serial homologies are identical? What other principal mechanisms might support the shift toward a new morphology? Further considerations will be addressed in the following paragraph.

To use a comparison from the world of classical music, Bach and Chopin use almost the same type und number of keys on the piano as Duke Ellington or other jazz pianists; they also use the same basic set of notes. There is hardly any difference in the pitches of the instruments, but there are significant differences in the harmonies and rhythms.

However, harmony only occurs when several notes sound together, and rhythm only occurs when notes follow each other in time. Chords, harmonies, and rhythm are thus analyzed on higher levels than those of single notes or tones. It is only on these levels that novelty appears. The works of individual composers become more distinguishable, not to mention the compositional form, including movements, themes, and secondary themes. Recognizing these requires observation on a higher level. Despite similarities in notes and even note sequences, as well as time and key signatures, no two works are identical.

Biologists disagree in their explanations of how evolutionary innovations occur. Some regard changes in gene regulation or switch combinations, gene duplications, gene shifts, or cis-elements as sufficient to explain the phenotype, while others are convinced that cells and cell assemblies, especially cell migration, cell adhesion, physical conditions, self-organization, and threshold effects in development are essential, in addition to the action of the environment on all levels. For these scientists, autonomous properties exist at all levels of organization in development. Innovation in evolution cannot be thoroughly understood without predictions about phenotypic plasticity and genetic accommodation. These frameworks and mechanisms play a supporting role in evolutionary novelty from the current perspective, and others are in the pipeline. How to explain innovation is, for some researchers (including myself), the predominant question in evolutionary theory. In Moczek et al. (2015), Günter Wagner is quoted as saying, "How novelty arises from within the confines of ancestral homology and how natural variation can lead to the evolution of complex, novel traits therefore remain some of the most intriguing and enduring questions in evolutionary biology." Life on Earth has consisted of novelty since its beginning. Every trait that exists today once resulted from an innovation. We should always be aware of this when we think of evolution.

Simply put, one might conclude that from the diversity of possible gene regulation mechanisms, nearly every phenotypic form in the morphospace can be produced, including novel ones. The focus of mainstream evolutionary biology is clear; it has traditionally been on mutations as the sole source of evolutionarily relevant phenotypic variation. Here, mutation is random with respect to its effects on the adapted phenotype; thus, natural selection is the only process capable of producing adaptive matches between the organism and its environment (Moczek et al. 2015). Innovation is thus a byproduct of continuous variation. Furthermore, the Modern Synthesis, at least in its earlier form, also implies that there is a simple relationship between mutations that initiate change and the phenotypic outcome (Peterson and Müller 2013).

From the evo-devo perspective, it is clear that natural selection alone cannot lead to the emergence of innovations. Selection is out of the question as the cause of the emergence of a novel structure because it requires existing ones (Müller 2010). The selection paradox, in this context, means that selection, in order to produce new traits, depends on inheritance in the population. However, that which is inherited cannot be new (Moczek 2008). It must therefore be explained how a new trait can come about and what is responsible for the innovation, if not selection.

Before that which is new or novel can be distinguished from that which already exists, the question of why a distinction between variation and innovation is necessary arises. The anticipated answer is that innovations provide a specific insight into development processes, and evo-devo aims to explore such innovations. Their development is clearly distinct from that of variations. This, in turn, has significant consequences for evolutionary theory.

The question of what constitutes an evolutionary innovation is a longstanding topic of discussion in biology. Each definition is incomplete and emphasizes a different point of view, such as the function of the new structure or phenotype or the absence of homology, i.e., the lack of correspondence with common predecessors. Consider this definition: "A morphological novelty is a structure that is neither homologous to any structure in the ancestral species nor homonomous to any other structure in the same organism." (Müller and Wagner 1991). Thus, feathers (Figure 3.18) were an evolutionary novelty when they first arose, but in the oldest extinct birds, which already possessed flightless predecessors with feathers, they were not. Feathers originally served a purpose entirely unrelated to flying. The function (ability to fly) is therefore less important here than the consideration of feathers as a novel phenotype. We are even less interested here in whether the innovation is adaptive since our primary concern from an evo-devo perspective is the developmental scenario (Peterson and Müller 2013). The insect wing (it probably underwent a functional change similar to that of the bird feather), turtle shell, first wing patterns of butterflies, or light-emitting mechanism of the firefly are all innovations of this nonhomologous type. An additional

Figure 3.18 Bird feather. An evolutionary innovation. Feathers were originally used for insulation, not for flying. Earlier in their evolution, they were simpler in construction. After many steps of modification, they achieved the recent asymmetrical shape of flight feathers.

finger or toe also falls into this category, even if four or five fingers or toes of the same type existed before its appearance. A new finger has no homologous counterpart because there was nothing in its place before the finger appeared. Therefore, it is seen as an innovation. Gerd B. Müller repeatedly points out that the definition of innovation cannot be applied to several organizational levels simultaneously. To do so would lead to misinterpretations, because we have just seen, for example, that the beetle horn is serially homologous to the wing, and that it is precisely this fact (existing homology) that attracted the attention of *Science* and gained front-page status. Should we dispense with homology altogether in defining innovation, as Moczek (Hu et al. 2019) suggests? We conclude in advance that the horn *phenotype* is novel, but the structure is not homologous. Müller points out that in such cases, innovation must be related to the phenotype alone, and that innovation on the phenotype level can not necessarily be defined genetically (e.g., Müller 2010, Peterson and Müller 2013). The novelty of a trait, besides the inevitable homologies, must be determined at the organizational levels below the phenotype (organs, tissues, cells, etc.) to be considered biologically significant. We will consider more examples of this in Chapter 4.

The discussion of innovation with examples from Müller (2010) will be limited to discrete new elements added to an existing body plan, although other types of innovation exist. For example, the long tooth of the narwhal (*Monodon monoceros*) appears as a fascinating and novel feature; nevertheless, it is only a modification of a preexisting tooth. The emergence of large body plans in the animal world during the Cambrian explosion approximately 540 million years ago is a separate category of innovation, which will not be discussed further here.

According to Moczek et al. (2015), evolutionary development relies on (a) of patterns and processes, from individual genes or genome elements for the evolution of new traits; (b) expansions of specific genomic domains, interactions of gene regulatory networks, or coordinated modifications of organ systems, such as those in the frog gut; and (c) co-option of multiple levels of biological organization, such as the wing patterns of butterflies, horns of beetles, shells of turtles, and others. This does not require novel genes or signaling pathways; rather, new functional phenotypes arise from differential combinations and redistributions of existing developmental modules. This distinction is explored in more detail in Kirschner and Gerhart (Section 3.5). However, it does not encompass epigenetic developmental components, such as those that may occur in intercellular signaling pathways to bring about novelty, or physical conditions. More on this is in the following section.

3.9.2 Conditions of Innovation Initiation

Three conditions for the emergence of phenotypic innovation are distinguished (Müller 2010):

1. The initiation conditions
2. The realizing conditions
3. The genetically permanent integration with its specific conditions.

What are the initiating conditions for the emergence of novelties in evolution? Based on several examples, West-Eberhard (2003) argues strongly that the most important initiators of evolutionary novelties are environmental conditions (Section 3.8). These may include a change in nutritional status, a change in climatic or other physical conditions, or the presence of a predator. In addition to environmental conditions, classical gene mutations or gene regulatory changes can also act as initiators of phenotypic novelties. The difference is that these do not affect the entire population and are usually not established permanently.

If the initiating cause of an innovation is originally nonspecific and general and, as in the case of environmental effects, typically occurs at the population level, then the conditions for the physical realization of a specific novelty must be sought in development (Müller and Newman 2005). Only development can be the medium for transferring environmental factors into the phenotype in a regulated manner.

3.9.3 Conditions of Innovation Realization

Once it has been determined who or what can initiate a change or innovation, it is necessary to explain how an innovation can be incorporated into the phenotype.

All the abovementioned prerequisites of innovations refer to developmental processes. The developmental system is genetic and epigenetic, interacting, self-regulating, and dynamic. It has two peculiarities, which the Modern Synthesis does not address. First, it can react to external stimuli and is thus environmentally dependent. Under this condition, an interacting process emerges in development; cells engage in bidirectional communication with other cells. They send information to the genome, and conversely, respond to genetic information. Cells as part of tissue aggregates further interact with those of neighboring tissues. All these pathways are subject to environmental stimuli. The sum of interactions of the system is shown in Figure 3.19. It cannot be said that a "genetic program" is executed during development. Cells are not simply executors of the instructions in DNA (Müller 2010). The same view was expressed by Kirschner and Gerhart (Section 3.5), Jablonka and Lamb (Section 3.6), and Noble (Section 3.7).

One element of novelty is that these responses can be nonlinear due to threshold effects following small external stimuli or genetic mutations. Thresholds cause unexpected jumps in otherwise linear changes. Sudden, major responses occur where one would expect continuity. Jumps are observed instead of gradual changes. However, these "quantum leaps" can trigger the phenotypic changes associated with innovations. I will

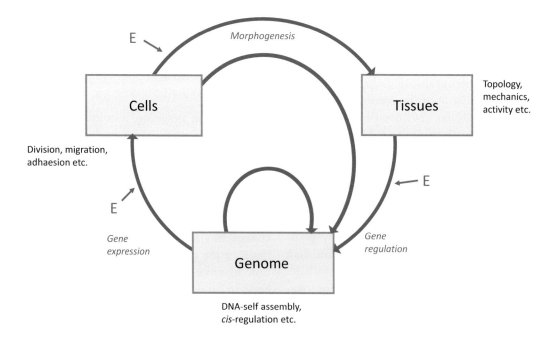

Figure 3.19 Dynamic interaction in development. Examples of autonomous properties of each level of phenotypic organization are displayed next to the boxes. U = environmental influences.

explain later, with an example, exactly how thresholds come about, how they can be modeled, and how they are visible during the evolutionary development of the cat's paw (Section 4.8).

The second novel element in embryonic development is the ability to self-organize. This is also discussed in Section 4.8 using the example of a Turing pattern at the cellular level of development. The initiating parameter(s) set in motion the self-organization of the entire system. This response by the system may generate a discontinuous output as an innovation in the form of, for example, a new skeletal element. The result is then presented as a turnkey to natural selection (Lange et al. 2014, Newman 2018).

According to Waddington (Section 3.1) and Andreas Wagner (Section 3.3), development buffers variations, and developmental paths are canalized; however, situations occur in the emergence of an innovation in which canalization reaches its limits and can no longer be maintained by the system. Development is thus decanalized (Müller 2010). Once again, that which follows the initiating disruptive factor can be discontinuous when thresholds of influencing variables are exceeded or undercut.

Such threshold mechanisms can also lead to the complete elimination of fingers or feet during embryonic development. One example is that of the Australian skink, a lizard species in which continuous evolutionary body elongation does not lead to gradualistic limb variation; rather, complete phalanges or whole toes always fall away in parallel with body elongation. Moreover, natural selection and adaptation do not explain the orderly omission of toes, specifically the first and the last. There is no advantage of not omitting a middle pair of toes. Rather, the developmental mechanisms in the limb provide the answer here. As early as 1985, through evo-devo experiments on amphibians, it was possible to predict which toe would be added or omitted and at which position in the limb when the embryonic tissue of the limb bud was manipulated. Thus, the last finger to develop in the embryo is the first to be eliminated. Observations in such experiments are consistent with those in natural populations (Müller 2010).

The system shown is genetic-epigenetic. Here, epigenetics does not mean epigenetics in connection with chromatin changes (Section 3.6). Rather, epigenetics at this point refers predominantly to the process of embryo formation, which has classically also been called epigenesis. This refers to the genetic, cellular, and environmental genesis in the developmental context.

In short, the developmental system encompasses both the autonomous capabilities of its components (cell behavior, cell assemblies, geometry, etc.) and the capacity for self-organization and nonlinear responses to local and global external conditions. Here, the explanatory capacity of evolutionary theory is extended to include the following:

- Feedback loops among development components and with the environment
- Nonadaptive and nongradual phenomena of phenotypic evolution
- Self-organizing capacity of development.

This is precisely the core of systemic eco-evo-devo research and an essential basis of the claimed theory extension of evo-devo.

The example of the beetle horn provides an illustration. Next, under the abovementioned realization conditions, in addition to the serial homology discovered by Moczek's team, further possible mechanisms, which were probably at play when the horn evolved to its present form, are possible in principle. The physical pressure applied when tissues drift toward the center of the body and meet there triggers various effects. Subsequently, several thresholds and mechanisms of self-organization may have played a role in producing the harmonic, curved, and tapering pointed horn. These mechanisms must be sought and elucidated, as they exist, for example, in the evolutionary development of the hand (Section 4.9).

3.9.4 Genetic Integration of Innovations

Understanding how an innovation is genetically and phenotypically integrated can help us to fully grasp what constitutes evo-devo and how strongly this extended school of thought can set itself apart from mainstream ideas.

A prerequisite for genetic safeguarding of new impulses is preexisting genetic variability (Nanjundiah 2003). Recall that Waddington spoke of canalization (Section 3.3.1 "Genetic Toolkit"), Wagner of robustness (Section 3.3). Genetic variability may not appear phenotypically under relatively constant environmental conditions because the environmental parameters are not conducive to its activation. As discovered by Waddington, it is true that only new environmental factor(s) unmask the hidden genetic/epigenetic plastic design potential. Thus, not everything in an organism that deviates from the pathway (masked mutations) becomes immediately visible; there are checkpoints at the molecular, cellular, and epigenetic levels. The protein HSP90 has already been mentioned in this context. It can not only represent a constraint but also open pathways along which a novelty can develop (Müller and Newman 2005).

Morphological variations can thus arise and subsequently be genetically stabilized, as described above. The change may be fixed by genetic mutations if the new phenotypic trait is advantageous and/or if the conditions that led to it (changes in environmental conditions or behavior) are sufficiently robust (Pigliucci 2008). Dependence on the initiating environmental factor is removed with genetic assimilation; the organism can become independent of it. Müller, Pigliucci, and West-Eberhard emphasize that genetic change in this case follows rather than precedes phenotypic change. *Genes are followers.*

In this context, the most important statement is that neither selection nor the genome alone controls the emergence of a novel phenotype. Rather, it is controlled by autonomous reactions of the development system to small disturbances. The response of the developmental system dictates the specific morphological product, in this case the innovation (Müller 2010). More recently, instead of possible autonomous responses, the discussion has turned to instances of agency and agents of the organism.

Evolutionary innovation is reportedly attributed to the absence of homology, i.e. lacking affinities to structures that already existed in the ancestral state (Müller and Wagner 1991). Researchers consequently associate this absence in terms of phenotypical homologies and not those related to the *process* of phenotypic formation (Peterson and Müller 2013). Currently, however, innovations are considered from a different evo-devo-perspective. Moczek and colleagues emphasize the need for homologous components to contribute to phenotypic innovation. These can be homologous gene regulatory networks as well as homologous source tissues. To this end, these researchers have developed the concept of an evolutionary innovation gradient. Such a gradient has "no true beginning, at most there may be key events along the way" (Linz et al. 2020). The purpose of the gradient is to express that homologies can occur in the course of the evolution of an innovation, for example in gene regulatory networks (remember the serial homology at the *Ontophagus*horn). However, the ancestral homologies bias the gradient to ensure that hotspots of innovation, such as a mutation, and sites of deep conservation accompany the evolutionary process (Linz et al. 2020). In this way, the concept of an innovation gradient is compatible with our intuitive thinking that new things always also consist of and emerge from existing components.

This complex theoretical connection of evolutionary innovation may be illustrated with a recent research finding: an inconspicuous point mutation may induce the appearance of one or more complete additional fingers or toes. Section 4.8 discusses this in more detail. Another example demonstrating the integration of evolutionary development and proving that a disturbance, be it environmental or genetic, does not result in chaos was provided by Moczek et al., who knocked out a head shape-associated gene in a beetle; this induced the formation of ectopic tissues in the middle region of the front of the head for a new eye. Moreover, nerve-like structures running to the brain were created. It was the first time that the silencing of an important gene had triggered such a constructive new development process. The authors called this surprising result "a remarkable example of the ability of developmental systems to channel massive perturbations toward orderly and functional outcomes" (Zattara et al. 2017). Moments that lead to such discoveries are surely among the lonely highlights of a scientist's life.

3.10 SUMMARY

Evolution and development are closely intertwined; therefore, it makes no sense to study them in isolation (Brakefield 2011). After 40 years, evolutionary developmental biology has reached a certain coming of age. It is seen by some as the modern cornerstone of traditional evolutionary theory. Evo-devo has provided unexpected discoveries for these researchers, developmental genes and gene regulatory networks. Evo-devo thus clarifies previously unsolved questions and new relationships in the phylogenetic tree of life.

For others, evo-devo is much more: a less adaptationist, more organism- and environment-centered platform on which an evolutionary theory emerges that transcends and leaves the Modern Synthesis behind. By moving beyond the gene-centered perspective of the Modern Synthesis, evo-devo sheds new light on the complexity of evolution, focusing more intensely on the phenotype. It is now seen as a site of integration of a whole range of interrelated mechanisms, from molecular and developmental to physiological and environmental. It follows that the patterns and processes of phenotypic change result from a combination of all these diverse, causal mechanisms, which play out at different scales, both organizational (from genes to environment) and temporal (such as development, life history, and evolution; Laubichler 2010). "Just as the Modern Synthesis was able to unify many aspects of biology in the mid-20th century, evo-devo is now positioned to transform and unify diverse aspects of biology, one of the primary scientific challenges of the 21st century" (Moczek et al. 2015).

In its conceptual, theoretical orientation, evo-devo can address the stumbling block that this discipline deals (only) with the individual organism, but the Modern Synthesis deals with the population and its adaptation. This incompatibility between the two theories makes it difficult for Manfred Laubichler (2010) to complete an integrative theory of phenotypic evolution. Alternatively, Stuart Newman, (2018) cites numerous examples of morphological forms that arose in isolation from successive cycles of adaptation. Once arisen, such traits become useful to organisms.

Alessandro Minelli (2015) cites several deficits of evo-devo. For example, the evo-devo research program is focused on the animal world (Metazoa). Plants, he says, are underrepresented. However, generalizations of findings from the animal kingdom cannot easily be transferred to other kingdoms. Minelli calls for the expansion of the research program to include unicellular organisms. He draws attention to the fact that the evolution of development extends beyond the view of developmental genetics. It is more than a sequence of changes that give rise to an adult multicellular animal or plant in complex steps that begin in the fertilized egg. He describes this view as a "naïve concept." Evo-devo will therefore evolve in a new direction. This should address fundamental problems around the question of how development evolved with its main features, from cell differentiation to complex, multicellular life cycles.

To date, evo-devo research has achieved its causal-mechanistic results using few model organisms. The necessary methods of genetic manipulation can be applied to them. These methods include genetically modified organisms, mutagenesis (creation of mutations in the genome to achieve the desired results), and cloning to produce genetically identical organisms. More and more quantitative data are available that can be incorporated into computer models and allow theories with predictive power. Only elaborate data comparison allows greater certainty about which gene regulatory networks are essential for a particular phenotype and which are free to vary in evolution (Moczek et al. 2015).

In Europe, the European Society for Evolutionary Developmental Biology (EED) was founded in 2006. It holds an international conference every 2 years with outstanding speakers and hundreds of presentations from all over the world (https://evodevo.eu). Not to be outdone by this successful program, the USA established its own society, the PanAmerican Society for Evolutionary Developmental Biology (EvoDevo PanAm) (http://www.evodevopanam.org). Their conferences alternate with those of the EED. The vast majority of the empirical research projects presented at these conferences aim to disseminate rather than stimulate theoretical discussion. Nevertheless, contributions to evolutionary developmental theory arise from them.

Chapter 4 presents selected results from empirical research, and at its conclusion (Section 4.9), formulates the implications of evo-devo theory and research for an EES. Chapter 5 introduces the theory of niche construction, one of the four research pillars of the Extended Synthesis, along with evo-devo, developmental plasticity, and inclusive inheritance. Finally, the contents of Chapters 3–5 flow together into Chapter 6 in the form of the large-scale EES project.

REFERENCES

Amundson R (2005) *The Changing Role of the Embryo in Evolutionary Thought. The Routs of EvoDevo.* Cambridge University Press, Cambridge, MA.

Bateson P, Gluckman P (2012) Plasticity and robustness in development and evolution. *Int J Epidemiol* 41:219–223.

Bonner JT (1982). Evolution and Development. *Report of the Dahlem Workshop on Evolution and Development Berlin 1981, May 10–15.* Springer, Berlin.

Brakefield PM (2011) Evo-Devo and accounting for Darwin's endless forms. *P Roy Soc B Bio* 366:2069–2075.

Carroll SB (2006) *Endless Forms Most Beautiful.* W.W. Norton, New York.

Carroll SB (2007) *The Making of the Fittest. DNA and the Ultimate Forensic Record of Evolution.* Norton, New York.

Danchin E, Charmantier A, Champagne FA, Mesoudi A, Pujol B, Blanchet S (2011) Beyond DNA: integrating inclusive inheritance into an extended theory of evolution. *Nat Rev Genet* 12:475–486.

Dawkins R (1976) *The Selfish Gene.* Oxford University Press, Oxford.

de Beer GR (1930) *Embryology and Evolution.* Clarendon Express, Oxford.

Dupré J (2010) It is not possible to reduce biological explanations to explanations in chemistry and/or physics. In: Ayala, FJ, Arp R (eds.) *Contemporary Debates in Philosophy of Biology.* Blackwell Publishing, Malden, 32–48.

Ehrenreich IM, Pfennig DW (2016) Genetic assimilation: a review of its potential proximate causes and evolutionary consequences. *Ann Bot-London* 117(5):769–779.

Ellis BJ, Bianchi JG, Griskevicius V, Frankenhuis WE (2017) Beyond risk and protective factors: an adaptation-based approach to resilience. *Perspect Psychol Sci* 12(4):1–27.

Flatt T (2005) The evolutionary genetics of canalization. *Q Rev Biol* 80(3):287–316.

Gapp K, Jawaid A, Sarkies P, Bohacek J, Pelczar P, Parados J, Farinelli L, Miska E, Masuy IM (2014) Implication of sperm RNAs in transgenerational inheritance of the effects of early trauma in mice. *Nat Neurosci* 17:667–669.

Gates RD, Edmunds PJ (1999) The physiological mechanisms of acclimatization in tropical reef corals. *Am Zool* 39:30–43.

Gilbert SF (2003) The reactive genome. In: Müller GB, Newman SA (eds.) *Origination of Organismal Form. Beyond the Gene in Development and Evolutionary Biology*. MIT Press, Cambridge, MA, 87–101.

Gilbert SF, Epel D (2009) *Ecological Development Biology. Integrating Epigenetics, Medicine and Evolution*. Sinauer, Sunderland.

Gould SJ (1977) *Ontogeny and Phylogeny*. Harvard University Press, Cambridge, MA.

Gould SJ, Lewontin R (1979) The spandrels of san marco and the Panglossian paradigm. A critique of the adaptionist programme. *Proc Roy Soc Lond B Biol Sci* 205:581–598.

Honnefelder L, Propping P (2001) *Was wissen wir, wenn wir das menschliche Genom kennen? Die Herausforderung der Humangenomforschung*. DuMont, Köln.

Hu Y, Linz DM, Moczek AP (2019) Beetle horns evolved from wing serial homologs. *Science* 366:1004–1007.

Jablonka E, Lamb MJ (2014) *Evolution in Four Dimensions. Genetic, Epigenetic, Behavioral, and Symbolic Variation in the History of Life*, 2nd ed. MIT Press, Cambridge, MA.

Karras GI, Yi S, Sahni N, Fischer M, Xie J, Vidal M, D'Andrea AD, Whitesell L, Lindquist S (2017) HSP90 shapes the consequences of human genetic variation. *Cell* 168(5):856–866.

Kirschner M, Gerhart J (2005) *The Plausibility of Life: Resolving Darwin's Dilemma*. Yale University Press, New Haven.

Laland KN, Uller T, Feldman M, Sterelny K, Müller GB, Moczek A, Jablonka E, Odling-Smee J (2015) The extended evolutionary synthesis: its structure, assumptions and predictions. *Proc R Soc B* 282:1019. https://doi.org/10.1098/rspb.2015.1019.

Lange A (2017) *Darwins Erbe im Umbau. Die Säulen der Erweiterten Synthese in der Evolutionstheorie*, 2. überarbeitete, aktualisierte Aufl. Königshausen & Neumann, Würzburg (eBook).

Lange A, Nemeschkal HL, Müller GB (2014) Biased polyphenism in polydactylous cats carrying a single point mutation: the Hemingway model for digit novelty. *Evol Biol* 41(2):262–275.

Lange A, Nemeschkal HL, Müller GB (2018) A threshold model for polydactyly. *Prog Biophysics Mol Bio* 137:1–11.

Laubichler M (2010) Evolutionary developmental biology offers a significant challenge to the Neo-Darwinian paradigm. In: Ayala FJ, Arp R (eds.) *Contemporary Debates in Philosophy of Biology*. Blackwell Publishing, Malden, 199–212.

Lind MI, Spagopoulou F (2018) Evolutionary consequences of epigenetic inheritance. *Heredity* 121:205–209.

Linz DM, Hu Y, Moczek AP (2020) From descent with modification to the origins of novelty. *Zoology* 143:125836.

Macagno ALM, Zattara EE, Ezeakudo O, Moczek AP, Ledón-Rettig CC (2018) Adaptive maternal behavioral plasticity and developmental programming mitigate the transgenerational effects of temperature in dung beetles. *Oikos* 127:1319–1329.

Maynard Smith J, Szathmáry E (1995) *The Major Transitions in Evolution*. Oxford University Press, Oxford.

McGinnis W, Levine MS, Hafen E, Kuroiwa A, Gehring WJ (1984) A conserved DNA sequence in homoeotic genes of the Drosophila Antennapedia and bithorax complexes. *Nature* 308:428–433.

Mills RE, Bennett EA, Iskow RC, Devine SE (2007) Which transposable elements are active in the human genome? *Trends Genet* 23(4):183–191.

Minelli A (2015) Grand challenges in evolutionary developmental biology. *Front Ecol Evol* 2:1–11.

Mitchell S (2009) *Unsimple Truth – Science, Complexity, and Policy*. University of Chicago Press, Chicago, IL.

Moczek AP (2008) On the origin of novelty in development and evolution. *BioEssays* 30:432–447.

Moczek AP, Parzer HF (2008) Rapid antagonistic coevolution between primary and secondary sexual characters in horned beetles. *Evolution* 62(9):423–428.

Moczek AP, Sultan S, Foster S, Ledón-Rettig C, Dworkin I, Nijhout HF, Abouheif E, Pfennig DW (2011) The role of developmental plasticity in evolutionary innovation. *Proc Biol Sci* 278(1719):2705–2713.

Moczek AP, Sears KE, Stollewerk A, Wittkopp PJ, Diggle P, Dworkin I, Ledon-Rettig C, Matus DQ, Roth S, Abouheif E, Brown FD, Chiu CH, Cohen CS, De Tomaso AW, Gilbert SF, Hall B, Love AC, Lyons DC, Sanger TJ, Smith J, Specht C, Vallejo-Marin M, Extavour CG (2015) The significance and scope of evolutionary developmental biology: a vision for the 21st century. *Evol Dev* 17(3):198–219.

Moczek AP, Sultan SE, Walsh D, Jernvall J, Gordon DM (2019) Agency in living systems: how organisms actively generate adaptation, resilience and innovation at multiple levels of organization. Proposal for a major grant from the John Templeton Foundation.

Monod J (1971) *Chance and Necessity: An Essay on the Natural Philosophy of Modern Biology by Jacques Monod*. Alfred A. Knopf, New York.

Morgan HD, Sutherland HG, Martin DI, Whitelaw E (1999) Epigenetic inheritance at the agouti locus in the mouse. *Nat Genet* 23:314–318.

Müller GB (2007) Evo-Devo. Extending the evolutionary synthesis. *Nat Rev Genet* 8(12):943–949.

Müller GB (2008) Evo-devo as a discipline. In: Minelli A, Fusco G (eds.) *Evolving Pathways: Key Themes in Evolutionary Developmental Biology*. Cambridge University Press, Cambridge, MA.

Müller GB (2010) Epigenetic innovation. In: Pigliucci M, Müller GB (eds.) *Evolution – The Extended Synthesis*. MIT Press, Cambridge, MA, 307–332.

Müller GB, Newman SA (2005) The innovation EvoDevo agenda. *J Exp Zool* 304B:487–503.

Müller GB, Wagner GP (1991) Novelty in evolution. Restructuring the concept. *Annu Rev Ecol Syst* 22(1):229–256.

Müller-Wille S, Rheinberger H-J (2009) *Das Gen im Zeitalter der Postgenomik. Eine wissenschaftstheoretische Bestandsaufnahme*. Suhrkamp, Berlin.

Nanjundiah V (2003) Phenotypic plasticity and evolution by genetic assimilation. In: Müller GB, Newman SA (eds.) *Origination of Organismal Form – Beyond the Gene in Development and Evolutionary Biology*. MIT-Press, Cambridge, MA, 245–263.

Newman SA (2018) Inherency. In: Nuno de la Rosa LN, Müller GB (eds.) *Evolutionary Developmental Biology. A Reference Guide*. Springer International Publishing, Cham.

Nijhout HF, Wray GA, Claire K, Teragawa CK (1986) Ontogeny, phylogeny and evolution of form: an algorithmic approach. *Syst Biol* 35 (4):445–457.

Noble D (2006) *The Music of Life. Biology beyond Genes*. Oxford University Press, Oxford.

Noble D (2015) Evolution beyond ne-Darwinism: a new conceptual framework. *J Exp Biol* 218: 7–13.

Noble D (2016) *Dance to the Tune of Life: Biological Relativity*. Cambridge University Press, Cambridge, MA.

Noble D (2017) Evolution viewed from physics, physiology and medicine. *Interface Focus* 7:20160159. https://doi.org/10.1098/rsfs.2016.0159.

Noble D (2019) Exosomes, gemmulues, pangenesis and darwin, revisited. In: Edelstein L, Smythies J, Quesenberry P, Noble D (eds.) *Exosomes in Health and Disease*. Academic Press, 487–502.

Noble D, Jablonka E, Joyner MJ, Müller GB, Omholt SW (2014) Evolution evolves: physiology returns to centre stage. *J Physiol* 592(11):2237–2244.

Nowotny H, Testa G (2009) *Die gläsernen Gene. Die Erfindung des Individuums im melokularen Zeitaltern*. Suhrkamp, Berlin.

Nüsslein-Volhard C (2004) *Das Werden des Lebens. Wie Gene die Entwicklung steuern*. Beck, München.

Pan Q, Shai O, Lee LJ, Frey BJ, Blencowe BJ (2008) Deep surveying of alternative splicing complexity in the human transcriptome by high-throughput sequencing. *Nat Genet* 40:1413–1415.

Peterson T, Müller GB (2013) What is evolutionary novelty? Process versus character based definitions. *J Exp Zool Part B* 320B:345–350.

Pigliucci M (2008) What, if anything, is an evolutionary novelty? *Philos Sci* 75:887–898.

Riedl R (1978) *Order in Living Systems: A Systems Analysis of Evolution*. Wiley, New York.

Roux W (1881) *Der Kampf der Theile im Organismus. Ein Beitrag zur Vervollständigung der mechanischen Zweckmäßigkeitslehre*. Verlag von Wilhelm Engelmann, Leipzig. https://archive.org/stream/derkampfdertheil00roux#page/34/mode/2up.

Shapiro JA (2022) *Evolution – A View from the 21st Century*. Fortified. Cognition Press. 2nd ed., Upper Saddle River, NJ.

Skinner MK, Mannikam M, Guerrero-Bosagna C (2010) Epigenetic transgenerational actions of environmental factors in disease etiology. *Trends Endochrinol Metab* 21:214–222.

van Steenwyk G, Roszkowski M, Manuella F, Franklin TB, Mansuy IM (2018) Transgenerational inheritance of behavioral and metabolic effects of paternal exposure to traumatic stress in early postnatal life: evidence in the 4th generation. *Environ Epigenetics* 4(2):1–8.

Waddington CH (1942) Canalization of development and the inheritance of acquired characters. *Nature* 150:563–565.

Waddington CH (1953) The genetic assimilation of an acquired character. *Evolution* 7:118–126.

Wagner A (1999) Redundant gene functions and natural selection. *J Evol Biol* 12:1–16.

Wagner A (2000) The role of pleiotropy, population size fluctuations, and fitness effects of mutations in the evolution of redundant gene functions. *Genetics* 154:1389–1401.

Wagner A (2005) Distributed robustness versus redundancy as causes of mutational robustness. *BioEssays* 27:176–188.

Wagner A (2008) Neutralism and selectionism: a network-based reconciliation. *Nat Rev Genet* 9:965–974.

Wagner A (2011) The molecular origins of evolutionary innovations. *Trends Genet* 27:397–410.

Wagner A (2014) *The Arrival of the Fittest: How Nature Innovates*. Penguin Random House, New York.

West-Eberhard MJ (2003) *Developmental Plasticity and Evolution*. Oxford University Press, Oxford.

Wieser W (1998) *Die Erfindung der Individualität oder die zwei Gesichter der Evolution*. Spektrum, Heidelberg.

Zattara EE, Macagno ALM, Busey HA, Moczek AP (2017) Development of functional ectopic compound eyes in scarabaeid beetles by knockdown of orthodenticle. *PNAS* 114(45):12021–12026.

TIPS AND RESOURCES FOR FURTHER READING AND CLICKING

The new perspective, focused on the phenotype is presented in: Brun-Usan M, Zimm R, Uller T (2022) Beyond genotype-phenotype maps: toward a phenotype-centered perspective on evolution. *BioEssays* 44(9):1–14.

A highly readable book, although it is limited to evo-devo in the context of gene regulation: Sean B. Carroll (2006) *Endless Forms Most Beautiful*. Norton, New York.

A fascinating textbook on evo-devo from an ecology perspective, that is, eco-evo-devo: Scott F. Gilbert and David Epel (2015) *Ecological*

Developmental Biology. The Environmental Regulation of Development, Health, and Evolution. Sinauer, Sunderland.

An informative compilation of topics/concepts in evo-devo that have not yet been explored (as of 2008) is available from Lewis I. Held, entitled 101 Unsolved Puzzles in Evo-Devo. https://www.sdbonline.org/sites/fly/lewheldquirk/puzzleq.htm.

Facilitated Variation in Three Minutes (YouTube). https://www.youtube.com/watch?vynEuJi0Umms.

Marc Kirschner Facilitated variation (YouTube). https://www.youtube.com/watch?vlbcpLPcXw9M.

Denis Noble gives an overall view of his fundamental critique of 20th century biology in a dialogue on his book Dance the Tune of Life (Nov. 16, 2016). https://thethinend.podbean.com/?sDenis.

Evolution Evolving: Armin Moczek. "On the origins of novelty and diversity in development and evolution: case studies on horned beetles" (YouTube). https://www.youtube.com/watch?v=K_Vuw Byeqmg.

Neuron time lapse video (exploratory behavior example) (YouTube). https://www.youtube.com/shorts/A9zLKmt2nHo.

Arrival of the Fittest with Andreas Wagner. https://www.youtube.com/watch?vaD4HUGVN6Ko

In March 2020, the journal *Development and Evolution* published a special issue on the topic *Development Bias and Evolution*, edited by Armin Moczek. Sixteen contributions by various authors illuminate the topic from a historical and philosophical perspective, cite case studies, and discuss developmental mechanisms and empirical tests among other things.

CHAPTER FOUR

Selected Evo-Devo Research Results

Chapter 3 explained the concept of evolutionary development with a focus on gene regulation, role of cells in facilitated variation, and novel role of the environment in the emergence of phenotypic plasticity and innovation. This chapter presents selected empirical evo-devo research achievements, including findings on phenotypic variations, such as the size of colorful spots on butterfly wings that can be explained in detail by gene regulation. For other phenotypes, such as supernumerary digits, which belong to higher-level evo-devo processes, self-organization at the cellular level is discussed. In general, two-dimensional color or structural patterns are easier to explain than three-dimensional patterns, such as supernumerary digits, bird beaks, or head shape in cichlids. Therefore, this chapter focuses on phenotypic variations that are difficult to explain via the Darwinian and neo-Darwinian evolutionary models. The discussed cases give rise to various predictions of the synthetic theory, such as discontinuous inheritance, developmental bias, and others, thereby supporting the previously presented epigenetic–systemic evo-devo theory. This chapter provides insight into the concept of the EES.

Important technical terms in this chapter (see glossary): canalization, developmental bias, development constraint, development plasticity, discontinuous variation, emergence, evo-devo, evolvability, genetic accommodation, genetic assimilation, Hox genes, morphogen, point mutation, polydactyly, self-organization, and threshold effect.

4.1 EVOLUTION OF BEAK SHAPE IN DARWIN'S FINCHES

Peter and Rosemary Grant made astonishing observations on Darwin's finch species inhabiting the Galápagos Islands (Abzhanov et al. 2004). They reported that a change in food supply lasting only a few generations led to a significant transformation of the beaks of these birds (Figure 4.1). The Grant couple spent 33 years on the Galápagos Islands studying Darwin's finches and discovered an extent of phenotypic variation that had previously been considered impossible in such a short time. The beak phenotype sometimes diverges rapidly when two reproductively isolated populations of the same species on an island specialize in seeds of different sizes (Abzhanov et al. 2004). Variation in beak size and shape also requires a change (of even a few millimeters) in the fit of the hornbill into the bones of the skull since the heads of finches do not grow to the same extent as the beaks. Therefore, changes in skull and beak proportions are accompanied by adaptations of the esophagus, trachea, and tongue because all these parts must be meticulously adjusted to each other.

A growth factor protein significantly involved in embryonic beak formation has been identified. Furthermore, the variable

Figure 4.1 Beak shape in Darwin's finches. Darwin's finches of the genus *geospiza* revealed a surprise. In a few generations, they could adapt their beak shapes and sizes based on changes in the food supply.

level of expression of this protein is correlated with the beak shape. Kirschner and Gerhart also mentioned that this protein, BMP4, is produced in embryonic neural crest cells, and when transgenically expressed in the neural crest of chickens, changed the beak shape as expected. The chickens developed wider and larger beaks than normal. Heterologous expression of other growth factors did not produce the same effect. Although the experimentally manipulated beak is altered in size or shape, it is nevertheless integrated into the anatomy of the bird's head. "There is no monstrous undesirable development" (Kirschner and Gerhart 2005).

Beak formation is a complex developmental process involving five nests of neural crest cells. The neural crest is the early embryonic structure from which the peripheral nervous system and other structures develop. The nests receive and respond to signals from facial cells at five locations. Therefore, traits affecting the neural crest cells alter beak growth in a coordinated manner. The neo-Darwinian theory of evolution should be able to explain how a sequence of random mutations and selections can lead to such an extensive, coordinated, phenotypic variation in only a few generations, which in fact requires the mutual interplay of several developmental parameters.

Kirschner and Gerhart call this process facilitated variation (Section 4.4). Variation is therefore not arbitrary. Rather, "conditionally facilitated variation influences one, functional output of phenotypic variation by an organism." There is compelling evidence that pathways for the coordinated development of the beak and head exist—functional, integrated adaptability that translates randomly distributed mutations into nonrandomly distributed phenotypic variations.

In this evo-devo example, the mode of action is as follows: a small cause (one or a few quantitative, regulatory protein changes) leads to a large effect (a functional change in beak shape), controlled by epigenetic processes of development, in particular by an extensive adaptive cell behavior of neural crest cells of the beak and the visual environment. From the well-researched knowledge of beak development and modification, it can be concluded that "rather large changes in beak size and shape could be accomplished with a few regulatory mutations, rather than a summation of a long series of small changes" (Kirschner and Gerhart 2005).

In this example, the trigger for changes in BMP4 levels during development has not been investigated. One possibility is the occurrence of random genetic mutations. However, it is more likely that developmental pathways respond to animal stress caused by an external factor such as ongoing fluctuations in the food supply.

4.2 EXPERIMENT 1: FISH CAN LEARN TO WALK

This section discusses how fish can learn to run ashore. In 2014, in an 8-month experiment with juvenile bichirs from tropical Africa (Polypterus senegalus), Canadian Emily Standen, investigated for the first time, the adaptation of bichirs to terrestrial conditions by completely depriving them of their aquatic habitat (Standen et al. 2014, Figure 4.2). Bichirs have a primitive lung and the ability to waddle on land. The trial, however, tested ways to improve their performance ashore. The animals adapted surprisingly quickly to the new conditions. Trained fish lifted their heads higher on land, moved their fins more efficiently, and slipped less frequently. The experimental animals not only survived but also thrived in the new environment. Their adaptations included changes in both musculature and bone structure. The test animals could run much better on dry land than the aquatic control animals. This high developmental plasticity allows evolutionary developmental biologists to draw conclusions about how sea creatures, such as the *Tiktaalik*, first went ashore 380 million years ago and how the transition from fins to extremities could lead to the gradual development of amphibians. In fact, the skeletal changes observed here reflect those observed in fossils describing the transition of vertebrates to terrestrial life.

The bichir experiment confirmed the hypothesis that animals have the plasticity to adapt their anatomy and behavior in response to environmental changes in a very short evolutionary time, even a single generation (Standen et al. 2014). In the long term, genetic mutations could support the conditions created by the new environmental situation and ensure inheritance of the adaptation. Therefore, the evolutionary pathway is not genetic mutation ⇒ environmental pressure ⇒ natural selection ⇒ adaptation in the population, but environmental change ⇒ permanent, environmentally dependent and nongenetically inherited phenotypic

Figure 4.2 Bichir on foot. An African bichir waddles on dry land in a laboratory test. Adjustments were made to muscle and bone structure in a short time. The experiment was intended to provide insights into whether the shore leave 400 million years ago could have stimulated phenotypic changes that subsequently became genetically fixed.

adaptation ⇒ supporting genetic mutations for accommodation/assimilation ⇒ genetic inheritance.

4.3 CICHLIDS WITH THICK LIPS AND LARGE BUMPS

Cichlids are a fascinating subject of evo-devo research. They are the third largest fish family, comprising 1,700 species, and thus, one of the most species-rich groups among vertebrates. In Africa, cichlids occur in several of the great lakes, including Lake Victoria, Lake Malawi, and Lake Tanganyika. There are several hundred cichlid species in each of these lakes, occupying various, often very small, niches. The habitat of a single species can be limited to a localized grouping of rocks. What interests us first are not their great color patterns, but the evolution of a new joint in a second jaw apparatus, and second, the striking similarities in outer head shape between cichlid species.

Unlike most freshwater fishes, which have a single jaw apparatus, cichlids have two jaws. The first jaw is homologous to ours. In the gullet behind it is a second jaw with its own jaw joint and teeth. Before swallowing, fish use this second jaw to grind food initially cut up by the first jaw apparatus. The basipharyngeal joint of this jaw (not the second jaw itself) is an example of an evolutionary novelty. Notably, this joint could not have been created solely by selection forces because natural selection, in the rudimentary sense, can only influence an existing phenotype (Section 3.9). At the University of Vienna, my former colleague Tim Peterson conducted precise biomechanical measurements using software-supported, mathematical finite element analysis to elucidate the mechanical forces that interact to form the new jaw joint in the pharynx of two types of cichlids. He demonstrated the individual forces that exert biomechanical pressure on the jaw system to trigger cartilage development for the new joint. These combined effects induce cartilage formation at a site between the base of the skull and the jaws of the throat where the new joint develops in cichlids. The example shows that physical forces contribute to phenotypic variation or innovation during embryonic development. In simple terms, the result can be summarized as follows: when two bones grow toward each other with sufficiently high pressure, a new joint is formed (Peterson and Müller 2018).

Let us now discuss the peculiarities of the head form of cichlids. In Lake Malawi and Lake Tanganyika, according to the synthetic theory, natural selection led to the conspicuous body shapes observed in cichlids, such as oversized lips (Figure 4.3b), short, robust lower jaws, bulging foreheads (Figure 4.3), and other characteristics. A neo-Darwinian evolutionary biologist would

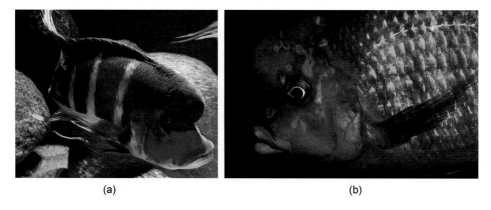

(a)   (b)

Figure 4.3 African cichlids. Tanganyika frontosa cichlid (*Cyphotilapia frontosa*, (a) and hump-head Mouthbreeder (*Cyrtocara moorii*) from Lake Malawi (b). Both species are endemic to their respective lakes. The conspicuous head shape was created by parallel evolution. No common ancestor bears the same characteristic.

use convergence—phenotypic similarity as a result of adaptation—to explain the presence of fish with similar traits in both lakes (Meyer and Stiassny 1999). According to this theory, similar environmental conditions exerted positive selection on random genetic mutations, which led to a comparable phenotype—a classic adaptation process. However, this explanation requires extraordinarily high similarity between lakes to account for the several parallel forms that have independently developed in each lake.

According to the Modern Synthesis, natural selection can exploit all physical possibilities to select the most suitable from an immeasurable spectrum, as detailed by Carroll (Section 3.3). However, from an evo-devo perspective, the full spectrum of physical possibilities is not available to the developmental program; the possible outcomes are limited by the developmental process itself.

Evo-devo researchers use the term *parallel variation* instead of *convergence* because the latter has already been claimed by proponents of the Modern Synthesis. Patterns of parallel evolution, such as the bulging heads of cichlids living in different lakes of the Rift Valley, provide a good visual representation of the concept at hand. Are phenotypic similarities due to comparable living conditions under which natural selection can act, or is the influence of biased development involved? Are there developmental pathways that influence how phenotypic variation should look? In this case, the course of evolution depends on the evolvability of the traits considered and how these traits conform to the fitness curve and requirements of natural selection. Evolvability is the capacity of a species for adaptive evolution. It can be small or large, and it can be oriented in certain directions. Therefore, according to the evo-devo perspective, natural selection and development work in concert. However, the relative proportions of both types of factors, extrinsic (selection) and intrinsic (development), remain unclear until more is known about the developmental link between the genotype and the phenotype. It can be hypothesized that adaptive evolution along the given developmental pathways orchestrates the development of phenotypic variation. Considerable empirical research is required to support or refute this hypothesis. Patterns of similar parallel morphological evolution in similar ecological environments are expected to occur. These patterns become predictable with precise knowledge of the development processes (Brakefield 2006, 2011). However, it is also clear that evolvability itself is an evolved and evolving product of evolution (Uller et al. 2018).

4.4 EXPERIMENT 2: BUTTERFLIES WITH EYES ON THEIR WINGS

Let us attempt to further solidify the subject of bias in evolutionary development. When there is a bias, certain phenotypes are easier, and thus, more likely to develop than others. Developmental mechanisms can limit or canalize evolutionary change. Thus, development follows a goal-less bias or direction because evolution is not a goal-oriented process.

The cichlid example from the preceding section does not provide much empirical evidence. One might argue that the parallel evolution of fish with bulging heads is not likely explained by natural selection alone. Such is not the case for butterflies, on which there is a wealth of data. The wings of butterflies with their countless colors and structural patterns are a goldmine for biologists. The reader might have already questioned how perfectly round eyespots developed on the wings of such beautiful creatures. Was the selection process readjusted thousands of times until the spots were truly round? Probably not. In the embryo, a morphogenic protein called *Wingless* is expressed in the center of the eyespots and diffuses outward in a concentric pattern, like a drop of cream in a coffee. *Wingless* was so named when its deletion in *Drosophila* led to wingless mutants. *Wingless* activates the expression of the target gene *Distal-less* (*Dll*) and, depending on its distance from the center, also increases or decreases the expression of other genes. *Dll* is a homeobox gene (Section 3.4) discovered by Carroll in the eye patches of butterfly wings. The outward diffusion of the morphogen only partly explains the circular form of the eyespot; the true mechanism is more complex. We will revisit morphogens when we discuss the emergence of fingers. At the heart of *Dll* expression in the eye spot is a signal center that directs the fate (differentiation) of all cells in its immediate environment. In the 1920s, Hans Spemann, a developmental biologist from Freiburg, Germany, discovered such an organizer in the early embryonic stage of tadpoles, where fundamental developmental decision-making takes place. Spemann was awarded the Nobel Prize for Medicine in 1935 for his discovery of a shape-forming gradient, which remains a guiding principle in developmental biology to this day.

The *Dll* gene was discovered several years ago in fruit flies, where it is involved in a completely different function from that in butterflies (as is *Wingless*)—the development of appendages, from legs to antennas to wings. In butterflies, it appears with a new function and switches; to use Carroll's words, "old genes learn some new tricks". In 1994, Sean B. Carroll's evo-devo discovery of the genetics of eyespots and their manipulation briefly brought him into the media spotlight. At the time, the question had already been asked: "If you can change butterflies in the laboratory, can you also change people?" The exciting story was published in the renowned *Science* magazine (Carroll et al. 1994).

For our purposes, we want to know whether it is possible to distinguish between natural selection and developmental bias in eyespots. The reader may know that the evolution of eyespots (including their sizes and colors) is characterized by an extraordinarily high degree of flexibility (evolvability), i.e., a high potential for independent variation. This concept applies to many other wing patterns and speaks more to a dominant role of natural selection than to developmental constraints (Beldade et al. 2002).

One experimental approach analyzed the factors influencing wing spot development using artificial selection. The experimental insect was the small, brown African butterfly, *Bicyclus anynana*, and a closely related Asian butterfly of the genus *Mycalesis*. Why study two species from different continents? The answer is to determine whether similar results for eyespots in laboratory experiments indicate similar developmental constraints and biases in the form of a coupling of the two eyespots on one wing.

Using artificial selection, i.e., breeding, individual eye spot size was targeted in both species. Therefore, the proportions of adjacent eyespots on the same wing received considerable attention. Artificial selection in different directions for the relative sizes of the two eyespots provided remarkable evidence of evolvability. The origin of the diagram in Figure 4.4 represents the absence of spots. The x-axis represents the direction of selection of the rear eye spot; the y-axis, that of the front eye spot on the same wing. The point in the top right corner thus represents both spots. In this region, breeding resulted in two large spots.

In more detail, the occupation of the morphological space for the relative size of the two dorsal forewing eyespots of the butterfly *Bicyclus anynana* is compared with the variations between species of this African genus and the closely related Asian genus *Mycalesis*. The four images of the wing in each corner of the morphospace are representative examples of the wing pattern after 25 generations of artificial selection in *B. anynana* in the direction of each of these corners of the morphospace, starting from the wild type for this species (star). Circles show the position of the average sizes of the same eyespots in different types of *Bicyclene* (closed symbols) and *Mycalesis* (open symbols). The dotted square includes species in which both eyespots are very small or absent and often difficult to measure.

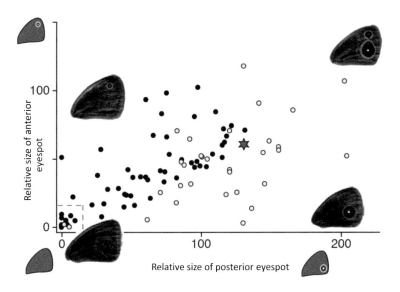

Figure 4.4 How do you like your butterfly? Do you prefer one or two eyespots each on the forewings? Better yet, would you choose a large one with a small one? Some things can be produced in the laboratory, but others cannot. Is development creating hurdles here? More on this in the text.

The overall phenotypes of the two species are distributed throughout the space. This is a clear indication of the high evolvability already mentioned, as well as natural selection. With highly independent variation, nature can flexibly adapt changing hunter-prey schemes and mating preferences in the best possible way.

Let us examine the distribution of the test results more closely. The individual points indicate average values. Some of the phenotypes fall near a straight line that crosses the origin. The imaginary line reflects positive developmental coupling of the two eyespots. In *Bicyclene* (closed symbols), there are no points far below this line. In other words, in addition to the confirmed influence of natural selection, this line or the frequency of points in its vicinity is seen as the bias of the interdependent sizes of eyespots. Thus, we have both, possible natural selection and possible developmental coupling (Brakefield 2006).

Among the approximately 80 natural species of the genus *Bicyclus*, there are numerous combinations of eyespots on the forewings of females, from no spots at all (bottom left on the diagram) to two on each wing (top right). There are also individuals with a single anterior spot (top left). However, there are no natural species with an exclusive posterior spot (bottom right), which also did not occur in the extinct *Bycyclus* species. Therefore, the absence in nature of a single posterior spot cannot be explained by its natural selection history, and one could not be produced in the breeding lines of *Bycyclus*. There are no closed dots right below the line in the diagram, but there are open dots representing the genus *Mycalesis*. The authors concluded that this trait can be bred in one species and not in the other; therefore, there is no constraint on the development of a sole posterior eyespot, but its absence is due to natural selection (Beldade et al. 2002).

However, if we examine the genus *Bicyclus* in isolation, the absence of a sole posterior spot in nature and via breeding (Table 4.1) indicates a possible developmental constraint that prevents its occurrence. This conclusion would only be inappropriate if the development of the trait in both species were so similar or even identical (due to their close

TABLE 4.1
A sole posterior eyespot on the butterfly wing

Characteristic	Genus	Breeding	Nature	Conclusion
Post. eyespot alone	*Mycalesis*	yes	no	natural selection
Post. eyespot alone	*Bicyclus*	no	no	constraint?

Is it prevented by natural selection or by developmental constraints? An alternative interpretation is possible.

relationship) that such a constraint would be unlikely. In this case, however, according to the assumptions of Brakefield, one would need to answer question of why the distribution of closed and open dots representing the two genera in Figure 4.4, bottom right, differs so widely. Each dot corresponds to the average value of several individual analyses performed over 25 generations (Brakefield 2006). The deviations are therefore statistically significant. An interesting working hypothesis would be that in this special case, a constraint suppresses the formation of a singular posterior eyespot. For this to be true, the development of the posterior spot would have to depend on that of the anterior one. Without the anterior spot, there would be no posterior spot and vice versa. Such a hypothesis could be used to test for causal dependencies in the developmental process. One could thus examine why the presence of a sole posterior spot is not pronounced. Nevertheless, there is still much research to be done on the evolution of eyespots. We will revisit *Bicyclus* in Section 6.4 when we discuss an Extended Synthesis project. Section 4.8 will touch another example of a trait that does not persist in nature—polydactyly.

In another experiment, David Houle and colleagues photographed and measured 50,000 fly wings (Houle et al. 2017) and found that some wing shapes were more common than others. The same was observed for mammalian teeth (Figure 4.5): some shapes that were expected to be common were not detected at all. At this point, the researchers questioned the existence of a connection between the directional developments observed in these experiments and the evolutionary diversification of the species. As

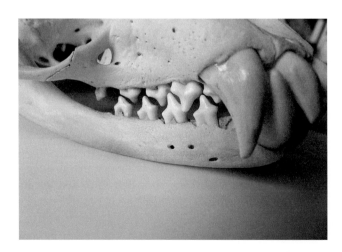

Figure 4.5 Skull of a Weddell Seal (*Leptonychotes weddellii*). The evolutionary diversity of tooth morphology in mammals is determined by the mechanism by which teeth develop in the embryo.

in the case of the *Bicyclus* butterfly, the non-occurrence of possible forms suggests a bias toward variation. The team had a good reason to suspect that evolution can be observed when development allows it, for example, through mechanisms of developmental bias.

The experiments described above revealed an apparent paradox: natural selection and the developmental bias may work against each other in the formation of morphological variants. Natural selection tends to maximize the number of possible options, while the developmental bias has a restrictive effect, making some forms more likely to occur than others. It could be argued on the basis of this apparent contradiction that the developmental bias in evolution may promote the ability to adapt and diversify. The only way to discover the reality is to represent the possible pathways in computer models with regulatory networks and to test various developmental and selection scenarios. These models could simulate the desired skewed phenotypic variations (cf. also Section 4.8). The fact that all phenotypes are produced by development implies that selection and directional development jointly determine the course of evolution.

The use of computer models helps to understand how biased development evolves and, conversely, the influence of biased development on evolution. Ultimately, according to Tobias Uller, this research could show how the evolution of the evolutionary process itself contributes to diversification and adaptation (Uller et al. 2018).

4.5 EXPERIMENT 3: GENETIC ASSIMILATION IN TOBACCO HORNWORM

Can a genome adapt retrospectively in such a way that a newly evolved phenotypic trait becomes genetically fixed *a posteriori*, i.e., hard-wired? An experiment was conducted in the USA by Yuichiro Suzuki and Fred Nijhout, (Suzuki and Nijhout 2006) on caterpillars of the tobacco hornworm butterfly (*Manduca sexta*). We refer to them here as species A. In nature, they occur as a green wildtype form (A) and a black mutant form (A′). A caterpillar of the related species B can develop a polyphenism that produces black larvae at an ambient temperature of 20°C and green larvae at 28°C. The aim of the experiment was to induce a species B-like dimorphism in species A (green) by exposing the larvae to a temperature stressor of 42°C for 6 hours. After 13 generations, using four breeding lines and one heat shock per generation and per line, the researchers produced a polyphenic mutant from A and A′, i.e., a species that, after discontinuation of the shocks for 13 or more subsequent generations, produces both black and green variants in the absence of the stressor (Figure 4.6), depending on the ambient temperature. The experiment aimed to produce a variant of species A with a phenotype as variable as that of the natural species B. The effect of the stressor on the first 13 generations of larvae was genetically consolidated, i.e., assimilated. The outcome of this experiment is comparable to that of Waddington on the veins of *Drosophila* wings (Section 3.1). Hence, genetic assimilation can be proven *a posteriori*, at least in the laboratory. Evolution in action!

To summarize, this example illustrates a different causal sequence from that of the Modern Synthesis: an environmental stressor leads to epigenetic variation and subsequent inheritance of a phenotype. This leads, with the appropriate mutations and their inheritance, to genetic assimilation. The environmental stressor is thus no longer required to maintain the new phenotype. This is the emphatic consequence of Mary Jane West-Eberhard's developmental plasticity (Section 3.8) and a pillar of the EES.

Further examples of genetic assimilation can be found in Section 3.8.

4.6 THE ALMOST NONCONSTRUCTABLE TURTLE SHELL

The turtle shell is not only a special feature in the animal world but also an anatomically

Figure 4.6 Frederic Nijhout, and tobacco hornworm caterpillars (*Manduca sexta*). The light-colored caterpillar on the left was treated with heat shocks in the laboratory. After prolonged exposure to the stressor, the caterpillar's offspring turned dark and continued to inherit the new color coat with heat shock treatment until the treatment was no longer required. Black was genetically assimilated.

complex structure. Widenings and outgrowths of the ribs are causally involved in its formation. The dorsal carapace represents a series of several evolutionary innovations via which the ribs formed the shell as we know it today (Rice et al. 2015). In the process—another remarkable event—bones emerged on the body surface. When asked the function of the turtle shell, you will say without hesitation that it serves to protect the animal. However, if we consider its gradual development via widening of the ribs, we may immediately view this idea critically. Over 50 million years, the intermediate shell forms did not form a carapace and thus offered no protection. Tylor Lyson of the Denver Museum of Nature and Science recently proposed a surprising new idea, postulating that broader ribs provided the animal with a stable base, and increased the stability of the spine facilitated digging. According to this hypothesis, the protective function arrived much later (Lyson et al. 2016). The evolutionary biologist uses the term exaptation to refer to a novel function of an existing trait. An analogous example is the bird feather, which originally served in thermoregulation and not flying.

More recently, the structure of the turtle shell, which has been considered enigmatic since the anatomical studies of the 19th century, has become a subject of renewed interest to evolutionary biologists. A few years ago, it was postulated that the facts of developmental biology suggest an origin of the shell through a macromutation (saltation) of the shoulder girdle, and thus, partially refute the synthetic evolution theory (Gilbert et al. 2001), at least insofar as it espouses gradualism as the only possible mechanism.

Explaining the origin of the turtle shell is akin to attempting to explain how a ship got into a bottle. If you place one hand around your shoulder and palpate the shoulder blade and collarbone with your fingers, you will notice that both the structures lie outside of your ribs. One would expect the shoulder girdle (composed of the scapula at the top and back and the clavicle at the front) to be located outside the ribs in the turtle, as in the body plan of all reptiles and mammals. It was therefore inexplicable how the shoulder girdle could enter the carapace, and thus become situated under the vertebral column and ribs via a gradual evolutionary process (Figure 4.7). Moreover, it is difficult to imagine viable intermediate forms. The current result amounts to a considerable skeletal reconstruction. Few other examples in

Figure 4.7 Development of the turtle shell. The shoulder girdle (arrow) is inside the carapace and thus inside the ribs in the turtle but outside the ribs in all other reptiles and mammals. Evo-devo investigates how this could have arisen during evolution.

animal phylogeny are more reminiscent of a *hopeful monster* than the shells of turtles. In fact, a saltation in the classical sense, was not likely required but neither were gradual variations per the Modern Synthesis, as summarized by Scott F. Gilbert (personal communication). Several more complex evo-devo reconstructions await discussion.

The process of turtle shell evolution might be imagined as follows: the decisive morphological trait is the ribs, which do not enclose the thoracic cavity as in all other vertebrates but grow into the upper (dorsal) body wall. There, they connect with skin ossifications, which presumably developed independently. During embryonic development, both formations are controlled by a cellular signaling pathway (the *Wnt* pathway), which is also involved, for example, in limb formation (Kuraku et al. 2005). The corresponding partial remodeling of preexisting developmental pathways (and their genes) is referred to as *cooption* and occurs frequently, although the details are not fully understood. The most likely explanation is the transformation of genetic switches (cis-regulatory sequences); a gene may be expressed in a new functional context without undergoing mutational change. During this transformation, the lateral body wall of the turtle embryo folded inwards, a process referred to as *evolutionary origami*. As a result, the shoulder girdle, which is otherwise located above the ribs, was displaced inward (the original position can be gleaned from the muscles attached to it; Nagashima et al. 2009).

This explanation is plausible; however, it has not gone unchallenged. In the meantime, the aforementioned paleontologist Tylor Lyson XE "Lyson, Tylor" stated that the position of the shoulder girdle within the ribs was the original position during the evolution of the amniotes (quadrupeds that can reproduce independently of water). Thus, the form present in, for example, mice and chickens arose later (Lyson and Joice 2012). At this juncture, a new question is raised: how did the shoulder girdle migrate from the inside to the outside of the rib cage? There are two possibly correct models, according to a team around Scott F. Gilbert, if a distinction is made between turtles with hard and soft shells (Rice et al. 2015).

The temporally parallel mechanism of abdominal carapace (plastron) development has not been fully elucidated. Presumably, it involves a cellular signaling pathway corresponding to the one that leads to ossification of the skull bones. Interestingly, a 220-million-year-old fossil of an aquatic turtle (*Odontochelys semitestacea*) discovered in China had a complete ventral carapace, but the dorsal carapace was missing (Li et al. 2008; Joyce et al. 2009). Meanwhile, 40 million years old turtle relicts with strongly broadened thoracic ribs (*Eunotosaurus africanus*) have been found in South Africa (Lyson et al. 2013). The family tree of turtles could be recontextualized in view of this finding.

4.7 HOW MANY LEGS DO MILLIPEDES HAVE?

The reader is certainly familiar with millipedes (*Myriapoda*) and centipedes (the class *Chilopoda* of millipedes), though it is less commonly known that 16,000 species of the former exist. Nevertheless, both names are a gross exaggeration. There are neither millipedes with 1,000 legs nor centipedes with 100. *Illacme plenipes*, a millipede species first described in 1926, has 750 legs on a 3-inch-long body, which makes it the species with the most feet among all the animals on earth. The centipedes (*Chilopoda*), which in reality have fewer than 100 legs, are more interesting from an evo-devo perspective. There are an estimated 8,000 species in this class, only 3,000 of which have been described. However, I would like to take a quick detour before we get back to the centipedes, which are of particular interest to us here.

Those who believe that among our zoological kin, brood care only exists in mammals are mistaken. Parental care is much more widespread in the animal kingdom. For example, the strawberry frog (*Oophaga pumilio*) is a poison dart frog barely the size of a human fingernail. Astonishingly, the father piggybacks the tadpoles one by one and carries them to a suitable aquatic habitat in the funnels of bromeliads (Figure 3.10). The father remembers each location where he has left a tadpole. If the tadpole has no food, the mother can lay an unfertilized egg, which provides the necessary calories for the young. Countless stories about self-sacrificing brood care could be told here, from ground-dwelling ringed tadpoles (*Siphonops annulatus*), which eat only the mother's skin for 2 months after hatching, literally grazing on it, to the deep-sea octopus (*Graneledone boreopacifica*), the female of which incubates her olive-sized eggs for an unbelievable 4 years and takes no food at all before dying after the young hatch. Let us return to the centipedes: among them is the order of earth lovers (*Geophilomorpha*), comprising more than 1,000 species alone, the females of which engage in brood care by lying backward around their egg ball and licking the eggs regularly to protect them from fungal attacks. Therefore, paternal care can be considered a particularly successful evolutionary trait that typically arose independently throughout the animal kingdom and continues to be highly prevalent.

All centipede species have one thing in common: an odd number of leg pairs, from 15 to 191. Biologists are challenged to determine how leg number in centipedes is so stringently controlled. The answer could be that an odd number of leg pairs is adaptive, and therefore, a result of natural selection. However, it is doubtful that 191 leg pairs make for a difference in fitness compared to 190.

Evo-devo provides another answer. There must be developmental constraints that prevent the incorporation of an even number of leg pairs into the body plan. The following explanation has been proposed. Segments corresponding to a pair of legs are always present in pairs, forming an indivisible module with two pairs of legs (that is, an even number of leg pairs). The first pair, however, is reserved for the poison claw, of which there is only one pair. The sum of leg pairs thus remains odd, no matter how many are added. The genetic basis for the double pairing of the legs on each segment has been studied in depth (Damen et al. 2009).

The odd number of leg pairings is a trend bias par excellence. The direction of evolutionary change (odd leg number) is influenced by the nonrandom structure of variation.

Even more can be learned from centipedes in the evo-devo context. For the moment, let us put aside the fact that leg segments on a centipede are a discontinuous trait. A sequential increase in the number of leg pairs by one double segment, i.e., four legs, is already a discrete variation discordant with the Modern Synthesis. Without knowing better, one might imagine the sequential evolution of even more pairs of centipede legs, not necessarily so. *Scolopendropsis duplicata* of the order of the giant tropical centipedes (Scolopendromorpha)), discovered in Brazil a few years ago, has 39 or 43 leg pairs (the females usually have a few legs more than the males). This is an anomaly because the sister species (*Scolopendropsis bahiensis*) has only 21 or 23 pairs of legs. Seven hundred other species belonging to the same genus also have 21 or 23 pairs. The two sister species with different numbers of leg pairs live in neighboring regions and are evolutionarily very young. Moreover, their common ancestral line has only 21 or 23 leg pairs. Studies have attempted to elucidate intermediate forms. However, in the same genus, there are no species with an intermediate number of leg pairs. Thus, there is no mechanism of selection or genetic drift working toward an intraspecific, basically uniform increase in the number of legs. The 39 or 43 leg pairs of *Scolopendropsis duplicata* can therefore be seen as a saltation, an evolutionary leap.

Genetically, this phenotypic jump may have been initiated by a point mutation. The mutation could have been the basis for a doubling of the number of segments. However, the constraint that prevents only a single new pair of legs from being created is probably genetically more complex, making it a more interesting challenge for evo-devo to explore in more detail. Regardless, this phenotype is a sensation. Such jumps could have occurred several times due to a simple doubling mechanism. The abovementioned neighboring earth lovers, with their maximum number of leg pairs of 191, could have jumped from an original 21/23 leg pairs to 39/43 and so on from there (Minelli et al. 2009).

Before we proceed from the discussion on centipedes, it is evident that the extent of variation in centipedes is of particular interest, not least because the variation takes place among closely related species. This is a clear example of what the German evolutionary biologist Richard Goldschmidt was referring to with his idea of *hopeful monsters*, for which he was rebuked by neo-Darwinists. William Bateson (Section 1.2) would certainly have liked to include *Scolopendropsis duplicata* in his main work.

4.8 SUPERNUMERARY DIGITS

The next example may be unfamiliar to most readers. Men usually frown when asked if they know the phenomenon. Women sometimes give answers such as: "Yes, I know that. My cousin's sister-in-law's daughter has that." In every medium-sized town, there are people who were born with six fingers, six toes, or both. The technical term for supernumerary digits is polydactyly (Figure 4.8). The current world record is held by a Chinese boy who inherited 15 fingers and 16 toes from his mother. The same variable phenotype occurs in dogs, cats, guinea pigs, horses, domestic pigs, and chickens. In humans, polydactyly is not as rare as one might think: on average, it occurs once in 10,000 births and much more often in Africa.

It is easy to explain why you may not have observed someone grabbing a mug with six fingers in a beer garden. In most cases, the extra finger is removed after birth, as is the extra toe to avoid the problem of overly

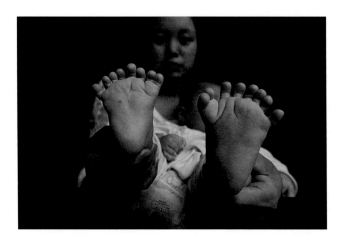

Figure 4.8 Polydactyly: a possible world record. Surplus fingers and toes are a common variation among numerous quadruped species. Here is a very rare case of a baby with 31 fingers and toes. Different numbers of digits develop in a single generation and are variably inherited depending on the mutation type. In such cases, siblings can have different numbers of fingers or toes. Polydactyly is a prime example of complex phenotypic variation. It is initiated by a single point mutation. However, genetic mutation alone cannot explain how new bones, joints, blood vessels, muscles, tendons, and nerves develop. That explanation is left to evo-devo.

tight shoes. Often, a tiny digit, with its still-soft bones, can simply be tied off after birth. Sometimes, a more complicated operation is necessary, as in the case of an incomplete doubling of the thumb, when bones, nerves, tendons, and muscles must be precisely separated in a difficult, nonroutine surgery, and the slightest damage to the important remaining thumb must not occur. Such an operation may have to be performed in one case on the first phalanx and in another on the end of the metacarpal bone at the base of the thumb. Polydactyly often forms much more beautifully on the outside of the hand, i.e., next to the little finger (postaxial). In contrast to supernumerary finger placement beside the thumb, which is somewhat annoying because of the unique thumb structure, there is space beside the pinky where a new finger can develop without restriction. The probability of an extra toe developing unencumbered is equally good on both sides of the human foot.

To a surgeon, polydactyly is a malformation, but from an evolutionary biologist's perspective, it is a highly interesting phenotypic variation. Moreover, the trait is complex, consisting of separate bones, joints, tendons, muscles, blood vessels, nerves of all skin layers, etc. The extra digit often functions without restriction. A person with a sixth finger can feel, touch, and grasp with it. A spontaneously created new finger can even have its own representation in the brain, with the help of which it is able to perform independent, coordinated movements just like the other fingers. Development, thus enables the control of additional body parts by the brain, a fascinating new discovery made in two young adults (Mehring et al. 2019).

Polydactyly is of interest to evolutionary biologists because it is, among other things, inherited. Once the autosomal dominant mutation is present, the trait remains in the line forever and can be inherited from either parent. The trait may not appear in the phenotype, even in the presence of the mutation, but a brother or sister will have an extra finger or toe, which will be passed on to future generations in a quasi-mysterious way, even in different numbers. In very rare cases, carriers develop certain syndromes, complicated congenital disease patterns, which may be accompanied by severe impairments. This occurs when polydactyly is caused by mutation of an important developmental gene, because such a gene may function in numerous organismal locations during embryonic growth and formation.

4.8.1 The Uniqueness of the Human Hand in the Animal Kingdom

Before we discuss the individual finger in more detail, let us examine the full evolutionary significance of the hand. Approximately one quarter of all the bones in the human skeleton can be found in the two hands (27 in each). These are held together tightly by joints in an intricate arrangement, intersecting ligaments, 33 muscles, three main nerve branches, connective tissue, thick to very fine blood vessels, and thousands of highly sensitive tactile sensors. Together, these structures "form the most filigree and versatile tactile and grasping tool that evolution has produced to date" (Böhme et al. 2019). (A possible exception to this statement might be the elephant's trunk with its 150,000 muscle bundles.) On each hand are five slender and delicate fingers. Attached to them are thin but strong tendons that not only extend into the forearm but are also connected to the muscular apparatus up to the shoulder, and of course, to the brain via nerve branches. If we did not use this intricate system every day as a matter of course, we could not likely conceive of it. Exceptional musicians such as Yuja Wang can play the piano so well that you doubt your sanity when you listen to her. In Glashütte, Germany, the watch center in Saxony, five specialists from the company A. Lange & Söhne spent 5,000 hours of manual precision work to restore the resurfaced, invaluable Grande Complication No. 42500 from 1902, a pocket watch consisting of 833 individual parts and one of the most complicated in the world. Some of the screws of this unique piece were only 0.05 mm in size. This indeed requires real tactile ability.

Madeleine Böhme, who recently discovered the spectacular, upright-walking *Danuvius guggenmosi* in the Bavarian Allgäu, and her coauthors describe the hand as an independent sensory organ (Böhme et al. 2019). In the dark, we can decide in a flash with our fingers, via their specific receptors in the brain, whether something is cold or warm, wood or stone, solid or fragile. The human hand with the highly advantageous specialized function of the opposable thumb has existed since the first tools of *Homo habilis* 1.8 million years ago and probably, according to Böhme, much longer. The ability to walk upright, which appeared much earlier than has long been assumed, freed the hands from walking and allowed them to perform countless new activities. The hand also contributed to the evolution of language. According to the linguist Michael Tomasello (Section 7.3), before the first spoken words, gestures were used for communication; if this is true, then the emergence of the hand was a decisive event in human cultural evolution.

Today, the development and evolution of the hand have become the object of computer models. We will become acquainted with one of these shortly. On the one hand, the simulations are astonishing and can give us insights into the rudimentary formation of the hand; on the other hand, the digital result is sparse compared to the natural reality.

4.8.2 Why Is Five the Default Number of Fingers?

Repeatedly, I am asked why so many animals have five toes on their limbs. Why is the default number not eight or ten? Whether in a mouse, elephant, whale, or sauropod (Figures 4.9 and 1.2), despite the huge differences in size, we usually find five toes. In less common cases, the default number is four, as in my cat's hind feet, three, as in bird wings, or one, as in horses. Pentadactyly (five-fingeredness) has been called an evolutionary enigma, a mystery. It has existed since after the Upper Devonian, almost 400 million years ago. The trait is persistent, stable,

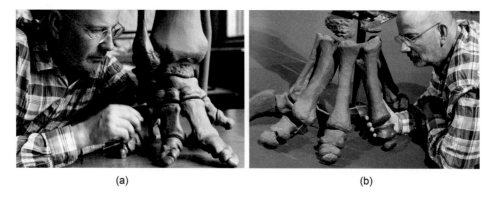

Figure 4.9 (a and b) Five fingers and toes for hundreds of millions of years. Pentadactyly (five-fingeredness) is an extremely robust homologous feature of the vertebrate hand. Depicted next to the author on the right is the hind foot of a *Diplodocus* from the Upper Jurassic with five toes and the same bone elements as in humans. On the left is the smaller, four-toed forefoot. The animal lived approximately 150 million years ago, grew up to 27 m long, and was thus one of the largest land creatures ever to inhabit the earth. The hind foot of *Diplodocus* was accordingly among the largest of all land animals' feet.

and robust; from small to large species, the pattern is the same.

Why then, are there not six digits on each limb? One is inclined to believe that an additional digit would confer an evolutionary advantage. Perhaps not, because someone with six fingers on each hand could not necessarily play the piano better. In any case, the panda has a thumb and thus six toes, even though the thumb is not a real finger but an extended carpal bone (Gould 1980). With it, he can grasp his bamboo twigs with great skill and eat them at his leisure. The number five per hand or foot must simply have proved to be a pretty outstanding solution during evolution. The deficiency in such statements is that they lack concreteness.

In Science, it is difficult to impossible to provide a positive answer to a negative question, i.e., why something does not exist. There is no empirical evidence to explain absence. One can ask why we do not have eyes in the back of our heads, like the jumping spider, but the answer cannot be provided. The positivism of Auguste Comte, who laid the foundation of our present positive science in the first half of the 19th century, determined that science investigates that which is present, visible, and observable before the eyes, and thus not what is invisible. Nevertheless, it is possible to query a nonexistent trait if causal reasons, in our case developmental constraints, are suspected of suppressing the trait. We have already asked such a question in the case of the nonexistence of a single posterior eyespot on the *Bicyclus* butterfly (Section 4.4). An even more impressive example is the recent laboratory experiment by Armin Moczek (Section 3.9), in which the switching off of a gene indicated that its normal function was to suppress the formation of a third eye in the horned beetle.. If the gene is knocked out, ectopic tissue for a new eye is formed, including the neural pathways to the brain (Zattara et al. 2017). Clifford Tabin, a Harvard expert on limb development, with reference to a Hox gene cluster, identified developmental constraints that do not permit more than five fingers (Tabin 1992). However, this explanation no longer remains convincing.

More recent studies have come closer to the question "why we have five fingers?" Developmental pattern formation in the embryonic tissue of the millimeter-sized bud that leads to the finger structure has been patiently explored by whole generations of scientists over several decades. From empirical observations made in the early 20th century to the empirical and increasingly theoretical works that began in the late 1960s, the molecular and cellular steps responsible for the patterning of fingers and toes have been elucidated using increasingly sophisticated computer models of self-organization. Nevertheless, the robust upper limit of the fingers is yet to be explained. Some parameters in the differential equations of Turing models are always more or less arbitrarily fixed. If any one of them is slightly varied under certain circumstances, the number of simulated fingers often changes very quickly. This is useful for the analysis of polydactyly but less so to prove the robustness of such a system (Lange et al. 2018). An entirely convincing explanation of five-finger robustness is yet to be offered, even after more than a century of research.

4.8.3 The Ancient History of Polydactyly

Scientists have always been fascinated by supernumerary fingers and other duplicated body structures. I was convinced that Darwin had been the first to mention surplus fingers and toes in humans and animals. He knew of people, dogs, and cats with extra toes, and he also knew that polydactyly could arise in a single generation. As already mentioned, Charles Darwin wrote casually about it, even though variation in digit number did not fit at all into his theory, according to which large phenotypic deviations arise cumulatively over long periods of time.

Surprisingly, Darwin was not the first to confront this topic. After painstaking research and lengthy discussions with ancient orientalists, my thesis advisor and I learned that the Assyrians recorded on cuneiform tablets approximately 4,000 years ago the fate of the daughter of a princess who was born with six fingers on each hand and six toes on each foot. Seeking to know more, Aristotle was the first person to deal with embryology. He was interested in the reason why certain duplications occur in the chicken egg. He imagined various liquids during fertilization that are not mixed as standard in the womb and therefore have an unwanted effect. That was a bold guess; however, his even more far-reaching idea was that the embryo was not present in miniature form during conception but first had to undergo a shape-forming process. It is said that he observed this in chicken eggs, which he opened on different days. His forward-thinking teaching was to be denied for centuries (Lange and Müller, 2017).

You are probably familiar with the Persian physician Avicenna Latinus from the film *The Medicus*. Avicenna was one of the brightest and certainly the most famous thinkers of the Arab world around 1000 AD. In a small, next to the famous *Canon of medicine* less known font, Avicenna chose the example of a supernumerary finger (*digitus superfluus*) to make it clear to his readers that rare events are never accidental or supernatural (Avicenna 1992). They always have a natural cause. This was a revolutionary thesis for a time when people in Europe had believed for centuries in supernatural forces for everything inexplicable. Not until the Enlightenment and Immanuel Kant did considerations like those of Avicenna gain acceptance in our culture.

In the early part of the modern era (17th–19th century), European scholars debated the origins of the embryo. One doctrine, known as preformationism, held the firm conviction that the embryo was fully and correctly formed as a miniature human being (homunculus) during fertilization and from that point on only needed to grow. Others supported Aristotle's view that the embryo underwent a lengthy, mechanistic design process. Unfortunately, following the invention of the microscope by Antonie van Leeuwenhoek, the discussion took an incorrect turn. Drawings appeared, in which a homunculus, a miniature human being, could be seen in the head of a sperm. Leeuwenhoek was the first person to see a human sperm. He also observed red blood cells, *Volvox algae*, fantastic planktonic organisms, and other life forms never before seen by human eyes. However, he could not have observed a homunculus in the head of the human sperm, because such a thing does not exist. Nevertheless, the preformation theory was bolstered by this story. Certainly, malformations such as polydactyly were not consistent with the (divinely inspired) preformation theory.

Francesco Marzolo, an Italian surgeon from Padua, stayed in Vienna in 1842 upon completion of his studies. There, he wrote a small Latin script entitled *De Sedigitis Dubia Physiologica*, which translates to "Physiological questions about hexadactyly." In this booklet, forgotten and never quoted until I rediscovered it on the old shelves of the Austrian National Library, Marzolo analyses an Italian family with hexadactyly (six fingers) over four generations. People with this trait did not always have the expected number of fingers and toes. They had between 20 and 24 fingers and toes per individual. Marzolo concluded that the trait could be inherited from the father, the mother, or both. Moreover, it could be passed from father to daughter or from mother to son. Today, such a trait is said to follow an autosomal pattern of inheritance. This realization on the part of Marzolo was revolutionary. It contradicted almost everything that had been taught about inheritance for centuries, namely, the material contained in the father's seed always determines the child's characteristics. Marzolo also observed that a trait could skip a generation and reappear in the next one.

The French universal scholar Pierre-Louis Moreau de Maupertuis had elucidated the inheritance of polydactyly almost one hundred years earlier in a study of a Berlin family and had justified it mathematically and statistically for the first time. However, Marzolo's statements were more concrete and far reaching. Gregor Mendel's work was still a long way in the future (Lange and Müller 2017).

Charles Darwin reported on 46 people with polydactyly with whom he had personally communicated. The term *polydactylism* appears 81 times in Darwin's complete works (darwin-online.org.uk). In his 1868 book *Varying of animals and plants in the state of domestication*, he admits to having faced significant hurdles when attempting to explain polydactyly. He wrote that he could not assign this variation to any rule or law. He also found that in the regular form (wild type), there is no rudimentary toe. Therefore, he assumed that all mammals, including humans, had a latent capacity to develop an additional finger. In *The variation of animals and plants* (Darwin 1875/1886), Darwin described polydactyly in dogs, in particular the Great Dane, as well as in cats. He observed inheritance of the trait over at least three generations.

As far as the tribal history is concerned, Darwin missed the mark. This happens occasionally, even to the greatest of minds. In 1868, Darwin was initially of the opinion that polydactyly must be an atavism, a relapse into the times of a many-toed ancestor. Other scientists, especially in Germany, including August Weismann, saw it as a deformity. A fierce quarrel broke out between the two men, and the controversial topic was debated over more than half a century in the magazines and books of the time. Darwin corrected his atavistic view, but by then it was already in all minds on the continent and was vehemently criticized. Not until 1922 was the idea considered outdated.

Around 1900, more than 1,000 cases of polydactyly in humans had received the scientific treatment, including by William Bateson (Section 1.2). Sewall Wright, cofounder of the Modern Synthesis, investigated 1,343 polydactylous guinea pigs in which the trait was highly variable. He hypothesized that the variation of polydactyly was under genetic and nongenetic (environmental) influences. He termed the latter *maternal effects*. He could prove, with the aid of statistics, that the prevalence of polydactyly decreases significantly with increasing maternal age. He also found that the prevalence of polydactyly was 50% higher in male guinea pigs born in winter than in those born in summer. It was also Wright who first pointed to the role of threshold effects during development to explain polydactyly (Lange and Müller 2017). We will return to these shortly.

4.8.4 Polydactyly and Genetics— Only Half the Story

In a 1968 laboratory experiment, polydactyly was produced artificially in chickens for the first time. Surprisingly, transplantation of embryonic tissue from the bud of the chicken wing into another bud led to a doubling of the toe number (Figure 4.10). This experiment yielded important insights into the development of the vertebrate limb. In 2008, Scottish geneticist Laura Lettice made a novel discovery when she succeeded in identifying a noncoding cis element in the DNA controlling the gene—*Sonic Hedgehog*—that is responsible for a certain form of polydactyly. Surprisingly, the cis element is not located directly next to the gene but approximately 1,000 DNA base pairs away. Moreover, it lies in the middle of another gene.

A few years earlier, noncoding DNA elements had been described as junk DNA, remnants of hundreds of millions of years

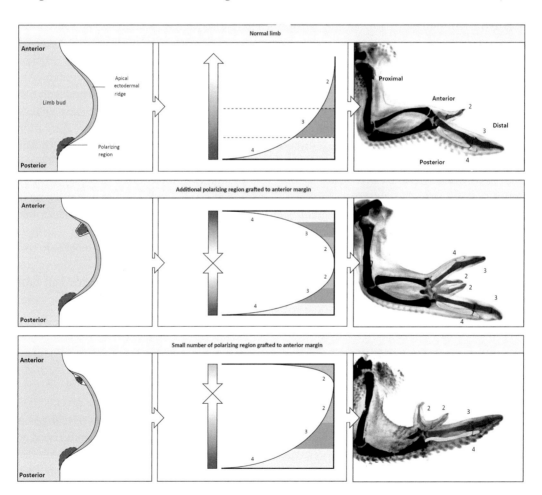

Figure 4.10 Experimental doubling of toe numbers in the chicken. In a spectacular laboratory experiment, the dark organizer region zone of polarizing activity at the posterior end of the chicken bud (above left) was removed and transplanted to the opposite side (anterior, below left) in another wing bud. The result was a doubling of the tow number (bottom right). Picture: Memorial University of Newfoundland.

of evolutionary change. The discovery of cis-regulatory elements shed light on the matter for the first time, and the term junk became recognized as a misnomer.

Despite its distance from the target gene, the cis element ZRS is a control element that directs the expression of *Shh*. This cis element can undergo a point mutation—the smallest possible error in heredity—independently of the *Shh* gene. The result is that new cell material is produced in the limb bud (cell proliferation). *Shh* and its protein SHH are jointly responsible for the development of the limb bud long before finger cartilage becomes visible. If something similar happens during early facial development, i.e., cell proliferation caused by a mutation in the *SHH* signaling pathway, the damage is considerably greater, and animals can be born with doubled facial parts, for example two beaks or three eyes. This extremely rare facial deformity is called diprosopus. Notably, even with such a dramatic deformity as this, indeed, even with a duplication of the upper esophagus, trachea, or visual pathways, chaos need not occur. The animals can breathe and eat, possibly through both canals. Thus, development coordinates decimalization insofar as possible. Pere Alberch, evo-devo pioneer of the early 1980s, knew of several such monstrous anomalies and would have enjoyed examining this case in more detail. Unfortunately, he died much too soon.

Back to the polydactyl hand: SHH is a diffusing protein, also called a morphogen (Section 4.4). If the bud is enlarged during morphogen diffusion, more fingers or toes develop. Curiously, these new cells do not accumulate on the outside, i.e., beyond the imaginary fifth finger or toe in humans, where *Shh* is normally expressed in an organizer region (zone of polarizing activity, ZPA), but on the inside, where the thumb or big toe appears. Thus, a new smaller organizer region spontaneously develops, which stimulates the accumulation of additional cells. In technical language, this process is called ectopic gene expression. The process is basically the same as in the partially transplanted region in Figure 4.10. The form of polydactyly that occurs on the inner side is called preaxial polydactyly. Ectopic gene expression is to evolutionary development as the Fosbury flop was to the high jump back then, when it was introduced. This new technique made it possible to improve the result by leaps and bounds in the truest sense of the word. In the high jump, a significantly higher bar could be surmounted; in development, new fingers represented a phenotypic novelty through ectopic expression. In both athletics and development, the technique must be implemented with precision. We will discover how this works for extra fingers.

So far, the clear connection between a newly discovered cis element, its exact location on the DNA, and its function in *Shh* expression, including ectopic gene expression and cell regeneration in the mutant, has been a brilliant new discovery. The result of a mutation in ZRS is one or more new fingers. But what exactly occurs during the formation of the new cells? Why does it result in one or more than one digit? And when does it not? Here, evo-devo comes into play with an extended view, which will be presented next. Certainly, the genetic background described up to this point is just as much an evo-devo research result as the subsequent events that occur in the bud. But genetics alone is not enough to explain polydactyly. We will now examine the other half, the epigenetic component, of polydactyly.

4.8.5 Hemingway's Cats

Let us return to the background of preaxial polydactyly. The carriers of this trait are sometimes referred to as Hemingway mutants. Ernest Hemingway owned about 60 cats and kept them in his beautiful colonial house in Key West, Florida. When I visited this house on a holiday in 1991, the offspring of his darlings, some of which were polydactyl, surrounded me in the garden. They were descendants of a cat gifted to Hemingway by a ship captain several years ago. Polydactyly was then inherited by this cat's offspring and subsequent generations. Cats with pre-axial polydactyly of this type are now called Hemingway mutants worldwide.

The domestic cat (*Felis catus*), which is descended not from the European wild cat but from the African wildcat (*Felis sylvestris lybica*) and was introduced to us by the Romans, has 18 toes—five on each front paw and four on each rear one. In the Hemingway mutant, polydactyly develops in the forefoot with a bifurcation, but in the hind foot, there are always one or two properly formed, separate new toes. In the similar Canadian mutant, even the new toe on the forefoot is complete.

We analyzed 485 polydactylous Maine Coon cats from the USA (Lange et al. 2014). The Maine Coon is a beautiful medium-longhaired cat. Polydactyly occurs relatively often in this breed. The reason is that the polydactylous British ancestors of this young cat race probably served on ships as mascots in the 17th century. It was probably believed that cats with extra toes, and thus larger paws, would make better mousers. In Boston and other ports on the east coast of the USA, the polys, as they are affectionately called, preferred to jump off deck after a long voyage. Over time, a rather large colony of polydactylous cats developed in Maine, where they reverted to a semiwild state as widely scattered farm cats. Today, almost all of these Hemingway mutants are registered as domestic cats in a database, which served as empirical material for my study on the evo-devo mechanisms of polydactyly.

4.8.6 Mysterious Numbers of Toes

An initial result of the study (Lange et al. 2014) was a frequency distribution of the number of toes in symmetrically polydactylous individuals (Figure 4.11). In this sense, symmetrical

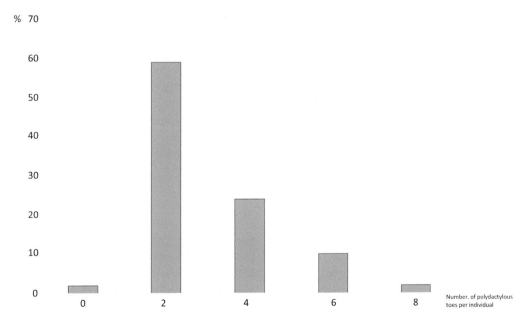

Figure 4.11 Distribution of toe numbers. A tally of the toe numbers of 317 symmetrical polydactylous cats revealed an unexpected frequency distribution of toe numbers per individual. The first bar at zero indicates how often the first toe is extended on the forelimbs. Here, a complete new toe does not yet occur. The most frequent occurrence was two additional toes; less frequent were four and six, and the rarest was eight. Such a pronounced polypheny, i.e., different phenotypes with the same genetic basis, was not expected from an identical genetic point mutation. The variation is biased in the sense that certain forms occur more frequently than others.

means that the number of additional toes is the same on both sides of the body. But there are also cases, albeit few, where the number of toes on the left side differs from that on the right. Stephen Jay Gould had already pointed out that development creates symmetry more easily than it does asymmetry. However, our case deals with a decanalization of development, to use Waddington's words. The mutation causes developmental chaos, which disrupts normal pattern formation. The result is sometimes more and sometimes fewer extra toes.

Unexpectedly, the obtained frequency distribution of additional toes indicates that two additional toes are more frequent than four, and four extra toes are more frequent than six or, seldomly, eight. Recall that a single point mutation leads to polydactyly. In the most simplistic scenario, the same phenotype would be expected from the same genetic material every time; however, this does not hold true in reality. It has long been known that phenotypic results can vary despite having the same molecular basis. However, this case was unique; there was a clear frequency distribution supported by statistics. Such a case, where different phenotypes arise from an identical genetic basis, is referred to as a polypheny. Since the polypheny in this case represents a random distribution, it is said to be biased toward the development of certain numbers of toes. A bias could also be detected in the difference in the number of polydactylous toes between the forefeet and hind feet, which also has a random statistical distribution with a maximum at two toes. In other words, polydactylous cats most often have two more supernumerary toes on their front paws than on their hind ones. In less frequent cases, the front and hind paws have the same number of polydactylous toes; more rarely, the front paws have two fewer, and so on (Lange et al. 2014).

The explanation for these distributions, for why the number of additional toes is neither disordered nor equally distributed, lies in the developmental program. In development, there are countless small, bistable, cell effects or reactions. The diffusion of morphogens, in this case *Sonic Hedgehog*, throughout the limb bud trigger signals in the surrounding cells of the mesenchyme. This is an early stage of cell differentiation before cartilage, where bone and other cell types emerge. Cell reactions in the mesenchyme drive forward the process of cartilage formation in the finger attachments. (The cartilage material only becomes ossified later.) In a bistable process, a cell either reacts to all signals or none at all. Thousands of bistable cell effects can be summed up, regardless of their specific character, and represented as an approximate normal distribution.

This example illustrates what Denis Noble described as the organism as a free agent. In concert with the genome, the organism determines the phenotypic pattern with the help of randomness. Higher organismic levels, in this case cells, use randomness to generate phenotypic order.

By contrast, the distribution of toe numbers described above is discreet and represented in a bar chart, which indicates its categorical nature. Each progression to additional toes is a discrete jump rather than a continuous increase. The number of additional toes concretely increases from two to four, i.e., the total number of toes increases from 20 to 22 or 24.

Combining the two distributions, continuous and discontinuous, produces an overall view of both, abstracted cell events and toe numbers. The main result of the investigation is that discontinuous jumps in toe number result from continuous cellular processes. There must therefore be threshold effects in development that trigger sudden events. This is by no means self-evident, but a close examination of the developmental process gives a clue. In the cell material that accumulates as a result of ectopic expression of *Shh* in the bud, precursor forms of half, quarter, or eighth toes are not evident. The cell material and cartilage precursors of a complete new toe are present; there are no partial toes.

Threshold value effects, as demonstrated here, provide key information for evo-devo and an important explanation for development processes. The gradualism-oriented Modern Synthesis, based solely on cumulative small changes, did not account for discontinuous, complex variation, threshold values, or developmental bias. The overall view of continuous development processes and discrete toe numbers is called the Hemingway model (Lange et al. 2014).

4.8.7 Computer Modeling of Polydactyly

To create a computer simulation of the Hemingway model, we digitized polydactyly (Lange et al. 2018). As a cellular automation, we used a Turing system with 10,000 (100×100) cells. Turing systems, named after the British mathematician Alan Turing (1912–1954), a true genius of the 20th century, originally modeled biochemical pattern formation processes, otherwise known as reaction-diffusion processes. In our case, however, pattern formation was transferred from the chemical to the cellular level. Interactions between neighboring and distant cells (which can activate or inhibit each other), in a complex process of self-organization involving only a few variables, form the pattern of fingers as they develop in the early phase of cartilage formation.

The simulation begins with a black screen dotted with a few white markers. The initially diffuse, then increasingly clear white pattern of the simulated toes slowly develops after a few minutes from millions of random cell signals. The idea that in the tiny tips of the limb buds of mouse or human embryos, which are barely 1 mm long, such a process of self-organization takes place in 48–72 hours and creates such wonderfully ordered fingers fascinated me anew hundreds of times when I worked with the Turing model. (On the sidelines of the EuroEvoDevo conference in Vienna, I once had the honor of explaining my simulation to Denis Duboule, the prominent geneticist who introduced Hox genes to the vertebrate landscape several years ago. Those are the beautiful moments in the life of a researcher).

In the model, a total of 20 million individual cell reactions are calculated in a single simulation run in 2,000 change steps (iterations). (My Mac reached its performance limit!). According to the principle of self-organization, in the initial equations of the model, there are no indications of the resulting pattern; thus, it cannot be predicted from them. Therefore, there are no commands such as "Generate five white vertical stripes in a rectangle." The pattern only appears in the complete simulation after 2,000 iterations and is independent of the initially randomly distributed, differentiated cells (chaotic initial distribution).

In further detail, Turing systems clarify that at the underlying biological (here, the genetic) level, the simulated phenotype (e.g., the stripe pattern) is not and cannot be explained. The pattern of the fingers is explained at the cellular level in the model and thus epigenetically. The detailed interactions among cells are determined by genes; however, their expression is predetermined in the model and is not addressed in the simulation. Genes are thus a necessary but not a sufficient condition for the emergence of the finger pattern. Another crucial element must therefore be added—the principle of self-organization. Above the gene level, a cell conglomerate organizes itself to determine the formation and number of fingers and toes. One might refer to this as an emergent behavior. Some people shy away from the term because "emergent" sounds as if the causes of an occurrence are not explainable (although they very much are). Turing processes provide a strictly mathematical explanation for various types of pattern formation, which, in a biological context, can be simulated at the biochemical cellular level but not at the lowest, genetic level of organization. In the case of the Turing system for polydactyly, pattern formation or patterning, as the experts call it, takes place at the level of mesenchymal cell accumulation in the hand bud.

Why go into this level of explanatory detail? The Turing system is part of a decades-long debate between geneticists and cell biologists over the question of how biological variation and patterns arise. In chapter 3, we became acquainted with two camps: on the one hand, the group of researchers who argue exclusively with genes and gene expression, and on the other hand, those who hold the cellular level and levels above it responsible for variation and pattern formation and relegate the power of genes to second place. Turing systems belong to models of the latter group. Rejected or deemed irrelevant for decades, they—and their creator—are now receiving the acclaim they deserve.

The model shown in Figure 4.12 is intended to demonstrate the hypothesis that toe number can be varied by changing a single variable in the computer equations. The variables in the equations stand for cell variables. We have not characterized them further. The model demonstrates a typical evo-devo statement: small cause—big effect. According to the principles of the Hemingway model, there are threshold values for the emergence of computer-simulated toes (white stripes): If the theoretical cell variable is increased in small, continuous steps, a completely new stripe appears from the threshold value of the variable, whereas there is no sign of it at a marginally lower value. For each additional simulated toe, there is a new threshold value.

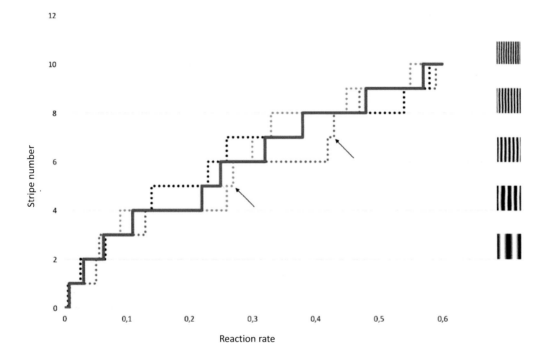

Figure 4.12 Hemingway model with threshold effects. Simulated toe numbers (y-axis, stripe number) as a function of a cell variable (x-axis, reaction rate). The continuous increase of a cell variable (reaction rate) leads to an abrupt or discontinuous increase in the toe number. The solid line shows the average toe numbers of three simulations (dotted lines). The interpretation of this simulation is that nonlinear threshold effects occur in the emergence of supernumerary fingers or toes during development: Thus, when a new toe emerges, it is immediately complete. The two arrows even indicate double jumps for two of the three simulations; here, two additional strips, symbolizing new toes, emerge at almost the same place.

In the diagram, the entire process looks like a staircase. Unexpectedly, as the value of the variable (the reaction rate) increases, the number of simulated toes increases nonlinearly; the staircase thus has rather uneven steps. Figure 4.12 shows the threshold values for 1–10 toes averaged over three simulations. We have termed the variable that determines the number of toes the *reaction rate*. It is a theoretical variable and generally signifies the propensity of cells to change to advance the cartilage formation process. In development, this means that an exact value of the variable will result in a new complete finger or toe.

The fact that the development of vertebrate extremities follows Turing processes has been discussed for decades and is rarely disputed any longer. There are numerous publications on this subject. The most recent of these also deal with polydactyly, and thus, with the evolutionary aspect of evo-devo (Raspopovic et al. 2014). Of course, in the evolutionary development of the hand, a long chain of gene expression events are involved from the beginning. Gene expression affects fine-tuning, such as determination of the length, width, and growth rate of the bud, length of the fingers, and position of the joints. Later, when the skeleton is in place, exploratory processes (Section 3.5) also form nerves, tendons, muscles, and blood vessels. Finally, the skin and nails or claws are formed. However, the pattern of fingers or toes is now considered a result of the Turing process. At the cellular level, the Turing process is a necessary condition for the formation of a pattern in the bone anlagen.

After a lecture on this topic, a listener recently asked me how the cell in the hand bud knew that it was to become part of a hand. In fact, it does not need to know. The self-organization mechanism alone ensures that the cells (through specific intercellular signals) form the rough pattern of the fingers or toes. Gene regulation provides the necessary cell differentiation and fine-tuning but not the pattern itself. As Denis Noble already noted, the heartbeat rhythm is not found at the level of the molecular components, and the parts of a system perform their task without the knowledge of the overall system (Section 3.7). The components are blind to each other; the same is true in finger development.

Polydactyly, introduced to the scientific community by William Bateson, is a good example of empirical and conceptual progress in biology. In at least one species, namely, the clawed frog *Xenopus tropicalis*, a well-known model organism from the African tropics, a sixth toe has been regularly recorded (Hayashi et al. 2015). However, evo-devo research on polydactyly is not primarily concerned with how often a variation occurs in reality but with the question of the embryonic mechanism that can generate the variation.

Today, 15 genetically different types of polydactyly are known. They involve Hox genes, *Sonic Hedgehog* and other genes on seven different chromosomes. The *Online Mendelian Inheritance in Man* (OMIM) database recorded a total of 500 entries for polydactyly in humans at the end of 2022. These include a number of syndromes, i.e., gene effects that are associated with polydactyly and other concomitant symptoms. The English term polydactyly appeared in more than 12,000 scientific publications as at the end of February 2022, according to the Yale University Library.

4.8.8 How Toe Numbers Challenge Evolutionary Theory?

The example of polydactyly was presented here in detail, as the results indicate a number of consequences for evolutionary theory and permit new predictions:

1. Radual evolution in small steps is not the sole yardstick for evolutionary events. The model shows how extensive, complex phenotypic variation can spontaneously develop in the embryo. In contrast to the previous view (chance, genetic single determination, gradualism), the new view deals with

discontinuous, epigenetic, namely cell-based evo-devo developmental mechanisms. Postmodern evolutionary theory must explain both continuous and discontinuous variations.

2. Variations can be biased in the sense that some forms (in our case, finger and toe numbers) occur more frequently than others. All optional forms (polypheny) are available for potential use.

3. Chance plays a subordinate role in evolution. Although it can initiate phenotypic changes, chance often does not play the leading role assigned to it by the standard theory in creating complex, discontinuous variations.

4. There is no single genetic determinant of complex phenotypic variation; further organizational levels above the genes must be included in the consideration in order to explain variation. This includes, among other things, cells and cell associations and the communication between them.

5. From an evo-devo perspective, natural selection loses its leading role in cases of complex variation, whereas the standard theory assigns it the role of permanently controlling numerous cumulative, individual steps. At the end of the process, however, natural selection has the final say.

6. Evolution can react faster via the threshold mechanism present here, and evolutionary processes are shortened.

The statement in point 3 seems to contradict the assertion of Denis Noble that chance should be seen in a new light, since organisms use blind chance to generate functional solutions, thus becoming actors. In fact, the two statements are not contradictory. In the Turing model of polydactyly, pattern formation arises precisely from innumerable random cell signals, in agreement with Noble.

The "random" mutation itself, which forms the basis of polydactyly, is however relatively insignificant in the overall process of structure formation, because the subsequent development process (form and number of fingers) is the new evo-devo insight and because it is here that the "music plays," to borrow a phrase from Noble. In this process, however, chance in the form of a mutation is required only once and not in the constant interplay with natural selection, as required by the Modern Synthesis.

The empirical research results mentioned in chapter 4 provide evidence of biased development, discontinuous variation, and environmental influences on variation. It remains a priority to investigate the extent to which these results occur in natural populations and withstand adaptation processes.

4.9 SUMMARY AND OUTLOOK

Table 4.2 shows a summary of the predictions of evolutionary progression that result from the evo-devo examples presented in Chapter 4. Discontinuities are noted if they can be proven to occur in one generation.

The demands arising from evolutionary developmental biology and from related disciplines for an Extended Synthesis with important modifications can be summarized as follows (Müller 2019a, b).

A system-oriented, interdisciplinary evo-devo theory is based on the following assumptions and observations:

1. Phenotypic changes depend on genetic and nongenetic factors, such as development, environment, and physics.

2. Genetic and phenotypic variations in populations are nonrandom.

3. Phenotypic modifications may precede genetic modifications.

4. Phenotypic changes can often not be gradual.

These assumptions allow the theory to predict the following:

TABLE 4.2
Predictions from the examples in Chapter 4

		Discontinuity	Nongenetic inheritance	Biased variation	Environmental induction	Genetic assimilation	Facilitated variation	Developmental constraints	Threshold effects
4.1	Beaks of Darwin's finches			x	x		x	x	
4.2	Shore leave of bichirs	x	x		x	x	x		
4.3a	New jaw joint bone in cichlids	x			x	x		x	x
4.3b	Head shape in cichlids				x			x	x
4.4	Eyes on butterfly wings	x			x		x		x
4.5	Tobacco hornworm	x	x		x	x	x		x
4.6	Turtle shell				x			x	x
4.7	Leg pairs in centipedes	x			x			x	x
4.8	Polydactyly	x			x		x	x	x

Evo-devo examples listed in Chapter 4 allow new predictions to be made about the evolutionary process in comparison to those of the Modern Synthesis.

1. Phenotypic variation tends to be systematic due to developmental constraints and facilitated variation.
2. Genetic variation has a stabilizing rather than a generating role.
3. Phenotypic innovation arises due to emergent and self-organizing characteristics of the development system.
4. Phenotypic evolution is discontinuous.

Let us preview Chapter 5 and the theory of niche construction, another major pillar of the Extended Synthesis alongside evo-devo. Chapter 6 will discuss how the Extended Synthesis is structured not only for niche construction and evo-devo but also for developmental plasticity and inclusive inheritance.

REFERENCES

Abzhanov A, Protas M, Grant BR, Grant PR, Tabin CJ (2004) Bmp4 and morphological variation of beaks in Darwins finches. *Science* 305(5689):1462–1465.

Avicenna L (1992) *Liber Primus Naturalium: Tractatus Primus, De Causis et Principiis Naturalium.* E. J. Brill, Leiden.

Beldade P, Koops K, Brakefield PM (2002) Developmental constraints versus flexibility in morphological evolution. *Nature* 416:844–847.

Böhme A, Braun R, Breier R (2019) *Wie wir Menschen wurden. Eine kriminalistische Spurensuche nach den Ursprüngen der Menschheit.* Heyne, München.

Brakefield PM (2006) Evo-Devo and constraints on selection. *Trends Ecol Evol* 12(7):362–368.

Brakefield PM (2011) Evo-Devo and accounting for Darwin's endless forms. *P Roy Soc B Bio* 366:2069–2075.

Carroll SB, Gates J, Keys D, Paddock SW, Panganiban GEF, Selegue J, Williams JA (1994) Pattern formation and eyespot determination in butterfly wings. *Science* 265: 109–114.

Damen WGM, Prpic M-N, Janssen R (2009) Embryonic development and the understanding of the adult body plan in myriapods. *Soil Org* 81(3):337–346.

Darwin C (1875/1886) The Variation of Animals and Plants under Domestication. John Murray, London.

Gilbert SF, Loredo GA, Brukmann A, Burke AC (2001) Morphogenesis of the turtle shell: the development of a novel structure in tetrapod evolution. Evol Dev 3:47–58.

Gould SJ (1980) The Panda's Thumb. More Reflections in Natural History. Norton, New York.

Hayashi S, Kobayashi T, Yano T, Kamiyama N, Egawa S, Seki R, Takizawa K, Okabe M, Yokoyama H, Kamura K (2015) Evidence for an amphibian sixth digit. Zool Lett 1:17. https://doi.org/10.1186/s40851-015-0019-y.

Houle D, Bolstadd GH, van der Linde K, Hansen TF (2017) Mutation predicts 40 million years of fly wing evolution. Nature 547:447–450.

Joyce WG, Lucas SG, Scheyer TM, Heckert AB, Hunt AP (2009) A thin shelled reptile from the Late Triassic of North America and the origin of the turtle shell. Proc Roy Soc Lond B 276:507–513.

Kirschner M, Gerhart J (2005) The Plausibility of Life: Resolving Darwin's Dilemma. Yale University Press, New Haven.

Kuraku S, Usuda R, Kuratani S (2005) Comprehensive survey of carapacial ridge-specific genes in turtle implies co-option of some regulatory genes in carapace evolution. Evol Dev 7(1):3–17.

Lange A, Müller GB (2017) Polydactyly in development, inheritance, and evolution. Q Rev Biol 92(1):1–38.

Lange A, Nemeschkal HL, Müller GB (2014) Biased polyphenism in polydactylous cats carrying a single point mutation: the Hemingway model for digit novelty. Evol Biol 41(2):262–275.

Lange A, Nemeschkal HL, Müller GB (2018) A threshold model for polydactyly. Prog Biophys Mol Bio 137:1–11.

Li C, Wu XC, Rieppel O, Wang L-T, Zhao L-J (2008) An ancestral turtle from the late Triassic of southwestern China. Nature 456(7221):497–501.

Lyson TR, Joice W (2012) Evolution of the turtle bauplan: the topological relationship of the scapula relative to the ribcage. Biol Lett. https://doi.org/10.1098/rsbl.2012.0462.

Lyson TR, Bever GS, Scheyer TM, Hsiang AY, Gauthier JA (2013) Evolutionary origin of the turtle shell. Curr Biol 23(12):113–119.

Lyson TR, Rubidge BS, Scheyer TM, de Queiroz K, Schachner ER, Smith RM, Botha-Brink J, Bever GS (2016) Fossorial origin of the turtle shell. Curr Biol 26:1887–1894.

Mehring C, Akselrod M, Blashford L, Mace M, Choi H, Blüher M, Buschhoff A-S, Pistohl T, Salomon R, Cheah A, Blanke O, Serino A, Burdet E (2019) Augmented manipulation ability in humans with six fingered hands. Nat Commun 10:2401.

Meyer A, Stiassny MLJ (1999) Buntbarsche–Meister der Anpassung. Spektrum Spezial https://www.spektrum.de/magazin/buntbarsche-meister-der-anpassung/825489.

Minelli A, Chagas Junior A, Edgecombe GD (2009) Saltational evolution of trunk segment number in centipedes. Evol Dev 11(3):318–322.

Müller GB (2019a) Evo-devo's contributions to the extended evolutionary synthesis. In: Nuno de la Rosa L, Müller G (eds.) Evolutionary Developmental Biology: A Reference Guide. Springer, Cham.

Müller GB (2019b) Evo-devo's challenge to the modern synthesis. In: Fusco G (eds.) Perspectives on Evolutionary Developmental Biology. Padova University Press, Padua.

Nagashima H, Sugahara F, Takechi M, Ericsson R, Kawashima-Ohya Y, Narita Y, Kuratani S (2009) Evolution of the turtle body plan by the folding and creation of new muscle connections. Science 325:193–196.

Peterson T, Müller GB (2018) Developmental finite element analysis of cichlid pharyngeal jaws: quantifying the generation of a key innovation. PLoS One 13(1):e0189985. https://doi.org/10.1371/journal.pone.0189985.

Raspopovic J, Marcon L, Russo L, Sharpe J (2014) Digit patterning is controlled by a BMP-Sox9-Wnt Turing network modulated by morphogen gradients. *Science* 345(6196):566–570.

Rice R, Riccio P, Gilbert SF, Cebra-Thomas J (2015) Emerging from the rib: resolving the turtle controversies. *J Exp Zool B Mol Dev Evol.* 324(3):208–220.

Standen EM, Du TY, Larsson HCE (2014) Developmental plasticity and the origin of tetrapods. *Nature* 513:54–58.

Suzuki Y, Nijhout HF (2006) Evolution of a polyphenism by genetic accomodation. *Science* 311(5761):650–652.

Tabin CJ (1992) Why we have (only) five fingers per hand: hox genes and the evolution of paired limbs. *Development* 116(2):289–296.

Uller T, Moczek AP, Watson RA, Brakefield PM, Laland KN (2018) Developmental bias and evolution: a regulatory network perspective. *Genetics* 209:949–966.

Zattara EE, Macagno ALM, Busey HA, Moczek AP (2017) Development of functional ectopic compound eyes in scarabaeid beetles by knockdown of orthodenticle. *PNAS* 114(45):12021–12026.

TIPS AND RESOURCES FOR FURTHER READING AND CLICKING

How fish can learn to walk. Land-raised bichirs provide insight into evolutionary pressures facing first vertebrates to live on land (*Nature*). https://www.nature.com/articles/nature.2014.15778.

CHAPTER FIVE

The Niche Construction Theory

The niche construction theory is another modern theory that has a novel perspective on evolution. Since 1988, Oxford evolutionary biologist emeritus John Odling-Smee has used the term "niche construction," to describe a process in which a species changes its environment causally rather than randomly. Species not only passively adapt to environmental changes but also actively change the environment. Based on abiotic (inanimate) and biotic (animate) factors, the environment and ecological niche of a species are part of the selection conditions for its evolutionary change as well as the evolution of other species. The resulting evolutionary change may even be systematically directional; evolutionary biologists refer to it as a *bias* (Odling-Smee et al. 2003). In the Extended Evolutionary Theory (EES), niche construction is one of the four central areas of research, along with evo-devo, inclusive inheritance, and developmental plasticity (Laland et al. 2015).

The idea originated with Darwin who, in his old age, studied earthworms, which reshape the soil in terms of not only its drainage capacity but also its chemical composition, thus promoting plant growth. Earthworms, from Darwin's point of view, are adapted to water but rather ill-equipped for terrestrial life. From a modern perspective, they must build their own simulated aquatic niche. However, this notion was not included in Darwin's theory as an independent evolutionary mechanism. It was Harvard professor Richard Lewontin, a self-proclaimed opponent of an all-too-strong adaptationistic dogma, who stated in a hypothesis in 1983 that adaptation to the environment is not exclusively passive. Thereafter, research on niche construction gained momentum. This chapter discusses niche construction in detail with examples.

Important technical terms in this chapter (see glossary): Developmental niche construction, ecological inheritance, gene-culture coevolution, niche construction, and reciprocity.

5.1 AN EVOLUTIONARY MECHANISM OF ITS OWN

The diversity of species on Earth cannot be explained without considering the massive early oxygenation of the atmosphere. Marine proliferation of bacterial algae (cyanobacteria), the originators of global photosynthesis, initiated the oxygenation process. Thus, their proliferation set in motion the evolutionary process that led to the biodiversity we see today. The oxygenated environmental niche then created the atmospheric conditions we know and depend on. In other words, the oxygenated niche is responsible for part of the selective environment that favors further evolution and further niches for numerous species.

Species thus create their own habitats, each in its unique way. For example, beavers change their environment on a large scale by building dams, thereby creating environmental conditions under which they continue to evolve (Figure 5.1). It is easy to imagine that the excellently adapted body and broad tail of the beaver evolved in its present form only after the beaver had consistently built its special niche of dam and water for several generations.

In addition to the ancient global oxygenation by algae and the evolutionarily young local beaver constructions, coral reefs, which enable the evolution of countless marine species, can serve as an intermediate example of niche construction. A niche is not at all uncommon in the context of evolution, although the term sounds like something that might be hidden. In fact, several species in the animal and plant kingdoms illustrate the principle of niche construction. However, there is a danger that if the concept of niche is applied too broadly, the idea of niche construction will be conflated with other ecological processes (Scott-Phillipps et al. 2014).

Odling-Smee and co-authors continued to expand his theory by citing numerous other examples, such as insect states that create specific physical and chemical environmental conditions, such as temperature, humidity, light intensity, etc., in their burrows (Laland et al. 2016). Offspring thrive under such biologically generated conditions. Odling-Smee also introduced the term "ecological inheritance" (Odling-Smee et al. 2003). He explicitly emphasized that heredity exists beyond genetic processes. Species behaviors in niches that represent environmental modifications do not require a strict genetic basis nor is it required that these behaviors optimize the fitness of the species. Both statements contradict the credos of neo-Darwinian theory. We will examine the reasoning in more detail below because it is not intuitively obvious.

The founders of the niche construction theory assumed that niche construction is an independent evolutionary mechanism or core evolutionary process alongside Darwin's natural selection. Thus, niche construction is adaptive because individuals with phenotypes that are better suited to the conditions of their constructs leave more offspring. Simultaneously, the genetic causes that lead to or can be associated with niche construction are not necessarily adaptive in terms of

Figure 5.1 Beaver dams–Masterful natural constructions. Beaver dams, this one in Grand Teton National Park, Wyoming, are often designed in a curved shape in the direction of flow. No easy feat for engineers with their water-level eyes. The dams usually have several dwellings above, but the entrances are always below the water level. The beavers can open the dam to allow flood waters to drain away. In this way, they regulate the water level. The largest known dam is in Canada; it is 850 m long. Beaver dams are a prime example of niche construction. The constructions have repercussions on the evolution of the beavers themselves and also on that of numerous other creatures living in the niche water world.

selective feedback from the niche. The causal influence of genes does not extend that far, despite the implications of Richard Dawkins. in his book *The Extended Phenotype* (1982).

Dawkins' theory. is simpler and more intuitive. We tend in principle to justify processes with causes, better yet, to reduce them to a coherent, causal chain, because a consistent, causal explanation is appealing. In several cases, it has a high subjective, persuasive value. However, our intuitive desire for coherence has nothing to do with the probability that a causally chained story is also true, as psychologist and Nobel Prize winner Daniel Kahneman explains (2012). A scenario involving two reciprocal evolutionary processes is more difficult to comprehend at first since the niche construction theory demands that we understand a complex cause–effect network in which there is precisely no continuous gene-centric cause–effect profile for adaptation. In the case of constructions by living beings, our thinking always strives in the direction of a clear, linear chain of causes, although such a chain is often less likely. Later, on the topic of prognoses, we will learn in more detail how our thinking often systematically deceives us and can produce cognitive errors (Chapter 8).

At this point, we must decide whether or not niche construction is an effective evolutionary mechanism. To explain the theory using different words, people who believe that the adaptability of genetic mutation and inheritance reaches so far that it can also causally codetermine the selective feedback of constructions, including adaptation, will say that niche construction is nothing new in principle but can be explained sufficiently well by the selection theory.

However, if people view reciprocity in this context and recognize the fact that evolutionary feedback from the construction (beaver dam, spider's web, or others) is not causally determined by the genes involved in the construction but by the construction itself and its biological and nonbiological environment, then they arrive at a different result. From this point of view, the construction, which up to this point was regarded as an effect (of the genes concerned), appears as a new, independent cause of further evolutionary processes.

In addition, the transmission of behavior during construction, as well as the transmission of biological and nonbiological changes in the constructed environment to subsequent generations, involves nongenetic inheritance mechanisms, those which Odling-Smee refers to as ecological inheritance. This includes all ecologically relevant components, such as the presence of dammed water in the case of the beaver, permanently altered temperature in the termite burrow, or microbiome in a bee colony, i.e., conditions that were changed by the construction in the first place, which reappear in each generation and are thus inherited.

The principle of natural selection still applies; however, niche construction is added as complementary, integral, and selective process. In summary, niche construction is its own evolutionary causality in the view of Odling-Smee and the co-creators of his theory. Causality interacts (reciprocally) with natural selection. Evolution, according to this new view, involves a network of causalities. The niche actively generates a novel, complex selection pressure that would not exist without niche construction.

Armin Moczek has kindly drawn my attention to three contexts that require emphasis to fully illustrate the implications of niche construction for modern evolutionary theory:

1. Before niche construction was proposed, natural selection was considered the only evolutionary process underlying the adaptation of an organism to its environment. The other three processes—mutation, migration, and genetic drift—do not lead to environmental adaptation. Niche construction thus includes another mechanism: organisms not only adapt to their environment over time but also make the

environment suitable for their existing traits. From this perspective, niche construction is key.

2. Niche construction further allows us to link microevolutionary and microecological processes with macroevolutionary ones. Niche construction applies to a range of situations, from dung beetles and beavers with local constructions to extensive coral reefs and cyanobacteria that alter the global atmosphere. This micro-macro connection with the same principle has not yet been achieved by conventional evolutionary theory with population genetics.

3. It should not be overlooked that niche construction can also be an important source of heritable variation that would otherwise be ignored. In some cases, these variations can be traced back to genes: some genotypes are simply better niche constructors than others. However, in other cases, the variation has less to do with genes than with the environment; for example, there may be certain resources in one population that are lacking in another, such as stones for tool making. Here, humans are perhaps the best example, such as when it comes to the domestication of plants and animals in different regions of the world. Jared Diamond (2017) reported in detail that in North America, domesticable animals were wiped out by hunters in the Stone Age. Moreover, the north-south orientation of the American continent prevented the spread of domesticable plants at temperate latitudes. The Isthmus of Panama, the deserts in the southern United States, and the tropical zones in South America continue to act as natural barriers to this day. In addition, people in the Middle East have been able to construct quite different niches from those available to Aborigines in Australia, for example, because of the availability of different grain and animal species. Several such geographic and climatic differences, given the same baseline scenarios for hunter-gatherers 13,000 years ago, resulted in corresponding effects on world history. With regard to the niche construction theory, it is crucial to note that even in these and other nongenetic cases, the constructed environment is heritable and influences the fitness of the descendants.

In Conversation with John Odling-Smee

John Odling-Smee was 84 years old when I spoke to him on the phone. He is almost blind; therefore, I was very glad that he agreed to be interviewed. The scientist was in his element and spoke with passion, such that I could not even ask some of my questions. Therefore, I have summarized some of his answers.

What was his path to rethinking traditional evolutionary theory?
When John Odling-Smee was born in 1935, the synthetic theory had not even been consolidated. He told me how he decided to study biology late in life, at the age of 28. The evolutionary origins of learning were his primary focus. When he started studying the songs of songbirds, boredom struck him; he wanted to know the true mechanisms at work in birds. For the first time, it occurred to him that there must be more at play than natural selection when he observed birds constantly changing their environment. It was the beginning of an intellectual development that eventually led him to the niche construction theory. However, there was still a long way to go.

Who influenced him the most?

Odling-Smee was the first to name Conrad Waddington as his greatest influencer. It was Waddington who put Odling-Smee on the trail with his book *The Strategy of the Genes* (1957). Odling-Smee described Waddington's view in retrospect as an "explosive system in which animals change their environment in the course of their lifelong development."

He continues: "Then I read Waddington's article *Paradigm for an Evolutionary Process* (1969) in a series of books. This paper received absolutely no attention. But I found there the idea that phenotypes operate on their environment." Odling-Smee was in contact with Waddington, who "got the same sort of opposition that we've been running into from the docks." He did not develop his ideas further right away.

A breath of fresh air came in 1982/1983. Odling-Smee visited Harvard University for a year. There he met Dick Lewontin, who he describes as an ecologist. The two shared many discussions, which motivated him immensely: "When most people said I was telling mad things anyway, I suddenly had to be taken seriously by these people."

When and under what circumstances did the idea of niche construction arise?

In 1988, he wrote for the first time about the construction of animals and introduced the term niche construction (Odling-Smee 1988). Simultaneously, Richard Dawkins. presented his theory of the extended phenotype (Dawkins 1982) to show that the expression of genes extends beyond physical phenotype into the environment. Odling-Smee was convinced that Dawkins did not think far enough and did not adequately name the consequences of his extended phenotype. Odling-Smee and his colleagues were then attacked by Dawkins in 2004 for going too far. Thereafter, the media had become aware and took up the issue, he recalls.

Odling-Smee shifted to the prime example of niche construction, beavers: "The genes underlying beaver dam construction that expands the beaver phenotype does not affect fitness resulting from dam construction feedback." This is a central tenet of niche construction that deserves all the attention it can get.

What is ecological inheritance?

Odling-Smee explained how he introduced the concept of ecological inheritance, using it to illustrate that in niche constructs, feedback not only occurs from coevolving organisms but also from inanimate environmental components. It quickly became clear that "evolutionary theory should not address the evolution of organisms but the interaction of organisms and their environment, and it means even more when organisms construct their own environment; however, this was not representable with the orthodox theory. There is the inheritance of naturally selected genes but also the inheritance of the environment to go with it."

What was the source of the niche construction theory?

"Organisms are selected based on whether they are capable of forming niche constructs," Odling-Smee points out. "I wanted to think about the active purpose of the phenotype as an agent." He acknowledges that everyone is currently aware of the existence of niche construction; however, "to ask if it is a separate process, is that just a question or is it more?" He promptly provides the answer, "Adaptation is a reciprocal process, not just a result of natural selection. Rather, niche construction alters the selection process."

Why do the protagonists of the Modern Synthesis have such a difficult time accepting niche construction as an independent process?

Odling-Smee's response steers purposefully toward the basic assumptions of the Modern Synthesis. He provides a historical context and explains how the Modern Synthesis arose from Darwin's and Mendel's ideas. "Assumptions were necessary to get it off the ground. But that probably could never have included everything that happens in evolution," namely, "the issues that we then took up." Novel concepts were then introduced, including that of epigenetics. They, together with niche construction, proved "incompatible with the basic assumptions of the Modern Synthesis visionaries," such as the "basic assumption that genetic inheritance is the only mode of inheritance in evolution—that's not true!" A clear statement.

Because of these inconsistencies, reluctance to accept the basic assumptions of the Modern Synthesis grew, which should have been revised, he continues. However, they became "frozen in time, so to speak, and became a mental kind of dogma and they shouldn't have done that. Your assumptions take this as they can. You need to revise it when they run into trouble with data."

I asked Odling-Smee to explain the difference between ecology and niche construction, as ecology also describes feedback mechanisms.

He clarifies that ecologists contributed the ecosystem engineering concept, according to which organisms remodel components of their environment but did not address evolution. Only "niche construction connects ecology and evolution." In several publications, Odling-Smee clarified the similarities and differences between the two disciplines.

As an evo-devo scientist, I would particularly like to hear from you how the two disciplines of evo-devo and niche construction, two of the pillars of the Extended Synthesis, can be connected, if at all.

Odling-Smee refers to ecological inheritance, which differs fundamentally from genetic inheritance and is more extensive. Ecological inheritance, he explains, depends on organisms bequeathing altered selective environments to their offspring by physically reshaping biological or nonbiological components of their environment. He points to a paper entitled *Niche Inheritance* that specifically addresses this issue (Odling-Smee 2010). There, he addresses the sought-after interaction of development and evolution when genetic inheritance is replaced by niche inheritance. Each individual inherits an initial organism-environment relationship or niche from its ancestors. Earlier evolutionary processes (evo-) affect later developmental processes (devo-). Conversely, the earlier evolution of individual organisms (devo-) may influence the subsequent evolution (evo-) of populations. (cf. on devo-evo Section 3.4 and the example of shallow tunnel depth in horned beetles., Section 3.8).

I hypothesize that medical/technological development today decouples humans from natural selection. How does he see this development? Can it be associated with niche construction?

His answer is straightforward: "Yes. I think it occurred to me. We screw up environments and exploit them. So, that's very new for me about it but it's part of the whole theory. If people want to talk about the positive side of it, destruction is always to it."

He adds, fittingly, that "cancer cells are another example of that culture" (and of niche construction).

At this point, I ask him to give a few more specific examples of gene-culture coevolution besides the well-known dairy farming.
Odling-Smee speaks of the controlled use of fire, which is unique to humans. No animal has ever brought fire under control, as far as he knows. He mentions the extensive network of human communication, of which music, language, and mathematics are unique forms. Technical developments are examples of gene-culture coevolution. When I speak of technical progress, he corrects me. He prefers the term *change* over *progress*.

My last question to him ventures a look into the future: we are entering a phase of development during which technologies in the form of artificial intelligence, robots, and the Internet could eventually surpass humans in general intelligence. In extreme cases, this could lead to our replacement as a biological species, a possibility foreseen by the likes of Nick Bostrom. If we accept that for a moment, would we be dealing with an evolutionary consequence of niche construction? What is your opinion on this?
"There is no reason," Odling-Smee replies, that we are not coming up with a sort of self-induced crisis. Species loss can be one of them, and that can also mean some form of artificial life." I add, regarding technology development, that we are not aware of what we are doing on our planet, and he confirms: "Humans are not the least bit aware of what is going on. They have to learn to understand their relationship with nature and the process of their evolution. Thus, we are not very adapted."

(I had the conversation abbreviated here on the phone in English with John Odling-Smee on July 11, 2019).

In conclusion, niche construction deals with the inheritance of acquired behaviors, the core idea of Lamarck's that had completely disappeared from the scene with the advent of the Modern Synthesis. It is time to revive this idea in the context of behavioral evolution and cultural inheritance.

5.2 DEVELOPMENTAL NICHE CONSTRUCTION—CASTLES AND PALACES FOR THE DESCENDANTS

An impressive example of innovation and niche construction in the history of animal evolution is the building of birds' nests. When a bird builds a nest, it creates selection in favor of the nest, which must be defended and kept in order. The new, diverse selection pressure affects both the parent birds and the behavior of the young that inhabit the nest. This adaptive response can also generate parallel evolution of independent species, such as those that specialize in stealing birds' eggs or cuckoo species that do not build their own nests but lay their eggs in other birds' nests. Nest building illustrates developmental niche construction because niche construction here takes place in a reciprocal relationship environment between parents and the developing young. Developmental niches cannot simply be seen as boundary conditions under which the development of the young passively takes place. Rather, developmental niche constructions, which are built anew by the parents of each generation with different resources and under modified environmental conditions, have an impact on the development and evolution of the offspring. They are thus both a cause and a consequence of evolution (Uller and Helanterä 2019).

We have already learned about the dung beetle, the females of which build deeper tunnels for egg laying within the dung ball on hot days, causing the beetles of several generations to grow to different sizes. Like the bird's nest, the dung ball is another typical developmental niche, in this case, for growing larvae. Using the genus *Onthophagus*, Chapter 6 will explore numerous other hypotheses in the context of the EES. For example, the females not only lay their eggs in the dung ball but also add a pedestal of their own feces as a supplement. The feces contain microorganisms, essential food for the young larvae. Not only that, but the larvae also remove the feces, leaving a new microbiome landscape in the dung ball along with their own food and feces. Like birds' nests, scientists view this process as an example of developmental niche construction.

In a systematic experiment, scientists removed the maternal feces, and thus the microbiome, and manipulated the niche constructed for beetle larvae in various ways or even prevented it completely. As expected, many changes in larval development were evident: larval sizes varied, and in two species, sexual dimorphism completely disappeared with respect to the differences in the shapes and sizes of legs between males and females (Schwab et al. 2017).

Like the previous examples, this one deals with feedback. Instead, of adapting their characteristics to the challenges of the environment, as in the usual case, larval organisms here systematically adapt their developmental niche, namely the dung ball including the maternal feces, to their needs. Conversely, these special developmental niche constructions have implications for their own development and evolution.

5.3 NICHE CONSTRUCTIONS OF THE HUMAN BEING

We have seen that niche construction existed before humankind entered the stage of evolution. However, this theory is of paramount importance in the context of human culture, which always concerns "built systems." In the beginning, they were simple artifacts in the form of tools. Today, they are vast, even world-spanning systems of energy production and distribution, transportation and financial systems, small businesses to global corporations, and worldwide communication networks such as the Internet. Culture is always observed in connection with its influence on nature. This no longer applies only to agriculture and the domestication of animals and plants but also applies to diverse global raw material extraction, all forms of production, the creation of transport routes, and progressive urbanization. Finally, elementary to the concept of culture are all techniques of writing, imaging, setting to music, and other media storage: "Culture means coding" (Klingan and Rosol 2019).

Humans are thus the champions among niche constructors. Our niche constructions are more powerful than those of any other species, and this is because of our cultural capabilities, according to Kevin Laland (2017). Human life takes place in myriad groups or meaning system constructions—villages, towns, schools, ping-pong clubs, or states (Wilson 2019). The problems with such a view are addressed below. By speaking of constructed systems, Wilson forms a bridge to niche construction; however, he does not expand on it. Human activity, including agriculture, forest clearing, and urbanization and megacities, has caused dramatic environmental changes. This has led to myriad changes in the way natural selection affects our species. Laland continues, "The more an organism controls and regulates its environment and that of its offspring, the greater is the advantage of transmitting cultural information across generations." Cultural niche construction thus becomes autocatalytic, meaning that more intensive environmental regulation leads to increasing homogeneity of the social environment. The same techniques are employed in neighborhoods, cities, and countries worldwide. Old and young alike engage in social learning from their parents and other adults.

A particularly impressive example of the feedback effects of human niche construction is rooted in several of the most significant genetic mutations to have occurred in human history. One such mutation first appeared somewhere between central Europe and the Balkans 7,500 years ago. It was followed by others elsewhere, such as in sub-Saharan Africa and also in Arabia. Its consequences have changed our lives like no other evolutionary breakthrough in the last 10,000 years. The mutation is the one that led to lactose tolerance in humans. In its wildtype form, the enzyme lactase, which is needed in the intestine to break down milk sugar (lactose), is no longer produced by the end of infancy. Upon reaching adulthood, mammals become lactose intolerant. This forces the offspring to find their own food and at the same time allows the mother, who is freed from breastfeeding to mate again. In parallel to a local mutation of lactose tolerance, which occurred independently several times in the regions of the Baltic Sea and other places, was the spread of dairy farming, a cultural development. This led to the formation of a niche construction, deliberately created by humans for themselves. This novel cultural environment led to an increase in the gene frequency of the mutant lactase gene in the population and thus to an evolutionary effect, and this in turn led to the even further spread of dairy farming (Feldman and Cavalli-Sforza 1989). Thus, in this exemplary gene-culture coevolution, the (in this case, man-made) conditions under which the mutation spreads in the population are more important than the circumstances under which it first occurred (Gerbault et al. 2011). Thus, the niche construction of dairy farming is both an evolutionary cause and an evolutionary consequence.

Here, the intertwining and mutual promotion of mutations and environmental conditions becomes clear (Figure 5.2). The dairy farming niche is a new environment created by humankind, and thus, a new selection basis for further human evolution. Thus, niches are not necessarily preexisting places in the natural environment that are occupied by an organism with characteristics appropriate for them. Rather, niches are selected and in several cases constructed by their inhabitants.

However, a 2018 publication presented surprising discoveries about dairy farming and lactase persistence. DNA analyses in Late Bronze Age populations of Mongolia revealed that people in East Asia practiced livestock farming although they were lactose intolerant. Analysis of the teeth of Mongolian individuals revealed high levels of milk consumption. Mongols are thought to have culturally adopted livestock management from Western steppe peoples approximately 3,300 years ago; however, unlike in Western people, no genetic exchange occurred in the Eastern population in favor of lactase persistence (Jeong et al. 2018). These discoveries are currently seen by the authors as contradicting the previously mentioned reasoning that lactase persistence is a strict selective requirement of dairy farming.

Numerous other examples of human niche construction could be listed here; the number of publications on this subject is growing rapidly. The use of fire is one such example. The cultural habit and niche of cooking meat and other food to improve its digestibility have, as a feedback effect, changed our biology from the immune system to the brain on a large scale. Another prime example is language, which helps determine our evolution and allows us to achieve cultural highs and, unfortunately, sometimes lows (Laland 2017).

At this point, let us briefly examine the niche construction theory as it applies to

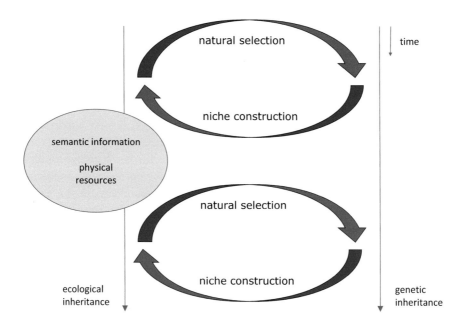

Figure 5.2 Niche construction. Evolutionary process with natural selection and niche construction. The environment generated by organisms (gray ellipse) arises as a counterpart to natural selection. The processes that cause organisms to modify their selective environment have a reciprocal causal relationship with natural selection in that particular environment. The niche construction here includes cultural knowledge (semantic information) and material culture (physical resources).

humans. Suppose we are silent listeners to a conversation between a humanities anthropologist and a representative of human niche construction, that is, a biologist. What may we expect from this discourse? The anthropologist has studied human niche construction and has come to the following conclusion: the niche construction theory represents fundamental new insights into evolution; however, its application to humankind constitutes the methodological error of combining the facts of two disciplines that do not mesh due to fundamental, methodological considerations.

The anthropologist continues: the history of humankind is one of the subjects, not of objects, that natural science must address and is characterized by the fact that people consciously act purposefully. However, the reasons for human action are individual or subjective and unique in each case and therefore cannot be placed in a causal context. In addition, the description by outsiders of human action based on reason is always subject to interpretation. The historian thus becomes "entangled in stories" (Schapp 2012). Human action deals with nonrepeatable and nonobjective facts and is thus unpredictable by nature. This does not mean that no historical predictions are made, be they those of Karl Marx, Oswald Spengler, or Samuel Huntington in *Clash of Civilizations*. However, the noncausality and nonpredictability in principle constitute the essence of historical science and its methodological distinction from the natural sciences.

On this basis, we cannot establish the causes of dairy farming introduction or human migration. There may be different, more or less understandable reasons for each of these events, but someone will inevitably interpret them. Different populations may also have different reasons for acting the same way at different times. Human actions and their consequences are therefore in a different category than nonhuman chains of events of an animate and inanimate nature. The latter can in principle be placed in a generally valid, lawful, necessary cause–effect context that permits predictions. This is the essence of the natural sciences and what methodologically distinguishes them from the social sciences and humanities, such as history. From this perspective, culture and nature are fundamentally different and must be strictly distinguished.

There are also multicausalities in nature, which may not be easy to sift through. However, in any case, we are dealing with different methodologies in human and non-human processes, which must not be confused under any circumstances, as niche construction implies. On the one hand is the category of human actions with (often individual) reasons for what they do; on the other hand is the category of causalities with cause–effect relationships in nature, specifically in evolution. Not considering different categories in the same theory is to commit a methodological error.

How does the biologist respond to this consideration? They acknowledge the methodological objection but argue against it that the reasons for human action do not matter. We do not need to know *why* humans penned and domesticated goats and cows a few thousand years ago. But we do see *that* they did, and that this behavior was dispersed over time. Second, in parallel, over longer periods of time, the heritable biological makeup of the population changed in terms of lactose tolerance. Milk tolerance increased where dairy farming was practiced. Culture and nature are converging; they cannot be seen in isolation, and indeed in many cases as "nature-cultures" they can no longer be separated from each other (Braidotti 2013).

To this the anthropologist replies, we may be dealing with a strong correlation on the part of the data material: increasing livestock farming correlates in various regions of the world with increasing lactose tolerance in the population. It is well known, however, that correlations are not yet cause-and-effect relationships; he does not have to explain that to the biologist. In any case, the data material is not sufficient to establish a cause–effect chain up to the reasons *why* humans have chosen livestock farming—nor does it need to, the biologist attests. He sticks to the data. In an empirical study, the data initially represent a correlation. The conclusion that the cultural activity of humankind can causally and hereditarily influence the biological makeup, even the gene pool of the population, must be possible in principle and must also be plausibly explained in every respect. This argument is acceptable to the biologist.

Both scientists are satisfied for the time being, even if the anthropologist needs time to reconsider. They desire to combine their results together in a logical way in the human context. To this end, their language must be synchronized. Further and repeated exchanges of views are required to clarify how they can use such discordant facts as genetic and cultural heredity in a biological consensus. The biologist, too, considers a reexamination of other niche construction cases in humans from a new angle. Cesarean sections are an example of a modern cultural behavior that can be seen as a niche construction. Their significant increase in emergencies in the Western world since the 1950s has been an intervention in natural selection and has caused a shift in the evolutionary balance. This equilibrium consists in the tension between the size of the maternal birth canal and the head circumference of the newborn. The two sizes have an opposing relationship. The disproportion between the two quantities, which is caused by the increase in cesarean births, was revealed by a sensational study at the University of Vienna (Mitteroecker et al. 2016). Furthermore, do cesarean births affect our basic biological makeup? What is inherited, and what is not? We will return to this explosive topic in Chapter 8. I thank Thomas Zwenger for discussing the differences in thinking in the natural sciences (the biologist) and humanities (the anthropologist) that I hope will eventually lead us to a unification of the two schools of thought.

In summary, human cultural activities direct mankind's evolution to the point of genetic adaptation (lactose tolerance, among others). Clearly, beyond dairy farming and the use of fire, humans have great potential in numerous fields to control, regulate, construct, and destroy their own environment. These capabilities generate a range of problems, including climate change, deforestation, and urbanization. Anthropologists with ecological training therefore find it easy to access human cultural niche construction with associated gene-culture coevolution (chap. 8).

5.4 IS THERE A FAULT LINE IN THE MODERN SYNTHESIS?

In 2014, criteria were presented for the first time to determine when a niche exists and when it influences evolution. These are (Matthews et al. 2014)

- The organism must significantly modify environmental conditions.
- Environmental conditions brought about by the organism must influence the selection pressure on the recipient organism.
- There must be an evolutionary response in at least one recipient population. The response is caused by the modified environment.

The first two criteria are sufficient for the presence of a niche; the third is a test of whether evolution by niche construction is present. The niche construction theory contains a complication of the previous theory of evolution that requires explanation. The neo-Darwinian theory views a unidirectional chain of successive events, reinforcing (cumulative) processes of mutation, selection, adaptation, mutation, selection, adaptation, and so on. Either the change in the environment itself does not depend on changes in the organism, or such a connection is not observed. By contrast, the niche construction theory does not consider environmental change over time a function of environmental factors alone but also dependent on the behavior of a species and its ancestors. Organisms, as illustrated, change their environment. Environmental and organismal changes over time, according to this view, are dependent on the environment and the organism, respectively.

Experts call this reciprocity or interdependence. Thus, "adaptations of organisms depend on natural selection modified by niche construction and on niche construction modified by natural selection" (Kendal et al. 2011).

In light of this interdependence, it is no longer clear whether natural selection is the primary factor in evolution or niche construction. Both interact with each other (Figure 5.2). This calls into question the idea that natural selection is the primary and only cause of evolutionary change. Causality can just as easily be seen in niche construction, or better yet, in both. Moreover, each is consequence of the other.

Organismic systems biology focuses on such feedback mechanisms (Section 3.7). I have emphasized interdependencies repeatedly in this book. Conventional evolutionary theory does not recognize such interdependencies; rather, it treats heredity and evolution as two separate issues, autonomous processes. The synthesis does not account for feedback between heredity and development. As a reminder, Waddington (Section 3.1), Kirschner and Gerhart (Section 3.5), Noble (Section 3.7), West-Eberhard (Section 3.8), Müller (Section 3.9), and last but not least, Moczek (Sections 3.9 and 6.5) elucidated such interdependencies. These scientists and numerous others have consistently drawn attention to the fact that heredity, development, environment, natural selection, and evolution constitute a complex, multicausal network.

However, this is precisely where supporters of niche construction theory see a possible fault line between the Modern Synthesis and the extended theory of evolution. The EES is not merely a supplement to the previous theory of evolution but changes its structure altogether (Uller and Helanterä 2019).

Of course, there are those who say that the niche construction theory fits seamlessly into the neo-Darwinian model and that it does not need its own theory. With mutation, selection, and adaptation, everything can be adequately described (Futuyma 2017). Let us assume for a moment that the niche construction theory does not intend to nullify the standard theory or any of its assumptions (cf. Uller and Helanterä 2019). However, these critics then miss something essential in terms of scientific methodology. They overlook the fact that fundamentally new information is being contributed in the form of assumptions and predictions about evolution. A new process is explained, which is not obvious from the old theory. Let us take a closer look at this with a simple example.

5.5 WHAT CONSTITUTES A NEW THEORY?

A theory can be thought of as an always-if-then statement: whenever assumption or condition A1 is given, then P1 (for prediction) holds. Of course, several assumptions A1, A2, A3, etc. can be mentioned in a theory, and several predictions P1, P2, P3, etc. can be made. In more complex theories, such as the theory of evolution, assumptions and predictions lead to a theory structure. Let us first examine the simplest case: if A1, then P1. A concrete example of this might be, whenever a stone is dropped (A1), it moves in a straight line toward the center of the earth (P1). This is a complete theory. Is it correct? Is it possible to imagine that it is not true in a certain situation or that it can be falsified? A gust of wind can change the trajectory of the stone. However, our aim is not to falsify the theory but to extend and improve it (cf. Uller and Helanterä 2019). To this end, we would extend our assumption A1. If we replace "stone" by "object," the generality or information content of the theory immediately increases. After all, P1 applies not only to a stone but to any material object. This is what science strives for: universally valid theories. The theory of evolution also strives to be such a generally valid theory. Evolution functions according to the neo-Darwinian scheme, simplified as mutation–selection–adaptation on repeat, *always* and for *all* life on earth. A uniform principle.

Let us now change the theory of the falling to whenever an object is dropped (A1) in the airless space (A2, new!), it moves rectilinearly in the direction of the center of the earth (P1) with Newtonian acceleration (P2). What is the relation between this new theory and the first one? But—and this is the crucial point—it does not provide as much information as the new one. The old theory tells us nothing about acceleration but only about the direction of the fall. Its informative content or informative value is lower. However, the new theory is extended and more valuable; it has more informational content. We learn something additional in the new theory, namely about acceleration. The new theory has extended the old one rather than disproving it. As David Sloan Wilson puts it, a new theory is not just a new interpretation of previous observations but opens doors to new observations, doors that were not even visible to the old theories (Wilson 2019).

The niche construction theory can be seen as such an extension. It does not necessarily falsify the previous one (mutation, selection, and passive adaptation), according to the well-known science theorist Karl Popper. The extension here is, for example, the active contribution of organisms to their own change with the help of their niche constructions. Other new contributions are the intrinsic mechanisms of the embryo for generating phenotypic variation, which we learned about in Chapters 3 and 4.

Let us now extend the discussion one step further. The statements of the niche construction theory, according to which there is an interdependence of heredity and development in the evolution of niche constructions (Section 5.3), bring predictions into play that cannot be derived from the previous theory at all, namely bias the niche construction processes. Consider the example of lactose tolerance (Section 5.3); a whole new culture emerges there and is inherited. This constitutes bias.

Transferring the same idea to our example theory of the falling object, we would now perhaps say: whenever an object is dropped in an airless space, it moves with Newtonian acceleration rectilinearly in the direction of the center of the earth and simultaneously with an earth rotation motion (P3). Including the earth's rotation in the theory introduces another assumption, namely that the earth rotates, and the stone accompanies the rotation (A3). This assumption is impactful and can be viewed as a fundamental reconsideration. It brings with it an equally new prediction and new informational content, namely that the object follows the earth's rotation (P3). One can say that the new assumption and prediction change the model drastically. They lead, analogously to the niche construction theory, to a new bias (object follows the earth's rotation) that was not included before. This bias could be seen as equivalent to that in the constructed niche.

We can now discuss whether the theory including the earth's rotation is a break line and therefore leads to a conceptually new theory or whether it is again an extension of the old one. Those who only need the absolute geo-position of the observer with the falling object will always argue the claim that the original theory is completely okay. He will say that the earth's rotation is of no importance, it is "nice," but necessary only in a few cases. Therefore, the inclusion of the earth's rotation does not constitute a fundamental change worthy of a new theory for them. Their colleague, whose profession deals with the calculation of the orbit of satellites, of course sees this differently and has only a smile for them.

We can summarize that the perspective or the environment of assumptions is crucial to a theory. From this point of view, both theories can possibly be valid side by side. However, those involved know exactly what and how much information is in each and for which cases it is applicable.

Of course, fault lines can also be seen in evo-devo theory (Chapter 3). For example, the independence of heredity and development (conventional view) is viewed critically compared to the assumption of interdependence (evo-devo view). Independence of the two would mean, according to earlier ideas, that genetic inheritance clearly informs the development of the phenotype. However, we have learned that this does not hold true. Furthermore, it is equally critical that the Modern Synthesis assumes genetic inheritance as the only form of inheritance, while evo-devo, as well as the theory of niche construction, incorporate inclusive or ecological inheritance.

No less relevant to evo-devo is the view of natural selection as the sole causal factor of evolution (conventional view) versus constructive mechanisms of development as causal factors of evolution (evo-devo view). More importantly, if interdependencies in the form of interactions of causalities are brought into play, i.e., if embryonic development and niche construction can each be evolutionary causes and effects (Section 5.2), then one can conclude that these views result in a whole new theoretical structure, for these aspects are not included in the synthetic theory of evolution. Depending on the evaluation of these different fault lines, there are possible incompatibilities with the standard theory of evolution. We will see, however, that the EES, as presented in Chapter 6, in contrast to some of its individual representatives, does not strive for such a departure from the synthetic theory of evolution today and justifies why it does so.

Recognition of a theory as novel is also a psychological feat. Daniel Kahneman has described the hurdles in the recognition of a novel theory with the term "theory-induced blindness." This is explained further at the end of this book in Chapter 9.

The niche construction theory, like the theory of evolutionary development, is integral to the Extended Synthesis. Chapter 6, which introduces the EES, will clarify both of them. With the introduction to evo-devo and to niche construction, the reader will be well prepared to follow the goals and reasoning of the Extended Synthesis in evolutionary theory. Chapters 7 and 8 will then focus on the dominant role of niche construction in the evolution of human thought and our culture. Chapter 9 will return to types of theories and clarify that the theory of the falling stone or object is a simpler theory than that of evolution and that one cannot therefore lump the two together without further ado. For the moment, however, the example is quite helpful.

5.6 SUMMARY

The niche construction theory is one of the four supporting elements of the EES. According to this theory, organisms actively shape and reshape their environment. Their own evolution then takes place in the environment they help to shape. Organisms are thus not passive recipients of selective influences. The niche construction theory recognizes reciprocal cause-and-effect mechanisms in evolution and views niche construction as a new, independent evolutionary selective mechanism in concert with natural selection. Among other things, the theory of gene-culture coevolution seems indispensable to a better understanding of the evolution of humans, whose actions are increasingly changing the globe.

REFERENCES

Braidotti R (2013) *The Posthuman*. Polity Press. Cambridge /UK and Malden, MA.

Dawkins R (1982) *The Extended Phenotype. The Genes as the Unit of Selection*. Oxford University Press, Oxford.

Diamond J (2017) *Guns, Germs, and Steel: The Fates of Human Societies*. Vintage, London.

Feldman MW, Cavalli-Sforza LL (1989) *Cultural Evolution: A Quantitative Approach*. Princeton University Press, Princeton, NJ.

Futuyma DJ (2017) Evolutionary biology today and the call for an extended synthesis. *Interface Focus* 7(5):20160145. https://doi.org/10.1098/rsfs.2016.0145.

Gerbault P, Liebert A, Itan Y, Powell A, Currat M, Burger J, Swallo DM, Thomas MG (2011) Evolution of lactase persistence: an example of human nicheconstruction. *Philos T Roy Soc B* 366:863–877.

Jeong C, Wilkin S, Amgalantugs T, Bouwman AS, Taylor WTT, Hagan RW, Bromage S, Tsolmon S, Trachsel C, Grossmann J, Littleton J, Makarewicz CA, Krigbaum J, Burri M, Scott A, Davaasambuu G, Wright J, Irmer F, MyagmarE, Boivin N, Robbeets M, Rühli FJ, Krause J, Frohlich B, Hendy J, Warinner C (2018) Bronze age population dynamics and the rise of dairy pastoralism on theeastern Eurasian steppe. *PNAS* 115(48):E11248–E11255.

Kahneman D (2012) *Thinking, Fast and Slow*. Farrar, Straus and Giroux, New York.

Kendal J, Tehrani JJ, Odling-Smee J (2011) Human niche construction in interdisciplinary focus. *Philos T Roy Soc B* 366:785–792.

Klingan K, Rosol C (2019) Technische Allgegenwart – ein Projekt. In: Klingan K, Rosol C (eds.) *Technosphäre*. Matthes & Seitz, Berlin, 12–25.

Laland KN (2017) *Darwin's Unfinished Symphony – How Culture Made the Human Mind*. Princeton University Press, Princeton, NJ.

Laland KN, Uller T, Feldman M, Sterelny K, Müller GB, Moczek A, Jablonka E, Odling-Smee J (2015) The extended evolutionary synthesis: its structure, assumptions and predictions. *Proc Roy Soc B* 282:1019.

Laland KN, Matthew B, Feldman MW (2016) An introduction to niche construction theory. *Evol Ecol* 30:191–202.

Matthews B, De Meester L, Jones CG, Ibelings C, Bouma TJ, Nuutinen V, van de Koppel J, Odling-Smee J (2014) Under niche construction: an operational bridge between ecology, evolution, and ecosystem science. *Ecol Monogr* 84(2):245–263.

Mitteroecker P, Huttegger SM, Fischer B, Pavlicev M (2016) Cliff-edge model of obstetric selection in humans. *PNAS* 113:14680–14685.

Odling-Smee FJ (1988) Niche constructing phenotypes. In: Plotkin HC (Hrsg) *The Role of Behavior in Evolution*. MIT Press, Cambridge, MA, 73–132.

Odling-Smee FJ (2010) Niche inheritance. In: Pigliucci M, Müller GB (Hrsg) *Evolution – The Extended Synthesis*. MIT Press, Cambridge, MA: 175–208.

Odling-Smee FJ, Laland KN, Feldman MW (2003) *Niche Construction: The Neglected Process in Evolution*. Princeton University Press, Princeton, NJ.

Schapp W (2012) *In Geschichten verstrickt. Zum Sein von Mensch und Ding*, 5th ed. Klostermann, Frankfurt.

Schwab DB, Casasa S, Moczek AP (2017) Evidence of developmental niche construction in dung beetles: effects on growth, scaling and reproductive success. *Ecol Lett* 188:679–692

Scott-Phillips TC, Laland KN, Shukar DM, Dickins TE, West SA (2014) The niche construction perspective: a critical appraisal. *Evolution* 68:1231–1243.

Uller T, Helanterä H (2019) Niche construction and conceptual change in evolutionary biology. *Brit J Phil Sci* 70:351–375.

Waddington CH (1957) *The Strategy of the Genes. A Discussion of Some Aspects of Theoretical Biology*. George Allen & Unwin, London. https://archive.org/details/in.ernet.dli.2015.547782/page/n1.

Waddington CH (1969) Paradigm for an evolutionary process. Biological theory. In: Waddington CH (ed.) *Towards a Theoretical Biology*, Vol. 2 Sketches. International Union of Biological Sciences & Edinburgh University Press, Edinburgh, 106–123.

Wilson DS (2019) *This View of Life. Completing the Darwinian Revolution*. Pantheon Books, New York.

TIPS AND RESOURCES FOR FURTHER READING AND CLICKING

Wells DA (2015) The extended phenotype(s): a comparison with the niche construction theory. *Biol Philos* 30:547–567.

CHAPTER SIX

Extended Evolutionary Synthesis

This chapter describes a project called the EES (https://extendedevolutionarysynthesis.com) and a follow-up project that surpasses it (Section 6.5). The EES project is not the official presentation of the EES theory. However, a total of 51 internationally known scientists from eight universities or academic institutions and various disciplines have contributed to it. Perspectives may vary among members of the same discipline. At this scale and with the budget approved, this coordinated program of empirical and theoretical research, launched in September 2016, is by far the largest, most representative, and dedicated to extending traditional evolutionary theory. By the end of 2019, the project had already yielded well over 100 publications in leading journals, with more to follow. In addition, there are relevant books, international workshops, and conferences (https://extendedevolutionarysynthesis.com).

The declared main goal of the project is to clarify the structure of EES theory and to strengthen empirical support for the statements it contains. The project title "Putting the Extended Synthesis to the Test" also expresses the latter intention. The project was financially supported by the US John Templeton Foundation with an amount in millions (https://extendedevolutionarysynthesis.com/-the-project/funders/). The individual disciplines are not equally weighted; for example, the project emphasizes niche design over evo-devo. In response to my inquiry, the project management confirmed that the EES project does not claim to be exhaustive in its selection of topics. For example, those responsible recognize threshold effects, self-organization, and the distinction between variation and innovation as important aspects, but these were not explicitly included in the scope of the current project.

The EES is described as a means of gaining a new understanding of evolution. It differs from the Modern Synthesis (MS), the standard model of evolutionary theory in use today (Chapter 2). However, the EES does not replace the MS; rather, it is repeatedly pointed out that its theses can be used in parallel with those of the MS to stimulate novel research in evolutionary biology (https://extendedevolutionarysynthesis.com/ about-the-ees/).

Two unifying concepts form the basis of the EES, namely, constructive development and reciprocal causation. The following key findings from four research areas are emphasized:

- Developmental bias
- Developmental plasticity
- Inclusive inheritance
- Niche construction

Important technical terms in this chapter (see glossary): Agency, agent, constructive evolution, developmental bias, developmental plasticity, evo-devo, inclusive inheritance, macroevolution, microevolution, niche construction, reciprocal causality

6.1 EMERGENCE OF THE EES PROJECT

The illustrations of the MS are, in the eyes of many, too general to explain evolution in the modern context. Thus, the EES is an alternative approach to the nature of development, construction of heredity, and causes of evolutionary change and adaptation. The article *The extended evolutionary synthesis: its structure, assumptions and predictions* (Laland et al. 2015), by a team of experts on various topics (evolutionary genetics, ecology, epigenetics, evolutionary developmental biology, and philosophy of science) was the first attempt to define the assumptions of the EES, describe its unifying ideas, and make some telling predictions.

The field of EES research involves an ever-increasing number of scientists, including members of the National Academy of Sciences, fellows of the Royal Society, fellows of the American Association for the Advancement of Science, members of other national academies, a past president of the European Society for Evolutionary Biology), the president of the EED Biology, former editors-in-chief of journals such as the *Journal of Evolutionary Biology*, *American Naturalist*, *Evolutionary Ecology*, and *Theoretical Population Biology*, and a winner of the Motoo Kimura Prize for Outstanding Contributions to Population Genetics. Other researchers who are not part of the current project are associated and support the effort (https://extendedevolutionarysynthesis.com/people/).

6.2 OBJECTIVES OF THE EES PROJECT

The objectives of the EES project are twofold (https://extendedevolutionarysynthesis.com/-the-project/summary-of-our-research/). The first is to test the EES predictions (Section 6.3) with an empirical research program, in other words, empirically substantiate them. For example, one might seek the relevance of certain predictions, concepts, or mechanisms in evolution, such as constructive evolution or developmental plasticity. Are the predictions in question exceptional or do they tend to be the rule in evolution? The project aims to provide solid underpinnings for the EES, stimulate new questions, develop critical tests, open new lines of research, and provide insights that extend beyond the traditional view of the MS.

The second objective is to clarify possible structural changes in the theory of evolution. We learned in Chapter 2 which form of the MS represents the principal causes of evolution. The scheme of the MS is greatly simplified: genetic mutation—natural selection—adaptation. These three principles are exclusively based on a genetic representation. Thus, the MS not only describes the evolution in terms of genes but also makes assumptions about the causal relationships in evolution by natural selection. For example, genetic mutations are the basis of trait selection. Alternatively, natural selection promotes the most suitable individuals for survival and reproduction. It is important to recognize that genetic representation is only one of several viewpoints and is not necessarily a true representation of nature in terms of the interplay of all the factors in evolution. There may be other descriptions of biological causes that are better suited to answering interesting questions about evolution. The EES project is devoted to such alternative causes. The overall structure of the EES is characterized by new assumptions, causalities, and predictions.

In particular, the above two key EES concepts, namely, constructive evolution and reciprocal causality, influence and change the above causal. Directed development, developmental bias, and epigenetic inheritance also impact the structure or causal relationships of evolutionary theory: not only do new causes appear, but cause–effect chains change and pivot.

Specifically, the subprojects of the EES project aim to (https://extendedevolutionarysynthesis.com/the-project/summary-of-our-research/)

- demonstrate the explanatory potential of the EES
- implement critical empirical tests of the key EES predictions

- develop new conceptual and formal-mathematical theories
- raise awareness of the importance of a conceptual framework and the promotion of pluralism.

Accordingly, the research provides

- clear assessments of the significance of controversial evolutionary processes (e.g., niche construction, nongenetic inheritance, etc.)
- clarification of the evolutionary significance of individual responses to the environment (plasticity)
- new theoretical approaches to complex genotype-phenotype relations
- clarification of the extent to which development processes explain long-term trends, parallel evolution, biological diversity, and evolvability.

6.3 NEW PREDICTIONS ABOUT EVOLUTION

The EES project arrives at different predictions about evolution than the MS. The predictions of both are compared in Table 6.1.

The EES project page emphasizes that all recognized causes of evolution (e.g., natural selection, genetic drift, mutation, etc.) and heredity (e.g., genes), as well as the multitude of empirical and theoretical findings that have been made in the field of evolutionary biology, are accepted in the project. It is emphasized that the EES does not reject the current theory or seek a revolution in the sense of Thomas Kuhn (Kuhn 1962) (https://extendedevolutionarysynthesis.com/synthesizing-arguments-and-the-extended-evolutionary-synthesis/). Rather, the EES seeks to complement the existing causal framework by recognizing additional causes of evolution (e.g., biased development) and heredity (e.g., inclusive inheritance) that fit completely within the framework of established causes.

6.4 BRIEF DESCRIPTION OF THE INDIVIDUAL RESEARCH PROJECTS

The EES program comprises 22 research projects (https://extendedevolutionarysynthesis.com/the-project/research-projects/) divided into four interrelated themes. Experimental and theoretical studies test EES hypotheses by comparing and evaluating predictions from traditional and EES perspectives. The summary of each topic presented here may seem a bit abstract, so please refer to the examples in chapter 4. Nevertheless, the topics discussed in this chapter can give an idea of how targeted work is being conducted on specific research questions to better understand evolution. Selected projects from this list were described in more detail earlier (Sections 3.8 and 5.2).

The first group of projects deals with conceptual issues in evolutionary theory. Historical analysis helps to understand the structure of evolutionary theory and how it has changed over time as well as the causes of and resistance to fundamental change. In a subproject involving philosophers and biologists, researchers are determining how the conceptual structure of evolutionary theory affects the answers to evolution's big questions: How are fitness, adaptation, and inheritance understood in traditional theory and from an EES perspective? By collaborating with theoretical biologists and philosophers, we are evaluating the explanatory power of models and their different assumptions about evolution and heredity (see Uller and Helanterä 2019).

The second project group consists of six individual projects dealing with evolutionary innovation. In them, research is conducted on how and via which mechanisms evolutionary development can produce innovations. For example, the interactions between short-term phenotypic plasticity and long-term genetic evolution are tested in the model. Conditions have been found that support the prediction that developmental plasticity influences evolution, and stabilizing genetic

TABLE 6.1
Traditional predictions and predictions stemming from the EES project

	Traditional predictions	EES predictions
1	Genetic change causes, and logically precedes, phenotypic change, in adaptive evolution	Phenotypic accommodation can precede, rather than follow, genetic change, in adaptive evolution
2	Genetic mutations, and hence novel phenotypes, will be random in direction and typically neutral or slightly disadvantageous	Novel phenotypic variants will frequently be directional and functional
3	Isolated mutations generating novel phenotypes will occur in a single individual	Novel, evolutionarily consequential, phenotypic variants will frequently be environmentally induced in multiple individuals
4	Adaptive evolution typically proceeds through the selection of mutations with small effects	Strikingly different novel phenotypes can occur, either through mutation of a major regulatory control gene expressed in a tissue-specific manner, or through facilitated variation
5	Repeated evolution in isolated populations is due to convergent selection	Repeated evolution in isolated populations may be due to convergent selection and/or developmental bias
6	Adaptive variants are propagated through selection	In addition to selection, adaptive variants are propagated through repeated environmental induction, nongenetic inheritance, learning and cultural transmission
7	Rapid phenotypic evolution requires strong selection on abundant genetic variation	Rapid phenotypic evolution can be frequent and can result from the simultaneous induction and selection of functional variants
8	Taxonomic diversity is explained by diversity in the selective environments	Taxonomic diversity will sometimes be better explained by features of developmental systems (evolvability, constraints) than features of environments
9	Heritable variation will be unbiased	Heritable variation will be systematically biased toward variants that are adaptive and well-integrated with existing aspects of the phenotype
10	Environmental states modified by organisms are not systematically different from environments that change through processes independent of organismal activity	Niche construction will be systematically biased toward environmental changes that are well suited to the constructor's phenotype, or that of its descendants, and enhance the constructor's, or its descendant's, fitness
11	Parallel evolution explained by convergent environmental conditions	Repeated evolution in isolated population may be due to niche construction
12	Ecosystem stability, productivity, and dynamics explained by competition and trophic interactions	Ecosystem stability, productivity, and dynamics critically dependent on niche construction/ecological inheritance.

All predictions are tested in the EES project, usually in several individual projects (https://extendedevolutionarysynthesis.com).

accommodation follows. Here, insight is also provided into how phylogenetic lines in animals and plants retain their ability to evolve while undergoing adaptation (see Rago et al. 2019). Another project quantifies the extent of developmental bias and how it affects evolutionary innovation (cf. Uller et al. 2018). In each case, research is being conducted on the core question of how innovation arises in evolution. This book has also posed this question repeatedly.

A project on *Onthophagus*, dung beetles is examining how evolutionary innovation, or the development of novel phenotypic variation, is influenced by developmental processes and ecological conditions, i.e., environmental conditions. It is testing the hypothesis that plasticity influences the evolution of morphological and life history traits (see Macagno et al. 2018, Section 3.8; Casasa and Moczek 2018; Pespeni et al. 2017). Using the examples of transitions from unicellular to multicellular organisms or the evolution of complex insects and human societies, another project is investigating the role of niche construction and nongenetic inheritance. Such transitions require complex group adaptations. To this end, researchers are attempting to trace the evolution of group adaptations using tools from computer science and mathematical biology.

In another experimental project, factors that promote the transition from unicellular to multicellular status are identified by comparatively analyzing unicellular and multicellular species of algae, fungi, and cyanobacteria (Figure 6.1). Using *Chlamydomonas*, a green alga that can occur in unicellular and multicellular states, as a model organism, scientists determined the contribution of directional evolution to multicellularity. In so doing, they found that several protozoan species exist under conditions known to enable multicellularity. In particular,

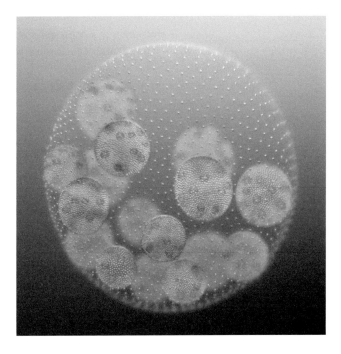

Figure 6.1 Multicellular green alga (*Volvox carteri*). *Volvox algae* are multicellular green algae two millimeters in size. Their individual cells resemble unicellular green algae. This makes them particularly suited to the study of the transition from unicellularity to multicellularity, one of the most important in the history of life on Earth. Before this transition, cells were uniform. Cell differentiation occurred several times in further steps during evolution. In *Volvox*, this was an evolutionarily recent event, as evidenced by its two cell types.

protozoa were discovered to have the role of a gelatin layer around the cell that allows them to pass through the gut unharmed by predators. This weakens the anti-predator thesis on the transition to multicellularity, which states that unicellular organism's transition to the multicellular form to protect themselves against predators, which may then no longer be able to prey on the multicellular organism. However, in this specific case, this is not absolutely necessary because of the protective gelatin layer on the unicellular organism. Therefore, the evolutionary emergence of complexity, which we must deal with during the transition to higher life forms, is in focus.

In the last project of this group, which deals with innovations, the role of developmental plasticity in the evolution of nest building behavior is investigated with the help of mathematical models, using the example of termite structure building (Figure 6.2). Constructive evolution and niche construction are contrasted with the traditional genetic view of evolution.

The third group of projects deals with inclusive inheritance in five subprojects. This includes symbionts, epigenetic variants, parental effects, and learned knowledge. This group investigates how and why symbiotic interactions between the host and the microbiota have evolved, with a focus on mutual niche construction. For example, one is interested in how gene-based adaptation exists without genome modification, such as host adaptation, by changing the microbiome composition. In one project, mathematical approaches are used to predict and interpret inheritance patterns of microbial diversity under different environmental conditions. This analysis is particularly relevant to medicine and also provides theoretical predictions relevant to other fields, such as pest control and conservation (see Kolodny et al. 2019).

Another key mode of inheritance in animals is social learning and teaching. Here, in a theoretical project, it was found that the learned skill of deception can evolve under numerous conditions and change during the life cycle. Deception is known to bird mothers who lure predators away from their young by feigning injury and a broken wing.

Current models of nongenetic inheritance are also being developed using state-of-the-art omics methods, i.e., genomic or proteomic methods (the proteome is the totality of all proteins), to document stress responses of the large water flea (*Daphnia magna*) (Figure 6.3) to toxic cyanobacteria. Researchers are interested in how these responses influence later generations through maternal effects and how natural selection subsequently influences inheritance as a function of environmental conditions (see Radersma et al. 2018).

Figure 6.2 Termite structures in Namibia. Termite structures are a typical example of niche construction.

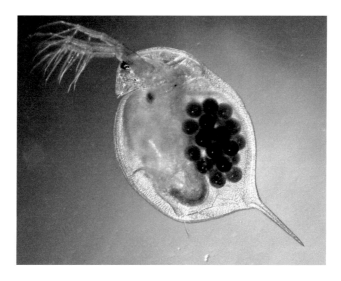

Figure 6.3 Large water flea. Depicted is a female with eggs.

Scientists refer to this as gene–culture–microbiome coevolution. The microbiome plays a crucial role in development and is often inherited genetically and epigenetically through the mother.

In a project on niche construction and microbiome function, *Onthophagus* dung beetles are used to investigate the role of the microbiome and larval conditioning in facilitating ecological specialization in dung beetle species and their response to novel anthropogenic, i.e., human-induced, challenges, such as antibiotics and toxins, and to natural environmental fluctuations. It will be tested whether novel members of the microbiome lead to the acquisition and stable, nongenetic inheritance of new functions (see Parker et al. 2019; Ledón-Rettig et al. 2018; Schwab et al. 2017; cf. Section 5.2; Schwab et al. 2016). The predictions of theoretical models designed in this study are empirically tested in *Onthophagus taurus*. In this horned beetle., the microbiome inherited via the dung ball is essential for digestion and growth, and host species diverge evolutionarily depending on their microbiota.

The fourth and last project group focuses on evolutionary diversification. Nine subprojects are dedicated to this area, and they test the core question: Can plasticity guide evolution? In the three-spined stickleback (*Gasterosteus aculeatus*), researchers investigated whether a high degree of ancestral plasticity is associated with an accelerated rate of genetic divergence. In addition, they examined whether patterns of genetic variation (consistent with the traditional expectation) or ancestral plasticity (consistent with the EES expectation) best explain the evolutionary change in saltwater and freshwater populations (see Foster et al. 2019). Regions of the genome were found to be selected in a freshwater environment, which can be taken as support for the thesis that the *Genes are followers* (Section 3.8). Finally, the result of genome-wide comparative studies should determine whether ancestral forms of plasticity have controlled evolutionary change during stickleback radiation.

A project on *Anolis lizards* (Figure 6.4) is testing the hypothesis that developmental responses to mechanical stress drive adaptive limb length diversification in lizards. One examines the role of plasticity in limb length divergence via (a) comparisons of phenotypic plasticity in animals reared under different conditions and (b) interspecific comparisons between phenotypically divergent species using comparative gene expression analysis of long bone growth zones (cf. Feiner et al. 2017, 2018). The result is that plastic traits, such as limb length, show higher evolvability.

Figure 6.4 Anolis (*Anolis carolinensis*). A species of the genus *Anolis* is used for an EES project to investigate how different mechanical stresses affect limb length.

A project is investigating the mutual causality between phenotypic plasticity and genetic evolution of eyespots in the African butterfly *Bicyclus anynana* (Section 4.4). The goal is to understand whether and how the evolution of plasticity and directional development influences evolutionary diversification as well as the extent of parallel evolution in butterflies in different geographic locations with similar environments (cf. Balmer et al. 2018; Nokelainen et al. 2018; van Bergen et al. 2017). This project confirms the EES prediction that bias can fundamentally shape diversification. However, it also showed that a bias can be overcome. Surprisingly, this case led to a "dramatic exploration of morphospace," or morphological possibilities (Brattström et al. 2020).

The next project investigates the dynamics of thermal adaptations in beetles and large and small dragonflies on micro- and macroevolutionary time scales. The relationships between thermal plasticity, behavioral thermoregulation, tolerance to different thermal conditions, and temperature-dependent morphological traits will be analyzed, as will the relationship between these traits and speciation and extinction rates. Plasticity is thus understood as a bridge between micro- and macroevolution.

Other projects focus on niche construction. For example, one project tests the hypothesis that in animals, artifact building (e.g., nests and spider webs) generates consistent and predictable selection pressures that lead to consistent responses to selection across species, and in turn, to predictable developmental trends and parallel patterns across traits. The project entails data collection on nest construction in birds and fish and web building in spiders, making predictions about expected selective responses to this niche construction activity, and testing of these predictions using comparative phylogenetic methods (see Laland et al. 2017). As predicted, a meta-analysis shows that components of the environment constructed by organisms (as distinct from nonconstructed ones) have different properties and they trigger their own evolutionary responses (Clark et al. 2019).

A project on the fauna of coral reefs is quantifying the niche construction of reef corals and assessing its impact on the local environment and biodiversity in the reef. Researchers are investigating the explanation of biodiversity in coral reefs and the role of different types of coral niche construction in this. The evolutionary effects of coral reef constructions on the biodiversity of the fishes living there are remarkable (Figure 6.5).

Figure 6.5 Coral reef. Coral reefs are diverse niche constructions that influence the evolution of several fish species and other reef inhabitants.

In a project on niche construction and evolutionary diversity in experimental marine microbial communities, researchers are investigating the effects of niche construction using synthetic miniature ecosystems subjected to experimental manipulations of various forms of niche construction. The functional responses of bacterial populations to complementary and conflicting activities of niche construction are examined, and evolutionary responses to niche construction are quantified (see Paterson et al. 2018).

Another project assesses the role of niche construction by ancestors in the emergence of macroevolutionary patterns. It considers previously neglected aspects such as the generation of new resources and morphological innovations made possible by niche construction. Scientists are developing a detailed theoretical model to understand the macroevolutionary dynamics of niche construction and ecosystem engineering. The predictions of this model are then tested using data from the fossil record (see Brush et al. 2018; Daniels et al. 2017). A final theoretical project will examine the importance of species interactions through predation, competition, and niche construction for ecosystem stability and diversity. The findings suggest that major transitions in evolution, such as from unicellular to multicellular organisms, cannot occur without niche construction and phenotypic plasticity.

6.5 A PROJECT BEYOND EES: INSTANCES OF AGENCY IN LIVING SYSTEMS

A new major project is already surpassing the EES. In 2019, a research proposal on *Agency in Living Systems* was approved by five universities in the US, Canada, and Finland (Moczek et al. 2019; https://agencyinlivingsystems.com). The project is led by Armin Moczek and evolutionary ecologist Sonia Sultan. Denis Walsh, a Canadian philosopher of biology, is responsible for the evolutionary theoretical underpinnings (https://agencyinlivingsystems.com/members/).

In the natural sciences, agents are entities whose evolved mechanisms operate in a self-regulating manner, contributing to their own permanence, maintenance, and function. They do not require intentions, goals, or knowledge to do so. The criteria for an agent are being objective and observable. An agent consists of systems that can respond to occurrences in a way that reliably produces operational, stable end states.

This project aims to recognize and describe organisms as living *agents* and not as passive biological objects as in traditional evolutionary theory. Rather, organisms as *agents* actively

In Conversation with Eva Jablonka

Professor Jablonka, you moved to Israel from Poland as a young girl with your parents in the 1950s. Please briefly describe what was it like then in the newly founded state. It must have been exciting for a child.

I was a child of 5 when I came to Israel. I hardly remember Poland, and I identified with Israel as it was, as I experienced it when I grew up and went to school in the 1960s. As I saw it, it was a country which was based on secular socialist ideas, on social justice, on solidarity. I had passionate love and loyalty for it. I was proud of it, I believed it would be a beacon of human rights. Unfortunately, it has not lived up to expectations. It is now a capitalist country with one of the greatest disparities between rich and poor in the Western world, which rules over territories we occupied over 50 years ago, and denies basic rights to the people who live there. It is also increasingly leaning toward religious fundamentalism. It is heartbreaking.

A scientist with your reputation could also have taught at one of the well-known US universities. You certainly had offers. Why did you stay in Israel?

There are many considerations in one's life, and I never seriously considered leaving Israel. My family was there. My friends, many colleagues and students, my language, my sense of cultural and social identity—even as I got increasingly critical of politics. I never felt intellectually constrained in Israel. On the contrary. It is one of the most intellectually vibrant places I have ever encountered. It is far less conservative than most of mainstream US academia.

You have studied nongenetic inheritance for decades. It took more than 30 years for Barbara McClintock's jumping genes, to be recognized. It seems that biology is slow to accept new perspectives on inheritance. Why is that?

Because inheritance is fundamental—practically, socially, and theoretically. Ideas about inheritance have impact on the way we think about politics, social justice, health and disease, and evolutionary history. The complex scientific history of heredity and evolution in the 20th century is a testimony to the fundamental and socially influential role of ideas about heredity on politics. Eugenics in the West and Lysenkoism in the USSR are known examples of the tragic misuse of ideas about heredity. Within biology, the way we understand heredity impinges on the way we understand development and evolution, so changing ideas about inheritance can change fundamental theoretical assumptions. This leads to understandable scrutiny and resistance: scientists will not abandon what they see as a well-tested mainstream approach in which they were socialized, for a novel one. Scientists have professional commitments, interests and habits of thinking, and their scientific style of thinking is embedded within a larger cultural system. All this can lead to resistance even when the evidence for a new theory is very good. It takes a long time for a style of thinking to change.

Biologists hesitate to accept epigenetics, whose contributions to evolution are considered insignificant. Do we have reliable empirical evidence that epigenetic inheritance has influenced evolution?

Marion Lamb and I have written so much about it. In our little book, *Inheritance Systems and the Extended Evolutionary Synthesis* (Jablonka and Lamb 2020) we give many examples, which we think present a very strong case. There is evidence from experimental lab and field studies that there is abundant heritable epigenetic variation in populations and that it is selectable; many different simulation and mathematical models show that within the range of parameters that have been experimentally established, epigenetic variation can affect population dynamics; processes such as hybridization and symbiogenesis, which are implicated in evolution above the species level, involve epigenetic mechanisms.

"Life is learning" said Konrad Lorenz. Karl Popper added that learning is the greatest activity of life. You emphasize the importance of sharing information through learning. Here I recognize a particularly *active* form of inheritance through which individuals or groups transmit information. Does learning play a central role in the evolution of all life on Earth?
Learning is fundamental to the biology of organisms. Simple, non-associative forms of learning exist in all living organisms and are essential for their survival. Neural learning is implicated in every aspect of animal evolution, and was probably one of the factors that drove the Cambrian explosion.

Your book insists that variation is not based on chance, neither in morphology nor in the behavior of living beings. Would you say that chance plays only a marginal role in evolution?
No, chance is fundamental. The physical world is a dynamic buzzing world with a huge amount of stochasticity. This is also the nature of the biological world. There is a huge amount of variability, especially at the molecular levels, part of it random, part semi-directed. But this stochasticity is harnessed by various evolved biological, developmental processes. If you introduce an unexpected, unfamiliar change in conditions, either environmental or genetic, the organism does not just accept it passively. It is trying to cope with it. Evolved mechanisms are recruited, and the organism employs them. So phenotypic variation is not random, it is developmentally regulated and developmentally selected. The processes underlying these nonrandom phenotypic adjustments is what West-Eberhard called phenotypic accommodation. These mechanisms operate at many levels, including the molecular level. Epigenetic variations can be random with respect to function, but can also be semi-directed because the epigenetic mechanisms that generate them are part of the response system of the organism. Importantly, there may be selection among the most successful accommodators and this may lead to genetic accommodation. So the point is: even when at the molecular level variation is totally random, phenotypic variation rarely is.

Cultural inheritance is a fairly new, exciting topic in biology. The theory of niche construction increasingly addresses cultural niches. How is culture related to evolution?
First, there is evolution of the cultural axis itself. Think about the evolution of human languages. Second cultural niche construction determines the kind of selection pressures which individuals will be subject to. For example, if you are living in a linguistic environment, there will be selection (at all levels—genetic, epigenetic, cultural) for

variations that make individuals thrive in it, most obviously, variations that directly affect the many aspects of linguistic ability and linguistic communication.

Although our communication methods have evolved from cuneiform to the Internet, and we will soon fly to Mars, we are still the same biological species.
I am not sure we are quite the same. Even if the genetic differences make no difference, epigenetically we certainly are not the same. Plasticity is a biological trait, and it has biological manifestations.

We will soon use CRISPR to customize our genome and may develop AI-based robots that challenge us. How far can we evolutionarily decouple from biology?
I don't think this will be a decoupling. It will be new evolutionary dynamics. The CRISPR technology is a biological technology. Future AI will present complex problems, I think. At some time in the future, AI may compel us to redefine our notion of life. But we are not yet there.

You are among evolutionary biologists such as Gerd Müller (Vienna), Denis Noble (Oxford), and Stuart Newman (New York) who support replacing Modern Synthesis with a new theory of evolution. Does the Extended Synthesis project compromise itself by addressing only extension rather than a new construction?
There is continuity between the MS and the EES, and the term "extended" highlights this. There is also a need for revision, which is not captured by this term. I am not sure what the ideal term is. Marion and I talked (tongue in cheek) about the "post-modern synthesis". The adjective "extended" was chosen in order to allow a dialogue with MS proponents so that they will realize that there is no wish to throw out the baby with the bathwater. It was a somewhat naïve assumption because the opposition did not appreciate it, maybe even regarded it as a sign of weakness.

Please allow me to ask two different but no less interesting questions: Together with the neurobiologist Simona Ginsburg, you recently published a book on the evolution of consciousness (Ginsburg and Jablonka 2019). That sounds exciting because we currently know almost nothing about consciousness. Denis Noble says that consciousness is not a neural object, and according to Thomas Nagel, consciousness even has an irreducible subjective character. Does that leave an objective view at all? How can you then write about the evolutionary origin of consciousness?
We think that a good biological understanding of consciousness is possible, and we profoundly disagree with Nagel. Consciousness is a mode of being of some living systems (we are not aware of non-living conscious entities). We think that an evolutionary approach focused on the transition from non-conscious to conscious organisms can be very informative, and do the same service to science as did the research program on the origin of life, which demystified life. This is the approach we employ.

The well-known neurobiologist and bee researcher Randolf Menzel said in a recent interview that "the bee knows who she is." Today forms of thinking in animals are known, but do you go so far as to ascribe animal's consciousness?

Yes, as we argue in detail in our book on consciousness, many (though by no means all) animals are conscious, subjectively experiencing colors and sounds, pains and pleasures. These include most vertebrates, some arthropods, such as the bee, and cephalopods (squid, cuttlefish, octopus)

The relationships to our animal ancestors are clearer now more than ever. What remains evolutionarily unique to man?
Humans live, as the philosopher Ernst Cassirer said, in the symbolic dimension. Our cognitive and emotional development and our evolution are molded and driven by the values of our symbolic cultures. We are profoundly different from other animals.

Professor Jablonka, thank you for this interview.

generate adaptation, resilience (stability), and innovation. Moreover, the corresponding mechanisms occur at different levels of biological organization. The processes involved can be described empirically and theoretically. The project is best described by the multimillion dollar research proposal, from which I drew the summary below. The factual content is already comprehensively explained by Denis Walsh in his book *Organisms, agency, and evolution* (Walsh 2015).

The project will identify agencies at different levels of organization. At the lowest level, these are gene regulatory networks; they can use cellular and environmental signals to dampen perturbations in developmental processes and also to promote new outputs along specific trait axes. At the higher level, cells and tissues interact dynamically in the developing embryo in such a way that integrated, potentially innovative functional structures emerge. At an even higher level, organisms in their entirety can modulate body proportions. This occurs via a plastic response to the environment, and in certain cases, through the epigenetic transmission to offspring. Finally, short-lived responses by individual social organisms can prompt a collective behavior, which then determines the functional properties of the entire colony. The various mechanistic levels thus span the entire spectrum from DNA variation to the phenotypic outcome that determines individual success and drives natural selection. Seen in this light, living systems can no longer be viewed and studied as mere objects of internal and external forces but rather as autonomous agencies with different instances of agency.

The research team of Armin Moczek emphatically point out that DNA sequences cannot "program" a phenotypic outcome, i.e., organisms do not come into being by reading out DNA alone, because countless other factors are involved at various levels. The gene-centric approach, according to which DNA is simply read out, misses the essence of biology. Instead, the researchers mentioned in this section focus on the active formative process, to overcome the self-imposed limits of previous understanding, and to make causal insights empirically and theoretically much more fully visible than before.

The project aims to answer the following two specific questions: What is an agency, and what is the science of agencies? Using various empirical approaches, researchers want to identify the mechanisms by which organisms (mammals, insects, and plants) can become agencies in the construction of their own new traits. Ultimately, this will result in a theory of organisms as purposive agencies and establish explanatory concepts and structures for a theory of instances of agency (as opposed to the previous object theory).

This project is extremely ambitious. To quote the objective and delimitation to the EES project from the project proposal kindly provided to me (Moczek et al. 2019),

> By addressing mechanisms from gene networks to colonies, and doing so across a broad range of organisms, the work presented here establishes an empirical and philosophical framework that includes the narrower foci of EES, but at the same time extends well beyond them to encompass all of biology rather than just evolution. Accordingly, the proposed work will ground the findings of EES in a coherent and robust theory applicable to all areas of biology and to all types of organisms. At the same time, based on rigorous philosophical concepts, our project will explore how recognizing organisms as evolutionary acting agents forces a change in how we study them and how we interpret scientific results. These conceptual challenges are also critical to the current EES project, but are beyond its intended scope. Our proposed work does not replicate or continue the EES project. It will, however, expand its impact by grounding its perspective in an important new way.

The project began in October 2019, and the scientific push includes numerous dissertations and publications in scientific journals by the end of 2022 (https://agencyinlivingsystems.com/publications/). In addition, there will be relevant cross-disciplinary books on *natural agency*. Overall, this will result in a coherent conceptual framework for numerous empirical processes across multiple biological lineages and all organismal levels.

In retrospect, we recognize that this research project is a continuation of earlier initiatives by Müller (Section 3.4), Kirschner and Gerhart (Section 3.5), Jablonka and Lamb (Section 3.6), Noble (Section 3.7), and others. The original evo-devo approach focusing on developmental constraints is now extended. There are several ways to determine the potential evolutionary activity of organisms. While constraints are thematically and conceptually passive in character, and evolvability implies neutrality, developmental bias shows a more orderly, active character. However, the formative and functional design potential of the organism and its organizational levels of organization comes to the fore even more strongly in terms such as facilitated variation or instances of agency and "organism as agent."

6.6 SUMMARY

Below, I summarize the predictions of the Extended Synthesis project (this chapter) with the results of evo-devo theory (Chapters 3 and 4) and niche construction theory (Chapter 5) in 22 points, organized here into seven themes now familiar to the reader.

The following list therefore contains research results on the EES project and on the other focal points of the Extended Synthesis.

1. Mutation
    - Mutation of DNA is one of several factors explaining phenotypic variation. Embryonic development provides the conditions for specific forms of phenotypic expression and thus ways to explain the pathway from genotype (the totality of an individual's genes) to phenotype (the totality of an individual's recognizable characteristics) (Pigliucci and Müller 2010).
    - Nonrandom genetic mutation also occurs. Genetic mutation is to be seen at most as an initiator of the phenotypic variation that subsequently arises in development.
    - On the one hand, chance (random mutation) loses importance (Jablonka and Lamb 2014); on the other hand, evolution actively uses stochasticity.

2. Natural selection
    - Natural selection is not solely responsible for generating

phenotypic variants; evolutionary development includes intrinsic mechanisms for producing phenotypic variation and innovation (Kirschner and Gerhart 2005; Müller 2017, and others). Natural selection thus loses sole control according to the new view. Its task is to fine-tune a trait that ontogeny has constructed and presented to it in a basic format (Lange et al. 2014; Newman 2018).

3. Phenotypic variation and evolutionary development
   - Phenotypic accommodation may precede rather than follow the genetic variation in adaptive evolution.
   - Phenotypic variation is, in several cases, biased and functional.
   - Strikingly different new phenotypes can result from mutation of a key regulatory control gene expressed in a tissue-specific manner, from facilitated variation, or from a variety of interacting developmental processes. In short, "phenotypic output is not a direct consequence of natural selection, but a consequence of the functional properties of the system" (Noble et al. 2014).
   - Development exhibits considerable plasticity, that is, it can produce more than one continuous or noncontinuous variable form of morphology, physiology, or behavior under different environmental conditions.
   - The explanation of phenotypic variation and innovation has recently begun to refer increasingly to instances of agency at different biological levels and to organismic agents. These terms also address the (physiological) function of the system as a whole (Moczek et al. 2019).

- Evolutionary developmental biology provides explanations for the pace of evolution. Its mechanisms allow rapid responses to selection, as well as mutational and environmental changes (Müller 2017). In particular, these are
   a. The principle of nonlinear threshold effects in the formation of discontinuous phenotypic traits (Meinhardt and Gierer 1974; Nijhout 2004; Tiedemann et al. 2012; Müller 2010; Lange et al. 2014, 2018). The possibility of the evolution of discontinuous traits must be a part of the evolutionary theory.
   b. The principle of self-organization (Turing 1952; Meinhardt and Gierer 1974; Müller 2010; Raspopovic et al. 2014; Lange et al. 2018; Newman 2018).
- Robustness (hidden, /masked genetic mutations) or channeling and genetic switches in gene regulatory networks that are easily changed create suitable conditions for phenotypic variation (Waddington 1942; Kirschner and Gerhart 2005; Wagner 2014).

4. Inheritance
   - In addition to genetic inheritance, inheritance mechanisms that must be incorporated into evolutionary theory also include nongenetic inheritance, learning, and cultural transmission. These forms are grouped with genetic inheritance under the term inclusive inheritance.
   - Individuals can acquire traits through reciprocal interaction with the environment and pass them on to subsequent generations.

5. Niche construction
   - Niche construction systematically tends to environmental changes adapted to the phenotype of the constructor or that of its successors (atmosphere with oxygen, coral reefs, beaver dams, cities, etc.).
   - The theory of niche construction views natural selection and niche construction as interacting causes of evolution and treats the adaptation of organisms as outcomes of both processes.
   - Niche construction is seen as a separate selective evolutionary process alongside natural selection.
   - Developmental niche constructions are built by parents for their offspring (nests, spider webs, and other dwellings). Parents thus modify their environment, and their offspring accommodate these constructions in their own development. Developmental niches represent specific selection conditions, modified by a particular construction, that help determine the evolution of the young.
6. Evolutionary diversification
   - Organismal diversity can be explained by the evolvability and constraints of the developmental system.
7. Conceptual topics/philosophy of biology
   - Biological systems have constructive elements in their development. They can change their own biology (biased development, niche construction).
   - According to the concept of reciprocal causality, evolution is based on a network of causality and feedback in which already selected organisms cause environmental changes; conversely, the modified environment (niche) causes changes in the organisms that modified it. Both processes are types of selection (Kendal et al. 2011).
   - Randomness is a basis for emergent order (Noble R and Noble D 2018).
   - Evolutionary theory is in transition from a predominantly population-statistical theory to a more mechanistic-causal theory (Müller 2010).

REFERENCES

Balmer AJ, Brakefield PM, Brattström O, van Bergen E (2018) Developmental plasticity for male secondary sexual traits in a group of polyphenic tropical butterflies. *Oikos* 127:1812–1821.

van Bergen E, Osbaldeston D, Kodandaramaiah U, Brattström O, Aduse-Poku K, Brakefield PM (2017) Conserved patterns of integrated developmental plasticity in a group of polyphenic tropical butterflies. *BMC Evol Biol* 17(1):59.

Brattström O, Aduse-Poku K, van Bergen E, Brakefield PM (2020) A release from developmental bias accelerates morphological diversification in butterfly eyespots. *PNAS* 117 (44):27474–27480.

Brush ER, Krakauer DC, Flack JC (2018) Conflicts of interest improve collective computation of adaptive social structures. *Sci Adv* 4(1):1–10.

Casasa S, Moczek AP (2018) The role of ancestral phenotypic plasticity in evolutionary diversification: population density effects in horned beetles. *Anim Behav* 137:53–61.

Clark A, Deffner D, Laland K, Odling-Smee J Endler J (2019) Niche construction affects the variability and strength of natural selection. *Am Nat* 195(1):16–30.

Daniels BC, Krakauer DC, Flack JC (2017) Control of finite critical behaviour in a small-scale social system. *Nat Comm* 8:14301.

Feiner N, Rago A, While GM, Uller T (2017) Signatures of selection in embryonic transcriptomes of lizards adapting in parallel to cool climate. *Evolution* 72(1):67–81.

Feiner N, Rago, A, While GM, Uller T (2018) Developmental plasticity in reptiles: insights from temperature-dependent gene expression in wall lizard embryos. *J Exp Zool* 329(6–7):351–361.

Ginsburg S, Jablonka E (2019) *The Evolution of the Sensitive Soul: Learning and the Origins of Consciousness*. MIT Press, Cambridge, MA.

Foster SA, O'Neil S, King RW, Baker JA (2019) Replicated evolutionary inhibition of a complex ancestral behaviour in an adaptive radiation. *Biol Lett* 15. https://doi.org/10.1098/rsbl.2018.0647.

Jablonka E, Lamb MJ (2014) *Evolution in four Dimensions. Genetic, Epigenetic, Behavioral, and Symbolic Variation in the History of Life*. 2nd ext. ed. MIT Press, Cambridge, MA.

Jablonka E, Lamb MJ (2020) *Inheritance Systems and the Extended Evolutionary Synthesis*. Cambridge University Press, Cambridge, MA.

Kendal J, Tehrani JJ, Odling-Smee J (2011) Human niche construction in interdisciplinary focus. *Philos T Roy Soc B* 366:785–792.

Kolodny O, Weinberg M, Reshef L, Harten L, Hefetz A, Gophna U, Feldman MW, Yovel Y (2019) Coordinated change at the colony level in fruit bat fur microbiomes through time. *Nat Ecol Evol* 3:116–124.

Kirschner M, Gerhart J (2005) *The Plausibility of Life: Resolving Darwin's Dilemma*. Yale University Press, New Haven.

Kuhn TS (1962) *The Structure of Scientific Revolution*. University of Chicago Press, Chicago, IL.

Laland KN, Uller T, Feldman M, Sterelny K, Müller GB, Moczek A, Jablonka E, Odling-Smee J (2015) The extended evolutionary synthesis: its structure, assumptions and predictions. *Proc Royal Soc B* 282:1019.

Laland KN, Odling-Smee JF, Endler J (2017) Niche construction, sources of selection and trait coevolution. *Interface Focus* 7(5). https://doi.org/10.1098/rsfs.2016.0147.

Lange A, Nemeschkal HL, Müller GB (2014) Biased polyphenism in polydactylous cats carrying a single point mutation: the Hemingway model for digit novelty. *Evol Biol* 41(2):262–275.

Lange A, Nemeschkal HL, Müller GB (2018) A threshold model for polydactyly. *Prog Biophys Mol Bio* 137:1–11.

Ledón-Rettig CC, Moczek AP, Ragsdale EJ (2018) Diplogastrellus nematodes are sexually transmitted mutualists that alter the bacterial and fungal communities of their beetle host. *PNAS* 115:10969–10701.

Macagno ALM, Zattara EE, Ezeakudo O, Moczek AP, Ledón-Rettig CC (2018) Adaptive maternal behavioral plasticity and developmental programming mitigate the transgenerational effects of temperature in dung beetles. *Oikos* 127:1319–1329.

Meinhardt H, Gierer A (1974) Application of a theory of biological Pattern Formation based on lateral inhibition. *J Cell Sci* 15:321–346.

Moczek AP, Sultan SE, Walsh D, Jernvall J, Gordon DM (2019) Agency in living systems: how organisms actively generate adaptation, resilience and innovation at multiple levels of organization. Proposal for a major grant from the John Templeton Foundation.

Müller GB (2010) Epigenetic innovation. In: Pigliucci M, Müller GB (eds.) *Evolution – The Extended Synthesis*. MIT Press, Cambridge, MA, 307–332.

Müller GB (2017) Why an extended evolutionary synthesis is necessary. *Interface Focus* 7:1–10

Newman SA (2018) Inherency. In: Nuno de la Rosa LN, Müller GB (eds.) *Evolutionary Developmental biology. A Reference Guide*. Springer International Publishing, Cham.

Nijhout HF (2004) Stochastic gene expression: dominance, thresholds and boundaries. In: Veita RA (ed.) *The Biology of Genetic Dominance*. Landes Bioscience, Georgetown, 2000–2013.

Noble R, Noble D (2018) Harnessing stochasticity: how do organisms make choices? *Chaos* 28:106309. https://doi.org/10.1063/1.5039668.

Noble D, Jablonka E, Joyner MJ, Müller GB, Omholt SW (2014) Evolution evolves: physiology returns to centre stage. *J Physiol* 592(11):2237–2244.

Nokelainen O, van Bergen E, Ripley BS, Brakefield PM (2018) Adaptation of a tropical butterfly to a temperate climate. *Biol J Linnean Soc* 123:279–289.

Parker ES, Dury GJ, Moczek AP (2019) Transgenerational developmental effects of species-specific, maternally transmitted microbiota in Onthophagus dung beetles. *Ecol Entomol* 44:274–282.

Paterson DM, Hope JA, Kenworthy J, Biles CL, Gerbersdorf SU (2018) Form, function and physics: the ecology of biogenic stabilisation. *J Soils Sediments* 18:3044–3054.

Pespeni H, Ladner JT, Moczek AP (2017) Signals of selection in conditionally expressed genes in the diversification of three horned beetle species. *J Evol Biol* 30:1644–1657.

Pigliucci M, Müller GB (eds.) (2010) *Evolution – The Extended Synthesis*. MIT Press, Cambridge, MA.

Radersma R, Hegg A, Noble DWA, Uller T (2018) Timing of maternal exposure to toxic cyanobacteria and offspring fitness in Daphnia magna: implications for the evolution of anticipatory maternal effects. *Ecol Evol* 8:1–10.

Rago A, Kouvaris K, Uller T, Watson RA (2019) How adaptive plasticity evolves when selected against. *PLoS Comput Biol* 15(3):1–20.

Raspopovic J, Marcon L, Russo L, Sharpe L (2014) Digit patterning is controlled by a BMP-Sox9-Wnt Turing network modulated by morphogen gradients. *Science* 345:566–570.

Schwab DB, Riggs HE, Newton ILG, Moczek AP (2016) Developmental and ecological benefits of the maternally transmitted microbiota in a dung beetle. *Am Nat* 188:679–692.

Schwab DB, Casasa S, Moczek AP (2017) Evidence of developmental niche construction in dung beetles: effects on growth, scaling and reproductive success. *Ecol Lett* 188(6):679–692.

Tiedemann HB, Schneltzer E, Zeiser S, Hoesel B, Beckers J, Przemeck GKH, de Angelis MH (2012) From dynamic expression patterns to boundary formation in the presomitic mesoderm. *PLoS Comput Biol* 8:1–18.

Turing A (1952) The chemical basis of morphogenesis. *P Roy Soc B Bio* 237:37–72.

Uller T, Helanterä H (2019) Niche construction and conceptual change in evolutionary biology. *Brit J Phil Sci* 70:351–375.

Uller T, Moczek AP, Watson RA, Brakefield PM, Laland KN (2018) Developmental bias and evolution: a regulatory network perspective. *Genetics* 209:949–966.

Waddington CH (1942) Canalisation of development and the inheritance of acquired characters. *Nature* 150:563–565.

Wagner A (2014) *The Arrival of the Fittest: How Nature Innovates*. Penguin Random House. New York.

Walsh D (2015) *Organisms, Agency, and Evolution*. Cambridge University Press, Cambridge.

TIPS AND RESOURCES FOR FURTHER READING

Baedke J, Fábregas-Tejeda A, Vergara-Silva F (2020) Does the extended evolutionary synthesis entail explanatory power? *Biol Philos* 35:20. https://link.springer.com/article/10.1007/s10539-020-9736-5.

Brattström O, Aduse-Poku K, van Bergen E, Brakefield PM (2020) A release from developmental bias accelerates morphological diversification in butterfly eyespots. *PNAS* 117(44):27474–27480: This is the best and the most pertinent example to explain how biased development determines evolution and how (in a butterfly genus) the bias can be overcome and diversification is made possible.

Brun-Usan M, Zoimm R, Uller T (2022) Beyond genotype-phenotype maps: toward a phenotype-centered perspective on evolution. *BioEssays* 44:2100225. https://doi.org/10.1002/bies.202100225: The authors in this study used development as a starting point, and represented it in a way that allows genetic, environmental, and epigenetic sources of phenotypic variation to be independent.

Cedar MA (2017) Domestication as a model system for the extended evolutionary synthesis. *Interface Focus* 7:20160133. http://dx.doi.org/10.1098/rsfs.2016.0133: An interesting contribution to EES in this study is the context provided here with domestication. The author calls for domestication to be considered a model system for testing the predictions of EES.

Clark A, Deffner D, Laland K, Odling-Smee J Endler J (2020) Niche construction affects the variability and strength of natural selection. *Am Nat* 195(1):16–30: A meta-analysis of selection gradients shows that environmental components constructed and non-constructed by organisms have different properties and elicit distinct evolutionary responses.

Gefaell J, Saborido C (2022) Incommensurability and the Extended Evolutionary Synthesis: taking Kuhn seriously. *Euro Jnl Phil Sci* 12, 24.

Kolodny O et al. (2018) Coordinated change at the colony level in fruit bat fur microbiomes through time. *Nat Ecol Evol* 3:116–124: This analysis of the microbiome of Egyptian fruit bats revealed that the meaningful ecological unit in the bat host microbiome is at the colony level, with coordinated intragroup microbial changes.

Kouvaris K et al. (2017) How evolution learns to generalize. Using the principles of learning theory to understand the evolution of developmental organization. *PLoS Comp Biol* 13:e1005358: This paper shows that evolution by natural selection, when operating on the basis of developmental parameters (not genes or traits), can produce effects previously thought to be impossible, which include the generation of pre-adaptation to novel environments and enhanced adaptation with experience.

Laland KN, Uller T, Feldman M, Sterelny K, Müller GB, Moczek AP, Jablonka E, Odling-Smee J (2015) The extended evolutionary synthesis: its structure, assumptions and predictions. *Proc Royal Soc B* 282:20151019: This study presents the current scientific consensus of the EES.

Ledon-Rettig C et al. (2018) Diplogastrellus nematodes are sexually transmitted mutualists that alter the bacterial and fungal communities of their beetle host. *Proc Natl Acad Sci* 115, 10696–10701: First publication to document the fitness relevance of interactions between hosts, microbiota, and symbiotic nematodes that are enabled through nongenetic inheritance and niche construction are enabled.

Noble DW et al. (2019) Plastic responses to novel environments are biased toward phenotype dimensions with high additive genetic variation. *Proc Natl Acad Sci* 116:13452–13461: This study provides important empirical support for plasticity-driven evolution and demonstrates that plastic responses have high evolutionary potential.

Pigliucci M, Muller GB (eds.) (2010) Chapter 1: Evolution - The extended synthesis. In: Pigliucci M, Muller GB (eds.) *Elements of an Extended Evolutionary Synthesis*. The MIT Press, Cambridge, MA, 3–17: This volume and introduction are the initial work of EES.

Uller T, Laland KN (eds.) (2019) *Evolutionary Causation. Biological and Philosophical Reflections*. MIT Press, Cambridge, MA: A short selection of nine important publications from the EES project: A comprehensive treatment of the concept of causality in evolutionary biology, making clear its central role in historical and contemporary debates.

Uller T et al. (2018) Developmental bias and evolution: a regulatory network perspective. *Genetics* 209:949–966: Reviews the evidence for directional evolution (bias), illustrates how it can be studied, and explains why it is not sufficient to characterize biased evolution as constraints.

Whitehead H et al. (2019) The reach of gene-culture coevolution in animals. PLoS Comp Biol 13(4):e1005358: This study reviews evidence that culture in animals can influence evolutionary processes in multiple ways, including triggering speciation, shaping gene flow, and promotion of coevolution.

CHAPTER SEVEN

Theories on the Evolution of Thinking

Can the standard theory of evolution adequately explain the extraordinary evolutionary growth of the human brain or neocortex (i.e., the part of the cerebral cortex with particular significance in primates)? Can it explain the peculiarities of human thinking, such as consciousness or empathy? This chapter addresses these questions from a modern point of view and, for this purpose, introduces two theories that do not belong to the extended synthesis but are related to it in their patterns of thinking. It will become clear that both theories, the theory of the social brain and Michael Tomasello's theory of the natural history of thinking, are congruent with our previous discussions of niche construction (Chapter 5) and have a higher propositional content than neo-Darwinian explanations. Moreover, the examples given here illustrate once more that the thinking of the EES is not limited to the evolution of organismic form.

Important technical terms in this chapter (see glossary): Consciousness, cooperation, Dunbar's number, intentionality, niche construction, ratchet effect, and theory of mind.

7.1 DARWIN'S VIEW OF THINKING

Twelve years after his epoch-making magnum opus *The Origin of Species*, Darwin published a book on human evolution in 1871. He had already announced this bold topic in the last sentence of his first book. The new work was entitled *The Descent of Man and Sexual Selection.* ||||In it, Darwin comments at length on the evolutionary origin of the mental capabilities of humans and animals. He states that we could not be convinced that our high mental capabilities evolved in stages if our mental powers were fundamentally different from those of animals. In terms of mental capabilities, he sees "no fundamental difference between man and the higher mammals in their mental faculties" but only slight variations that have developed gradually. He then discusses a variety of examples from the animal world, specifically, attention, memory, imagination, and reason. He goes on to comment on imitation, comparison, and choice, emphasizing their various gradations. He states that from the behavioral concomitants of action, we can infer whether it is taken by instinct, reason, or a mere association of ideas. He also addresses in detail the (conflicting) views of his contemporaries on the evolution of thinking, such as, no animal has the capacity for abstraction, ability to form general concepts, or ego-consciousness, and no animal uses tools or language.

Darwin invalidates all these views, which was a remarkable accomplishment during his time. He concludes with the following remark: If the abovementioned abilities of animals, which he sees very differently, are themselves capable of development (evolution), then it is not improbable that complicated abilities such as abstraction, ego-consciousness, and others have developed from simpler ones. To the question of why the intellect of the monkey is not so far developed as that of humans, he answers that only ignorance of the successive evolutionary stages prevents him from providing a more precise answer to this. In summary, Darwin's intention is to present the differences in the intellectual abilities of humans and animals as gradual and not principled. Otherwise, his theory would be inconsistent, and he could not derive our cognitive abilities from those of animals (Darwin 1871). More skeptical than Darwin was his contemporary Alfred Russel Wallace. To him, the mental abilities of humans, language, art, and morals appeared so unique that he did not dare to make natural selection responsible for them. In this respect, he remained attached to his religious origin. After Darwin's death, toward the end of the 19th century, psychologists began to develop the concept that humans are the product of their evolutionary past, not only in their physical form but also in their behavior and culture.

Behaviorism and other concepts interrupted this line of reasoning until the middle of the 20th century. According to behavioral theory, the mind cannot be a subject of scientific discussion because it eludes direct observation. Above all, according to the abbreviated conviction of behaviorism, one cannot excavate the contents of thought, and thus, cannot say anything about one's own evolutionary history. Archaeology has long adhered to this principle and has excluded several questions from the study of human evolution, such as emotions and intentions. However, today, few doubt that humans are a legitimate object of research for anthropologists, biologists, and psychologists, not only in their actions and cultural embeddedness but also in their thinking. Today, the majority of these are convinced that, in addition to our cultural achievements, our mental abilities can also be explained on the basis of our phylogenesis.

The focus is on the following two well-known central evolutionary theses: (1).

Evolution by natural selection and adaptation is the only known natural process that can produce a complex structure like the human mind. The features of thinking that we possess today are adaptive because they were advantageous to our ancestors in the natural selection process. (2). Evolution acts over the long term. Our minds are therefore shaped by the long-term challenges that humans have faced in their natural environment. We must be aware of this: during the greatest evolutionary time span, our hominin ancestors were hunter-gatherers, but during this long evolutionary period, humans and their ancestors have gone through significant stages of change.

However, in the second half of the 20th century, the inclusion of human culture and the idea of cooperation gained importance as pillars of the study of human behavioral evolution. The framework of the neo-Darwinian synthetic evolutionary theory, long limited to Darwin's *Survival of the Fittest* (i.e., survival of the best adapted or most able to reproduce), was thus greatly expanded (Chapters 3–6).

7.2 THEORY OF THE SOCIAL BRAIN

How does modern science view the evolution of thinking? The theory of the social brain or social brain theory postulates that the social environment and group size of a species promote the evolution of the brain via an increase in its size and thus also the evolution of the thought process. In short, humans have strengthened the selection pressure on their increasing brain size in the group itself. In the last 2 million years, a climate-conditioned strengthened selection pressure inevitably led, by evolutionary favor, to increasing group size, whereby again larger brains with still more complex thinking ability were selected. Brain size would thus be a constraint, a limiting factor, for the social group size of a species. I will analyze this in a little more detail below.

The change in human thinking that this theory outlines is described in terms of so-called orders of intentionality. Mentalization is the ability to suspect and understand what another is thinking, or the ability to recognize that others may have views of their own. This has been elaborated in the *Theory of mind* (Gamble et al. 2015). Mentalization is more than empathy and involves cognitive understanding of the other in addition to feeling what the other is feeling.

Six orders of intentionality are distinguished. In the first order, I am aware of myself; I can recognize myself in the mirror. Besides us, some small apes, other mammals, and some birds have this ability. In the second order, I additionally believe something about what my counterpart believes. For example, I believe that my girlfriend loves me. A 5-year-old child can already do this, but our early ancestors with small brains as well as the apes could also do this. In the third order, it becomes more difficult. Here, I think about what my counterpart believes concerning a third person, although I do not believe it. For example, she believes that he loves her, but I do not, or you believe your boyfriend is convinced that you will not go on vacation without him, but I do not. Processing such thoughts is a demanding task that undoubtedly requires the capacity for language. A language can originate as a system of gestures. Species that utilized such systems also had the ability to produce compound tools. On the next levels, orders four to six, the concept of language increases in abstraction. Such thought processes are reflected exclusively in narratives and myths rather than oral communication. Only humans or Neanderthals with a highly developed language are or were therefore capable of thinking on higher order levels. Species at this level can not only master technology but also mentally process situations such as the prolonged absence of another and even death, which is also necessary in a group society. At this point, it is important that human beings with mental intentionality of the fifth or sixth order can permanently deal with more conspecifics in the social environment than their ancestors with intentionality of the second or third order could.

But how is it possible to know at present what our ancestors as a group were mentally capable of when their social behavior cannot be excavated? Brain sizes, which can be read from found fossils, are assigned to average group sizes. These represent possible individual social ways of life. Increasing group sizes are accompanied by group social requirements of increasing complexity:group size, for example, the ability to empathize, strategize, and manage stress. Laughter (Figure 7.1)

Figure 7.1 Laughter. Laughter, music, and sports help people to interact while living in a social group. The release of endorphins promotes stress reduction.

is one such evolved form of stress management in humans.

Let us first address the following questions: Can complex social behavior in humans be reduced to a common denominator? Can a single quantity, such as a species-specific index, be used to simplify comparisons of brain size and social structure? In 1992, Oxford psychologist Robin Dunbar developed a quantity known as Dunbar's number, which may be helpful in this regard. With his hypothesis, the British researcher has been cited nearly 100,000 times (a record) in scientifically relevant magazines as of 2019. Dunbar's number is the average upper limit of the number of people with whom an individual can maintain stable social relationships. This refers to relationships in which an individual knows who each person is and their relationships with others. Dunbar once explained his number as "the number of people you would not feel embarrassed about joining uninvited for a drink if you happened to bump into them in a bar" (Dunbar 1998).

Dunbar's number predicts a group size of approximately 150 individuals from an average brain size of more than 900 cm^3 in human evolutionary history. The Dunbar numbers determined for humans vary between 100 and 250, but generally, 150 is used as the limiting size. The crucial connection is that Dunbar's number is a direct function of the relative size of the neocortex, and the neocortex size in turn limits the size of the social group. Incidentally, Dunbar did not first determine this number empirically, for example, by statistically analyzing the number of people we are in close contact with on average and the brain size that fits proportionally. Such a study could be performed by any biology student with a basic knowledge of statistics. Rather, Dunbar's ingenious achievement was to linearly extrapolate Dunbar's number for humans from the brain and group sizes of other primates. Only afterward did he and others empirically verify the number thus determined for humans several times.

Dunbar thus provided a uniform standard, an index that clarified important relationships. On this basis, the social brain hypothesis could be further improved and developed into a theory. The correlations in primates are summarized schematically in Table 7.1:

Dunbar placed a high value on proving that his number remains stable over evolutionarily long periods of time. Whether the figure also applies to online social media, such as Facebook, XING, mobile address directories, or others, has been the subject of numerous scientific studies in which it has been repeatedly confirmed. In any case, it is important to monitor the implications of megacities and the coexistence of tens of millions of people for human communicative abilities and where our cognitive limits may be exceeded. In retrospect, it is true that our ancestors with small brains lived in smaller groups, mainly because of their limited brain sizes. The six order levels of intentionality and the corresponding Dunbar's numbers show the preconditions under which social thinking could develop evolutionarily (Gamble et al. 2015).

What do we learn when we examine the subject in the context of niche construction? The Australian philosopher Kim Sterelny emphasizes that the externalist explanation

TABLE 7.1
Relationship between brain size and social structure

Brain size	Group size	Dunbar's number	Order of intentionality	Social structure
Small	Small	Small	Low	Simple
Large	Large	Large	High	Complex

The core statement of the social brain theory is that social structure and group size evolutionarily drive brain size.

(i.e., the mutation–selection–adaptation approach) is insufficient to explain the evolution of social intelligence and for the "explosion of cooperation" (Sterelny 2007). According to the theory of niche construction, species modify their own environments. They create their own specific niches in which new selection processes emerge that further drive the evolution of the species. The niche created by the species is a determinant of its own evolution. According to Sterelny, "Hominin evolution is hominin response to selective environments that earlier hominins have made" (Sterelny 2007).

The social brain theory equivalently states that the early first formed small groups, and later, larger ones. The function of group formation was to increase the probability of individual survival. In such groups, tools were used for the first time, fire was built, and language was developed. The social group or niche is the decisive selection factor, for under the conditions of the social group, especially the group size, the brain could grow. The average size of the social group is correlated with that of the brain. The size of the social group niche drives that of the brain, but the increasing brain size again changes the niche by allowing even larger groups to form with more complex social structures; in larger groups, new mental demands on group members, such as those mentioned above, reappear. The niche thus continuously determines and changes its own evolution. We obtain a higher propositional content with this explanation than with the one-dimensional approach of mutation, selection, and adaptation. The theory of social brain evolution extends the synthetic theory, placing it on a new, broader foundation.

Some researchers argue that the relationship between social structure and brain size is too simplistic to explain the evolution of higher intelligence. This gave rise to the hypothesis that natural selection must favor general intelligence rather than a specific ability (social intelligence). Brain size alone, as we know, does not initially equal intelligence. Darwin's and Einstein's brains were no larger than mine, yet my intelligence does not match theirs. It was therefore necessary to define what constitutes the general intelligence that characterizes humans. According to this definition, general intelligence is our ability to reason, plan, solve problems, think abstractly, design complex ideas, learn quickly, and learn from experience (Laland 2017). A group led by Kevin Laland summarized measures of general intelligence, namely, "social (social learning, tactical deception), technical (tool use, innovation), and ecological (extractive foraging, diet breadth) components of intelligence". These metrics were applied to 62 primate species to determine if they could be ranked on a scale of general intelligence. But caution must be taken when interpreting the results, as they represent nothing more than correlations. As with social brain theory, excellent statistical matches can be obtained. However, the fact that the observed factors are also causally related does not follow logically from this but must first be deduced with solid justifications.

First, the group succeeded in arranging all 62 primate species on an ascending scale of general intelligence. As expected, the arrangement also correlated with the cross-species increase in brain size. However, the researchers argued that it was the "battery of hypotheses," which they summarized as "general intelligence," that drove the evolution of the brain and that a single factor, namely social structure, could not satisfactorily explain causally the phenomenon. They stated that social intelligence (i.e., the above criteria of social learning and tactical deception) dominates the evolution of the brain. Simultaneously, they emphasized the surprising result that using only the social intelligence/group size factor in primate experiments in the laboratory revealed a different scaling of primates than when all five factors were considered (Laland 2017). The experiment provided additional evidence that everything in nature is multifactorial and thus complex. Overall, the evolution of the social brain is an excellent example of human niche construction and more complex than the neo-Darwinian model assumes, namely with selective feedback from the self-created niche of the social group.

7.3 TOMASELLO'S NATURAL HISTORY OF THINKING

Michael Tomasello, a former American psychologist at the Max Planck Institute for Evolutionary Anthropology in Leipzig, presents in his book *A Natural History of Human Thinking* (2014) the evolution of thinking as a pathway from nonhuman apes via early humans to modern humans. In doing so, he distinguishes (in contrast to the previously mentioned orders of intentionality) three evolutionary stages of thinking, namely, individual intentionality in the great apes as well as—under the umbrella term shared intentionality—the two forms of joint intentionality in early humans and collective intentionality in modern humans (Figure 7.2). Intentionality refers to the self-regulating, cognitive way of dealing with things. To summarize Tomasello's viewpoint, humans think by means of shared intentionality in order to cooperate with each other, whereas the great apes behave largely individualistically. According to Tomasello, language, developed by way of symbolic pointing gestures is the keystone of human cognition and thinking, not its foundation. Given the levels of intentionality that can easily be interpreted as human niches, it is surprising that Tomasello himself does not make the connection to the theory of niche construction.

Cooperation is a novel concept in evolution that contrasts with the *Survival of the Fittest* and must be considered equally important.

Great apes can think, and they do so in terms of individual intentionality. They possess three key thinking skills, the first of which is schematic, cognitive representation. They can use this to create imagery and know that a leopard can climb trees. With the second key competency, great apes can draw nonverbal, causal, and intentional conclusions, an ability that is traditionally not ascribed to animals. In a well-known experiment, an object is hidden behind a viewing screen. The monkey expects to find it there, but when he sees it being taken away and replaced by another one, he does not expect to find it again behind the screen. The fact that apes can understand the goals of other apes also implies causal inference. As a third key competence, great apes can monitor their own behavior, i.e., monitor the outcome and elements of a decision-making process. For example, they know when they do not have sufficient information to make an appropriate behavioral decision. According to Tomasello, this first form, individual intentionality and instrumental rationality, applies to nonhuman apes during the period after the ancestors of modern humans diverged from those of chimpanzees to become the australopithecines. Such beings are primarily competitive and always think in the service of competition. Unable to speak, great apes can cognitively represent the world and have a certain awareness of their own actions. The discontinuity in primate evolution that sees humans as the only thinking beings (and only by means of language) is thus no longer tenable.

Gareth Evans' generality condition can be used to evaluate an animal's capacity to think. According to Evans' definition, thinking occurs when each potential subject of a thought can be combined with different predicates, and likewise, each potential predicate can be combined with different subjects. Both can be achieved linguistically and non-linguistically. An example of a thought subject with different predicates is the idea that a great ape thinks a leopard can run fast, climb a tree, chase monkeys, and also eat them. In the second case—a predicate with different subjects—a monkey knows that leopards can climb trees, but so can snakes and small monkeys. According to this criterion, which addresses only representational ability and does not use the ability to reason and reflect on one's own actions, which is cited for great apes, it may be at least partially true that great apes can think (Tomasello 2014).

According to Tomasello, early and modern humans have developed a second and a third level of more complex cognition, which he summarizes as shared intentionality and includes both joint and collective intentionality (Figure 7.2).

The joint intentionality of early humans, who did not yet have a conventional language, comprised communal activities, such as foraging, with shared attention and common goals, a collective intentionality with individual roles and perspectives. Communication occurs through natural gestures of pointing, so-called iconic gestures or gesture games. Early humans transform the individual intentionality of apes into shared intentionality through cooperation. The cooperation partners become interdependent: the survival of the individual depends on how the cooperation partner judges him. Communication is strongly oriented toward a "me and you"

scenario and not yet to a larger, anonymous group. One participant has an interest in helping the other to play his role. To do this, he must supply the latter with information that is interesting to him. The conclusions drawn via such a thought process are no longer individual but are socially recursive, i.e., the intentions of the partner are reflected alternately and repeatedly. Great apes are not capable of such a cooperative form of thinking. They do not make joint decisions and, consequently, cannot reflect on them together.

Finally, on the highest level, collective intentionality, cooperative thinking has evolved to the point where it takes place in a group-oriented culture. Here, knowledge and skills are transmitted cumulatively across generations. Great apes, unlike humans, are not motivated to inform others about things or to share information with them. In human thinking, this motive leads to what is known as the ratchet effect. In the ratchet effect, successful cultural adaptations to local conditions are preserved across generations and their knowledge is even extended. This kind of thinking can result in stable, cumulative, cultural evolution. Modern humans exhibit a

Individual intentionality	Joint intentionality	Collective Intentionality
• schematic cognitive representation • nonverbal causal intention (conclusions) • introspection • behavior: individual, competitive	• joint attention • common goals • but: individual roles • gestures, sign language	• group cohesion • accumulative transfer of knowledge (ratchet effect) • language • objective norms for the group • behavior: "we" instead of "I"

Great Ape   Australopithecus   H. habilis   H. erectus   H. sapiens

Figure 7.2 Intentionalities according to Tomasello. Individual intentionality in apes and australopithecines. Shared intentionality in the forms of joint and collective intentionality in the human genus (*Homo*). Only humans developed a collective society and the ability to accumulate knowledge from generation to generation (ratchet effect).

stronger ratchet effect than early humans and apes (Tomasello 2014). For example, they disseminate technical information at a high rate in the same generation and pass it on to the next one.

Within the framework of collective intentionality and sociocultural thinking, individuals use language to create objective norms for the group (regulations, laws, traditions). These norms can be reconsidered and justified by any individual to convince others of them. The group procures normative conventions and standards and can reflect on them using objectified criteria. Shared intentionality is seen as a key innovation or system change in evolution. In cultural evolution, the group itself can become a unit of natural selection (group selection).

As a further evolved ability, humans appear to have unique possession of a pronounced episodic memory, which enables them to clearly assign events to the past, present, and future in a cognitive manner and to distinguish between them. We can undertake mental time travel in both directions, combined with the ability to design nested mental scenarios, e.g., to plan and execute complex technical or artistic projects.

The theory of the evolution of thinking as developed by Tomasello does not explicitly mention niche construction. However, it is obvious that his accounts can be viewed as analogous to the social brain theory in the context of the niche construction theory. The idea of niche construction supports Tomasello's theory. Thus, individuals who mastered collective intentionality created a niche for the group, where the abilities of the individuals are reflected in the cultural–social behavior of the group. The niche codetermines collective intentionality, and thus, the evolutionary course via the feedback of selection forces from this niche, such as pressure toward an even larger group. This view can be found in quite a few scientific articles on Tomasello. In any case, Tomasello's argument for epigenetic, symbolic inheritance including the ratchet effect, with cultural group selection and with the addition of niche construction is a brilliant way to challenge neo-Darwinian thinking.

7.4 CONSCIOUSNESS—A QUESTIONABLE SCIENTIFIC OBJECT

Can Michael Tomasello and the proponents of the theory of the mind understand individual human consciousness in evolutionary terms? Can we therefore expect an answer to the question of what consciousness is? In fact, the aforementioned theories do not claim that we can; we cannot explain consciousness. Tomasello speaks of various forms of intentionality and describes these intentionalities as having properties related to consciousness. This is legitimate, of course, but it does not mean that he also has an answer to what individual consciousness and the human mind are with analyzed contents or categories of thought. After all, Tomasello uses intentionality as a trick to create a bridge between cognitive content and consciousness. However, the true nature of consciousness is beyond his knowledge as well as ours. Human consciousness is a result of the evolution of the brain. Self-representation enhances evolutionary fitness and is therefore within the framework of evolutionary theory. Can it be, nevertheless, that the theory of evolution, indeed the natural sciences as a whole, cannot fully explain the phenomenon?

The neurobiologist Joachim Bauer talks about the self-systems in the cerebral cortex and describes how we—only in resonance with reference persons, primarily the mother—become what we are in our early childhood development. However, Bauer always makes a sharp distinction between the neuronal correlates of the self in the brain and the actual consciousness or self. In other words, from the blank slate of the newborn brain, a self-structure in the brain is only gradually formed in daily contact with the mother. Thus, according to Bauer, in the interpersonal relationship, in Bauer's words, the brain makes biology out of spirit, but of course it is not a spirit (Bauer 2019). Biophysicist and molecular biologist Alfred Gierer maintains that cognitive processes conform to the laws of physics. The same laws that apply to a machine also apply to the brain, but it by no means follows that we can also tap into all the hidden features of the brain. Gierer therefore concludes that the brain–mind relationship may not be fully decodable (1998). Thus, he believes the relationship cannot be completely reduced to the physical level. (This includes, in addition to consciousness, the issue of free will, but I will not discuss this further here). However, the aforementioned considerations do not prevent science from approaching the relationship between the brain and the mind with a variety of methods.

Not only do we know a little about the world; but more importantly, we are aware of our knowledge. In addition to the representation of the external world, we possess a representation of the internal world of our own state. This meta-level raises a whole series of nontrivial problems.

If the biologist takes external signs, i.e., behavioral patterns, as given and implies that the observed behavior can only be understood if something is behind it in the way we directly consciously experience it in ourselves, then he can thereby try to conclude that an elementary ego feeling also exists in *other* people and is thus an objective given. He can then assume this ego as a product of evolution that serves survival. We can follow this line of inquiry to investigate how consciousness arose in our tribal history, a task that requires cooperation across several disciplines. Work in ontogenesis (individual development) is just as much a part of this as work in phylogenesis (tribal history), psychology, and behavioral research. However, future findings in artificial intelligence can also lead to further advancements.

One can also approach the concept of self differently, more critically. For this purpose, let us make a short foray into philosophy to determine what science, and especially the theory of evolution, can and cannot do to answer questions of consciousness and the mind.

As a rule, science deals with an object of investigation, of which several, if not numerous objects of the same kind are available for analysis. But in the case of consciousness, I can only be aware of myself. Since consciousness does not have the object character demanded by the natural sciences, according to dualistic thinkers, it is not scientifically objectifiable. For representatives from the humanities, consciousness is therefore not only until today, but in principle not objectively recognizable, thus for different scientific analyses of the same kind. Consciousness, according to this view, is something fundamentally different from the entities that we can perceive in the physical world. Denis Noble, whom we met in Section 3.7 as a systems biologist, also states that the self cannot be found anywhere because "the self is not a neural object" (Noble 2006). Accordingly, a dualism between mind and matter emerges here. Both mind and matter are principally different for dualists and not recognizable in the same way. We will deal with this dualism a little more closely in a moment.

Other researchers from the natural science faction, by contrast, are intuitively convinced: the self, consciousness, and the mind must be explainable from the material, neuronal components of the brain as a biological organ, i.e., from its physics and its physical processes, and only from these. This materialistic view is the dominant monistic doctrine in the natural sciences, dating back to Aristotle, a paradigm also called monism, physicalism, or naturalism. Dualists accuse the monistic theories of suffering from an "explanatory gap." By this they mean the fundamental difficulty of imagining how insights into the activity of simple neurons can help us understand what happens qualitatively in the brain, i.e., how we feel pain or think about ourselves. A bridge between the two aspects is lacking. It is also important to keep in mind that neuroscience is heading in a reductionist direction with such considerations, whereas in this book, we have repeatedly experienced new insights brought about by taking a nonreductionist approach.

In summary, it might be difficult or even impossible to try to explain the human being as the sum of individual parts. The brain is indeed an objectively describable organ in space and time, and the physiology of this organ is therefore also the object of object-scientific investigation. However, only the one who is in the examined brain as ego-consciousness can say whether the observable processes in the brain also have a meaning.

Intermediate forms between the dualism and monism sketched very simply here are abundant. The topic is known as the "mind-body problem" and has been for centuries; the literature on this topic is extensive. It is even spoken of as a "swarm of naturalisms" (Demmerling 2008). Of course, in the modern literature, some take the position that the mind-body problem is bogus, and thus discussions of it are obsolete. However, I am constantly aware of my own consciousness but not that of others, no matter how close they are to me. I can only make assumptions about their consciousness on the basis of their behavior. For example, I may know that I love a woman. I am aware of it. But whether she really loves me, I can only assume. I may be able to deduce it with reasonable confidence from her behavior, but I can never know it scientifically-objectively. The same goes for the taste of chocolate. I cannot possibly know another person's perception of it.

By way of another example, our conscience is part of consciousness, and as such, also inaccessible at its core. Nevertheless, Eckart Voland and Renate Voland (2016) have outlined a theory on the evolution of conscience. They must rely on two reliable external manifestations of conscience that we know: guilt and shame. Both emotions can be observed when we have a guilty conscience, but conscience itself cannot. The authors deduce from the manifestations how the evolution of the human conscience can be explained adaptively, that is, in accordance with evolutionary theory. Only we humans have a conscience, according to the two authors, which is considered as a courageous theory that deserves to be discussed (not only) in scientific circles.

Philosophers, such as Thomas Nagel in his *Mind and Cosmos* (2012), continue to emphasize the problem of dualism. Nagel perhaps put it best when he stated that "We won't have an adequate general conception of the world until we can explain how, when a lot of physical elements are put together in the right way, they form not just a functioning biological organism but a conscious being" (Nagel 1987).

Today, the radical concept of monism or naturalism, with its undeniable successes in science and technology, has prevailed in the natural sciences. For the monist, there is nothing but things, events, and facts of which we have objective knowledge. "The world disintegrates into facts," as Ludwig Wittgenstein stated uncompromisingly and strikingly in terms of linguistic philosophy and was thus quoted countless times. However, if humans establish connections by assigning real things to each other in a meaningful way, then they methodically use subjective knowledge that only attains validity in communication. A cup as a cultural object is only a cup when a group reaches a communicative agreement on its function, namely what one can do with it. This function is not one of its many physical properties (such as heat resistance or fragility), but the function of a cup is the construct of meaning or purpose that its makers give it: a container for drinking. A rock crystal or a waterfall, on the other hand, was not created for such a purpose and thus has none. It can still be valuable and beautiful to us, but that is different from being purposefully created.

Monists criticize the dualistic perspective above all for the fact that the focus on the purpose, and thus, on the subjective opens the door to intellectual and theoretical proliferation. Everything could be thought of subjectively. However, philosophers have always been aware of this danger. Since the Enlightenment, they have pointed out the necessity of critical, rational control argumentation. Thus, there is no room here for creationism and similar intellectual aberrations.

In its course, evolution once reached a point where such connections of meaning began to appear in the form of artifacts. Biological evolution has thus taken a unique path with the human being, which stands out from everything that existed before. Human existence demands an expanded way of thinking. No longer did the descriptive how-being of objects or processes alone occupy humans, but they also began to ask *why*. Why hold a hand ax? Why sew a fabric? Why did Beethoven compose the Moonlight Sonata? Only a being conscious of itself can ask such questions and give subjective answers to them, because it can conceptualize purpose. But we human beings can conceive of purpose only as a conscious idea. Without consciousness, no purpose exists for us. So much for the philosophical view of humankind.

Until the 1960s, the difference was crystal clear: humans can think, but animals cannot. Humans have awareness, and animals do not. Humans make and use tools, and animals do not. The difference between humans and animals has always seemed fundamental. Today, we know that also other mammals and birds have some form of consciousness. The mirror test is one possible way to verify this (Figure 7.3). Likewise, we know that we can observe a form of purpose-rational behavior in, for example, chimpanzees, which use and work with tools. Chimpanzees also adopt orphaned monkey children. At least as impressive is the behavior of a group of New Zealand keas (*Nestor notabilis*). They cooperate perfectly as a team to collect food in one box. The New Caledonian crow (*Corvus moneduloides*) and the Hawaiian crow (*Corvus hawaiiensis*) not only use but also make themselves fine poking instruments, i.e., tools (Rutz et al. 2016; Figure 7.4). They may even use several different tools in succession to reach the next most useful one and eventually, the coveted food. When crows can distinguish between a short and a long tool for their purpose, they are said to have the ability to think abstractly.

Figure 7.3 Cleaner fish. Animals can think and do so on a gradient of levels. In 2019, the blue-striped cleaner wrasse (*Labroides dimidiatus*) passed the mirror test. Among other things, it tried to scrape off a label on its body that it could only see in the mirror. This was interpreted as the presence of basic ego-consciousness. However, the study has since been interpreted differently (Kohda et al. 2019). Outside of mammals and birds, this fish would be the first species to demonstrate a similar behavior.

Figure 7.4 Hawaiian crow (*Corvus hawaiiensis*)—master of tool use. These birds can adaptively, i.e., not only by following the example of selected individuals, select tools in a purposeful manner and use them sequentially. Researchers speak of sequential intelligence and abstract reasoning.

But can animals also spontaneously find a novel use for a tool? In fact, this has also been demonstrated. A Eurasian jay (*Garrulus glandarius*) was placed in front of a tall, narrow vessel half filled with water. The bird could not reach the worms in the water with his beak. Stones and corks were placed next to the jar for him. Immediately, he began to throw the stones into the glass, causing the water level to rise. He also tried this with a cork. However, it floated, which caused the bird to subsequently use only stones. After a short time, he was able to reach the worms in the glass. The experiment revealed in the bird, the intelligence and forward planning of a 7-year-old child. More and more animals are being discovered that use tools in surprising ways. For example, a night heron (*Nycticorax nycticorax*) at the zoo was observed not eating pieces of bread thrown to it by visitors but instead placing them on the water, whereupon fish took the bait. The heron got what it wanted. However, when a big carp approached, he took the bread again, apparently so as not to waste it.

In fact, in addition to the use of tools, a dozen other criteria were found to prove human uniqueness: upright gait, speech, brain size, empathy, laughter (Figure 7.1), symbolic thinking, and many more. However, none of these alone was able to prove our special position in the long run. But wait! What about music and rhythm? Reinhard Piechocki has recently devoted himself to the study of *Homo synchronicus*. He points out that only humans are able to move in unison to a given beat and rhythm in pairs or even in groups: in synchronous drumming, dancing, clapping, beating, singing, shouting, and music-making (Piechocki 2019).

However, in general, the following applies: during evolution, humankind has increasingly developed characteristic features that build on each other. We have long been called upon to rethink what makes us human. As the famous paleoanthropologist Louis Leakey said, referring to the findings of the world-famous chimpanzee researcher Jane Goodall: "We must redefine tool, redefine man, or accept chimpanzees as human" (janegoodall.org). By the same token, we may also redefine "consciousness," "purpose," and "syntax," and likewise revisit the question "what for?".

Some readers may be inclined to object here and attest that one should not exaggerate; nevertheless, the difference between humans and animals remains intuitively obvious. We are about to fly to Mars, and we can build a global Internet, humanoid robots (Chapter 8), and countless other things. Humans can also think in dizzyingly abstract ways. Despite the truthfulness of this statement, it does not address the topic at hand. The objective is to trace the subtle and principled beginnings of these differences, the starting points from which complex forms developed in evolution. Where are they to be discerned? Is there or is there not at any point a sharp dividing line between humans and animal? Are there or are there not attributes that belong exclusively to us? The answer is: there was no singular moment when humans came into being. As for the number of declared differences, in the past, there were dozens. Not so today, as one by one, the differences between humans and animals are blurring. Evolutionary biologists tirelessly (and successfully) trace pathway after pathway to our animal ancestors and place us on a deeply rooted genealogy. We are animals. Finally, Richard Leakey, famed paleoanthropologist and son of the aforementioned father and equally renowned mother Mary Leakey, put it succinctly in a historic 2011 conversation with his colleague Donald Johanson (the discoverer of "Lucy"). When asked what makes us human, he replied, "We are apes!" We are, Leakey said, just a slightly more complicated species of ape that walks on two legs. The differences evolved gradually. (Leakey/Johanson, YouTube).

How do we want to exclude that the above animals act purposefully, i.e., that they know why they do something? The expert speaks cautiously of sequential intelligence in crows. Not so long ago, the possibility of intelligence in birds had been completely ruled out; for 100 years, people were convinced that birds had no neocortex, i.e., that part of the cerebral cortex that is the seat of coordination, planning, thinking, and our ego personality. Onur Güntürkün of the Ruhr University in Bochum, Germany, disproved this erroneous view after years of debate on birds with his guild. According to this, birds have a brain structure that converges on the neocortex. Not only can they think; some even exhibit ego-consciousness. Some primates and intelligent birds thus provide evidence that the supposedly unique, consciously purpose-oriented actions of humans have their evolutionary roots in the animal kingdom after all.

To empirically conclude that consciousness exists in the abovementioned and other animals (Figure 7.3), we must collect analogous phenomena in them. If we do this, then a conscientious neurobiologist like Randolf Menzel, after decades of research on bees, can cautiously reach the conclusion that "The bee knows who it is" (Menzel 2019). Menzel, of course, knows perfectly well that he can only infer the bee's behavior and cognitive properties from the perspective of his own human thought world; it appears to him as though the bee knows who it is. In this way, we gradually and very carefully build bridges to animals with which we are closely or even very distantly related.

Let us return to the purposes of which we are aware. The German philosopher Peter Janich speaks of a change of perspective. The human being changes from object to originator, author, purpose setter, criteria selector, and method realizer. The human being understood in this way is "not a definitional product of a biological taxonomy or any other natural science, but is grasped by an open list of attributions that evolves over the course of cultural history" (Janich 2008). An "open list of attributions" may be taken to mean that the term "human" is not objectively definable. This open list is then the product of free human judgment and thus not a scientific term. You, dear reader, are a different person after reading this book, at best, similar to the one you were before. In this sense, you would also not be a scientifically tangible object.

Such questions about meaning and purpose or significance initially overtax the explanatory possibilities of biology and all other empirical sciences. To date, biology cannot be used to understand the mental ability of humankind. It has no access to the spiritual origin of the concepts set by human reason, such as freedom, morality, culture, hate, love, happiness, beauty, enthusiasm, or music. However, it may not need to, for it can nevertheless objectify numerous manifestations of these spiritual qualities and examine their evolutionary, selective aspects. However, the essence of the subject "spirit" remains an open question for all of us, because it is an indispensable part of our existence.

Let us assume that we would obtain a linguistically and semantically precise grip on evolution or heredity, the heartbeat or the physiology of a muscle, and that we would know exactly on all biological levels how the embryo forms in the womb and develops its myriad of biological functions. (In chapter 9, I will discuss why much of this will probably never be fully graspable anyway because of the complexity of the issues and processes involved). However, even against this background, we could not rule out the possibility that the nature of the inquiring mind developed by a growing being is a subject of an entirely different class of its own.

Summarizing philosophically once again, the spirit could also be described simply as the primordial human ability to distinguish the natural from the artificial, i.e., created by oneself. The artificial can be an artifact (Figure 7.5) or a mental achievement, for instance, in the form of a novel or a symphony. The spirit is thus humans' distinguishing ability to express themselves symbolically in a pure object world and to communicate symbolic meanings with their peers.

At present, research on consciousness stands on an uncertain foundation. What constitutes us as human beings will occupy us for a very long time, perhaps even forever; it was important to clarify this here. The discussion about this topic must not be understood as a relapse into irrationality. Rather, Alfred Gierer points out that there can be nonformalizable presuppositions beyond objectifiable knowledge and that cognitive ability also depends, not in the least, on the cognitive apparatus (Gierer 1998).

Figure 7.5 Artifact. The Rosalind Franklin Mars Rover from Airbus, named after the British biochemist who played a decisive role in the discovery of the double helix structure of DNA. The rover is a state-of-the-art artifact and will be used to search for traces of past life on Mars in 2028. It was built to serve this purpose, putting it in line with humankind's oldest artifacts, such as hand axes. They all served or serve specific purposes. The depicted high-tech artifact, a top product of modern human engineering and intellectual achievements, thus symbolizes the millennia-old cultural inheritance of human knowledge.

As it stands today, dualism has fallen by the wayside in the science race, so to speak. The natural sciences have won. Nevertheless, we cannot rationally decide whether dualism or monism is the true view of the world. We can empirically prove neither human free will nor similar human abilities unambiguously (Loh 2019). The "true" view of the world is decided—at least in our time—historically-pragmatically and thus not philosophically or logically.

7.5 SUMMARY

The placement of the evolution of thinking in a social context is a milestone in the theory of the evolution of thinking. The size of the social group drove the evolution of the brain and the emergence of the mind, not vice versa. This is true even if there is feedback at play. When social brain theory and Tomasello's natural history of thinking are linked to the concept of niche construction, both appear in a new light. Both theories represent human niche constructions, in which groups of people create complex social assemblages. Consequently, according to the niche construction theory, these constructions release their own selective forces, which lead to group adaptation. Only the various niches in different group sizes and intentionalities made it possible to go through the evolutionary stages we know in human history, in other words, only the theory of niche construction allows an overall view of the two theories presented with their independent, interacting selection processes. The self, consciousness, and the mind are, from the natural–scientific–monistic point of view, products of neuronal elements and processes, which we can probably explain in the future. From the philosophical–dualistic point of view, they are not scientifically objectifiable; some would say that we cannot therefore understand them in principle. From this point of view, the self or consciousness eludes scientific access. Nevertheless, we can assign patterns of appearance and behavior to it and investigate its evolutionary origin.

REFERENCES

Note: Citations from German-language books and articles that are not published in English are translated by the author.

Bauer J (2019) *Wie wir werden, wer wir sind. Die Entstehung des menschlichen Selbst durch Resonanz.* Karl Blessing Verlag, München.

Darwin C (1871) *The Descent of Man, and Selection in Relation to Sex.* John Murray, London.

Demmerling C (2008) Welcher Naturalismus? Von der Naturwissenschaft zum praktischen Naturalismus. In: Janisch P (ed.) *Naturalismus und Menschenbild.* Felix Meiner, Hamburg.

Dunbar (1998) *Grooming, Gossip, and the Evolution of Language* (1st Harvard University Press paperback ed.) Harvard University, Cambridge, MA.

Gamble C, Gowlett J, Dunbar R (2015) *Thinking Big – How the Evolution of Social Life Shaped the Human Mind.* Thames and Hudson, London.

Gierer A (1998) *Im Spiegel der Natur erkennen wir uns selbst.* Rowohlt, Reinbek.

Goodall J https://www.janegoodall.org/our-story/our-legacy-of-science/.

Janich P (2008) Naturwissenschaft vom Menschen versus Philosophie. In: Janisch P (ed.) *Naturalismus und Menschenbild.* Felix Meiner, Hamburg.

Kohda M, Hotta T, Takeyama T, Awata S, Tanaka H, Asai J-Y et al. (2019) If a fish can pass the mark test, what are the implications for consciousness and self-awareness testing in animals? *PLoS Biol* 17(2):e3000021. https://doi.org/10.1371/journal.pbio.3000021.

Laland KN (2017) *Darwin's Unfinished Symphony – How Culture Made the Human Mind.* Princeton University Press, Princeton, NJ.

Leakey R, Johanson D (2011) Human Evolution and Why It Matters: A Conversation with Leakey and Johanson. YouTube. https://www.youtube.com/watch?v=pBZ8o-lmAsg.

Loh J (2019) *Trans- und Posthumanismus zur Einführung.* Junius, Hamburg.

Menzel R (2019) Die Biene weiß, wer sie ist. In: Klein S (ed.) *Wir werden uns in Roboter verlieben. Gespräche mit Wissenschaftlern.* Fischer, Frankfurt, 64–79.

Nagel T (1987) *What Does It All Mean?* Oxford University Press, Oxford.

Nagel T (2012) *Mind and Cosmos: Why Materialist Neo-Darwinian Conception of Nature Is Almost Certainly False.* Oxford University Press, Oxford.

Noble D (2006) *The Music of Life. Biology beyond Genes.* Oxford University Press, Oxford.

Piechocki R (2019) *Rhythmus und Ekstase – Zur Natur- und Kulturgeschichte des Homo synchronicus. Teil 1. Naturwissenschaftliche Rundschau* 72(857):544–556. Part 2 and 3 in 858 (2019) und 859 (2020).

Rutz C, Klump BC, Komarczyk L, Leighton R, Kramer J, Wischnewski S, Sugasawa S, Morrissey MB, James R, St. Clair JH, Switzer RA, Masuda BM (2016) Discovery of species-wide tool use in the Hawaiian crow. Nature 537:403–407.

Sterelny K (2007) Social intelligence, human intelligence and niche construction. *Philos Trans R Soc Lond B Biol Sci* 362(1480):719–730.

Tomasello M (2014) *A Natural History of Human Thinking.* Harvard University Press, Cambridge MA.

Voland E, Voland R (2016) *Evolution des Gewissens. Strategien zwischen Egoismus und Gehorsam.* Hirzel, Stuttgart.

TIPS AND RESOURCES FOR FURTHER READING

All the above books are recommended for further reading.

Other recommended readings are as follows:

Laland KN (2017) *Darwin's Unfinished Symphony—How Culture Made the Human Mind.* Princeton University Press, Princeton, NJ. Evolution drives culture and culture drives evolution. The extensive research on this topic is made accessible in a captivating and easy-to-understand manner.

Richerson PJ, Boyd R (2005) *Not by Genes Alone. Alone—How Culture Transformed Human Evolution.* The University of Chicago Press, Chicago, IL. "Culture is biology" is the message of the two authors. The core statement of their book is not easy to understand, but the book is indispensable for a modern understanding of the evolution of culture.

Sachser N (2022) *Much Like Us. What Science Reveals about Thoughts, Feelings, and Behaviour of Animals.* Cambridge University Press, Cambridge.

Aware: Glimpse of Consciousness (Purchase video). https://aware-film.com. In this video, scientists are discussing about consciousness. Watching this video has become a great experience for me personally. It illustrates how mysterious consciousness is to scientists and individuals who aim to understand and define it. The video is embedded in fantastic natural imagery. The subjectivity of consciousness is addressed here in contrast to the objectivity of scientific research.

CHAPTER EIGHT

The Evolution of Humankind in Our (Non)Biological Future

In this book's conclusion, I would like to discuss the future of *Homo sapiens*, which is, at best, still determined to a limited extent by biology. Impressions of humankind's possible future suggest that a theory of evolution restricted to genetic inheritance or even biology is no longer appropriate. We will realize that our culture and biology are rapidly merging, resulting in gene-culture coevolution. Culture is a product of evolution, but humans have been intervening with their evolution for millennia via their inventions.

According to the concept of gene-culture coevolution, the cultural and technological components of evolution will become equally important as the genetic component. The genetic component of evolution with respect to prenatal selection has gained importance with the application of, genome editing for treating genetic diseases, cloning, acquiring explicitly desired traits, etc. (Brosius 2003). The importance of the cultural and technological component of evolution has increased exponentially with the development of medical technology, sensors, robotics, and artificial intelligence. This chapter emphasizes developments in these areas and their possible consequences with respect to evolution.

Evolutionary theory can and must conceive humankind in a scenario where natural selection is increasingly replaced by competition in the field of technology—there is even discussion on our replacement as a biological species by machines. Evolutionary theory must be able to address this. Culture, technology, genetic manipulation, artificial intelligence, and robotics are and will be human niche constructions with major impacts on our future. Niche construction theory provides a theoretical framework for mapping our technoevolutionary future. Theories addressing adaptive machine behavior, hybrid human–machine behavior, and human–machine coevolution must be integrated; the necessity of such integration is discussed in the following section.

Important technical terms in this chapter: (see glossary) artificial intelligence, *genome editing*, nanobots, niche construction, superintelligence, synthetic biology, technological singularity, transhumanism, and maladaptation.

If you ask members of society whether they believe humans will continue to evolve, you may be met with doubtful looks. Often, today's human being is seen as a kind of final result. However, we, similar to all living beings, are embedded in continuous evolutionary processes. We are the result of millions of years of tinkering, far from being perfect in any respect because perfection does not exist in evolution. In historical times alone, which is but a brief moment in evolutionary history, we have undergone noticeable biological changes. Mutations are known to have occurred in the past millennia and have since been adapted. The blood group B, which originated in the mountains of the Himalayas, was added to the possible antigenic phenotypes. There was a reduction in the wisdom teeth as the ability to chew became less important for survival than it was for early nomads. Malaria resistance (sickle cell anemia) is also a recent mutation in humans. The most significant mutation in this series is lactose tolerance. It arose several times independently on Earth and made way for a new, valuable food for adult humans, who, like other adult mammals, were originally intolerant to milk. As a result of this change, animal husbandry came into being; this, in turn, promoted the spread of the mutation (Section 5.3). As a consequence, more and more people were able to survive with this advantageous trait, a typical gene–culture niche construction (Gerbault et al. 2011; Chapter 5).

8.1 HUMANKIND TAKES CONTROL OF ITS OWN EVOLUTION

The situation of modern humans is a new one, as we have taken our evolution into our own hands. The days when life was shaped exclusively by the hard-falling forces of evolution are over (Doudna and Sternberg 2017). We are purposefully changing ourselves, replacing natural evolution with technology. In the process, the distinction between natural and artificial is becoming increasingly difficult, and a nature–culture interaction is ongoing (Klingan and Rosol 2019).

If I had been born 50 years earlier and had been affected as a child by the same autoimmune metabolic disease that I have today, I would have lived only a few months from the time of diagnosis. According to Darwinian theory, I would have been naturally selected, so to speak, to leave no offspring. In the current era of synthetic, short-acting insulins, long after the first application of insulin in humans by Canadians Frederick Banting and Charles Best, I live without difficulty and have three children. My own and my children's DNA will remain in the human gene pool, as they will hopefully have children of their own, with or without (hopefully without) diabetes. Natural selection has been overridden. This is only one example among innumerable others of the fact that humans are increasingly evading natural selection. We are now an artificially selected species; almost anyone can be retained in the gene pool in highly developed societies. *Survival of the Fittest*, the central Darwinian adaptation process, obviously no longer plays a significant role. In other words, nearly all participants are made fit in our time.

Unlike evolution in the Darwinian sense, evolutionary manipulation by humans proceeds with a purpose. Synthetic biology, genetic engineering, and modern medical technology are well-known examples of this and are featured and debated in the media. Transhumanism and posthumanism discuss these efforts of humans to "improve" themselves intellectually, physically, and psychologically and to free themselves from their biological bonds, so to speak. Transhumanism affirms techniques for the improvement of humankind. The categorical dualisms of Western culture—the real versus the virtual world, human/animal, human/machine, nature/culture, and even male/female—are increasingly being deconstructed; staunch representatives of emerging posthumanism want to overcome and dissolve these dualisms altogether. "More is better" becomes the implicit credo of transhumanism (Loh 2019). Abstractions become practice, for example, as Japan, in August 2019, became the first country in the world to grant permission to implant human-induced pluripotent stem cells in animal embryos and let them grow human organs until birth. The results are chimeras, hybrids of animals and humans. Their organs could be a promising alternative in view of the disastrously low willingness to donate organs. Remarkably, compared to natural evolution, the new, technologically driven evolution is proceeding at a rate approaching the speed of light.

Transhumanism knows numerous currents, and some of them are philosophically underpinned; however, we shall not undertake a comprehensive discussion of them here. The original idea dates back to German philosopher Friedrich Wilhelm Nietzsche's famous vision of the superman in his book *Thus Spoke Zarathustra*. It should be emphasized above all that transhumanism is based on evolutionary processes. It calls for technological support of evolution for a collective and comprehensive transformation of our species. Accordingly, evolutionary biologist Julian Huxley, an early proponent of transhumanism, also speaks of evolutionary humanism (Loh 2019). "Humans are no longer seen as the crown of the natural world or as entities categorically distinct from purely natural ones." Rather, transhumanist thought binds humans into an evolutionary process. Within this, they are themselves an evolutionary species that differ incrementally from other species (Sorgner 2016). Following the idea of perpetual progress, transhumanism arrives at the conviction that the improvement and optimization of humans to posthuman beings will never reach a conclusion (Loh 2019). This includes, at the same time, the statement of the permanent incompleteness of evolution.

8.1.1 Synthetic Biology and Artificial Life

Let us highlight a few examples. Synthetic biology is a modern and lucrative growth market. Bioengineers are specifically working on modifying biological systems, thereby creating novel, modular, synthetic-chemical units known as BioBricks. This confers various novel functions on biomodules made of cells and tissues. In the simplest case, they can be manufactured with increased heat and cold tolerance or with surfaces more suitable for different purposes. There have long been databases on the Internet through which anyone can order BioBricks online, and students can use them to experiment with genes. Biology is following a path similar to that taken by chemistry more than 100 years ago. Chemistry, too, was initially analytical and descriptive and subsequently led to the production of hundreds of thousands of new synthetic substances, many of which we hold in our hands every day.

In addition to golden rice, which became known in 2000 and was developed to prevent vitamin A deficiency, research is being conducted into synthetic vaccines, among other things. Insulin that is produced synthetically with the aid of recombinant DNA (Humalog) is identical to human insulin. It is produced in huge bacterial colonies and was a revolution in diabetes care at its introduction in 1996. Today, it is a common and rarely talked about market product. Currently, researchers in Cambridge, Massachusetts, are developing a drug for urea cycle disorder that can be administered orally, either in the form of tablets or in the form of suspensions. Urea cycle disorder is a genetic disorder in which affected individuals are deficient in one of the enzymes involved in urea cycle in the liver; as such, nitrogenous compounds in food are not converted to urea, which is excreted in the urine, and toxic ammonia is formed. Ammonia accumulates in blood and can cause severe brain damage and even death. The drug consists of transformed *Escherichia coli* that absorb large quantities of ammonia when intestinal oxygen concentration decreases. Probiotics comprising live microorganisms with environment-dependent responses can be used for treating various hereditary and malignant diseases or as diagnostic tools (Lu et al. 2016).

The age-old idea of creating artificial life has reached its highest point in research thus far. In 2016, the US genetic researcher Craig Venter succeeded for the first time in creating an artificial single-celled organism in the laboratory, the first "artificial life," as he announced in the media worldwide (Service 2016). To accomplish this, the DNA of a *Mycoplasma* bacterium was reduced to a minimum in the laboratory, so that it consisted only of the components necessary for its naked survival and reduplication. To call this artificial life, however, would be making a rather lofty statement. In fact, it was necessary to insert the reduced and thus artificial DNA produced by the laboratory experiment into the cell body of a *Mycoplasma* bacterium. Life, even that of a single-celled organism, was reduced to DNA by Venter and colleagues (Gibson et al. 2010), but life is more than mere DNA.

Venter's approach is referred to as the top-down approach. An alternative to this is the bottom-up strategy, where one concentrates on a specific property of the cell and first tries to form this property synthetically, i.e., from inanimate building blocks, in the laboratory. For example, the German biophysicist Petra Schwille and her team succeeded for the first time in producing a synthetic cell membrane, one of the elementary functional units of the cell, without a cell nucleus or cytoplasmic contents (Kretschmer and Schwille 2016). The goal was to cause this compartment, which resembles a simple soap bubble, to divide. Cell division (mitosis) is a basic biological function. It was achieved in the laboratory using a molecular oscillation mechanism that locates the center of the membrane and divides the compartment into two equal parts. Thus, one of numerous, highly complex cellular functions was synthetically reproduced, a first step on the arduous path to an artificial cell. Nevertheless, the road to a cell constructed from scratch is still long and untraveled. Fundamental questions about cell biology remain unanswered to this day: how would an artificial cell increase its cytoplasm before it divides? How does synthetic machinery work that copies DNA on its own and passes on the genetic information?

Perhaps the most spectacular current approach to creating artificial life is the xenobot, a biological mini-robot reported in the renowned journal PNAS in January 2020 (Kriegman et al. 2020). Researchers at the University of Vermont in the US have developed a method to automatically construct complete biological machines from scratch. How was this accomplished? The robot, which is less than a millimeter long, should be able to move inside the human body in a targeted manner, repair itself, and in the future, also transport substances such as drugs or microplastics. In addition to skin cells, pluripotent stem cells from the embryo of the African clawed frog *Xenopus laevis*, which form heart muscles, were used as the basic building blocks constructing this robot—as an advance into new avenues of research. The rhythmic movements of these cells achieve the desired propulsion of the xenobot. Furthermore, the researchers simulated the optimal interaction of the cells and the geometric shape of the xenobot in a supercomputer for several months using a specially-built AI system. This system was able to virtually run through countless cell combinations and finally select those that were best suited for certain tasks, such as directed locomotion. The algorithm of the AI system thus represents nothing other than the natural selection process. The xenobot was then built in the laboratory from a few hundred cells; in the future, however, this will also be performed automatically. Let us leave the question of whether the xenobot meets the requirements of real life open for the moment. In any case, it does not represent the developmental end of synthetic biology.

In summary, the vision of artificial life research is to create custom microorganisms with novel, useful properties, such as bacteria that produce biofuels, recycle plastic waste, or deliver medical agents. Nevertheless, according to Joachim Schummer, the idea of artificial life is ultimately based on an outdated, strictly causal, deterministic fundamental conviction according to which life, on the one hand, can be completely and unambiguously broken down into functional modules, and on the other hand, all these components are determined by one or a combination of several gene sequences (Schummer 2011).

8.1.2 Genome Editing—Interventions in the Germ Line

The discovery of a new method of targeted gene manipulation called CRISPR/Cas, genome editing, or gene scissors, represents a breakthrough in genetic engineering and medicine that made the world sit up and take notice. In one fell swoop, genetic engineering leaped from the Stone Age into the computer age. With this mechanism, which became known in 2012, DNA from plants and animals, including humans, can be specifically cleaved and modified. In the process, foreign genes can be inserted, removed, or switched off. The method opens the way for new possibilities for genetic manipulation—while being more cost-effective and efficient than previous methods—and has become a social issue in the USA. The high relevance of CRISPR holds true despite the unexpected hurdles that have emerged. Of course, CRISPR is not a completely error-free method either, and its use can, in rare cases, produce new, unfavorable mutations if, for example, despite all caution, the cuts are not made precisely.

All the questions that had already been asked earlier in genetic engineering were back on the table at once for a short time: can lethal genetic diseases be eliminated with the new method? Does it open the door to genetic manipulations aimed at improving desired characteristics? Can we use it to make ourselves more intelligent, more beautiful, or more efficient? What consequences will we see in animal and plant foods? The discussion culminated in November 2018 when a young Chinese doctor used the CRISPR technique to perform an intervention on human twin embryos to make them immune to HIV. The international scientific community immediately reacted with a spontaneous, unanimous rejection of this procedure. On December 8, 2018, German geneticist and Nobel laureate Christiane Nüsslein-Volhard placed an appeal in the *Frankfurter Allgemeine Zeitung* entitled Hands off our genes! In the article, she drew attention to the risks that even the smallest manipulations of the human genome entail. The consequences of this and similar interventions are completely unforeseeable, she said. According to the researcher, we know next to nothing about the role that genes play in their complicated interplay in humans, but also in other species, both during development and later in life. Thus, a change that appears beneficial at one point may unexpectedly prove harmful at another.

Similar to Nüsslein-Volhard, Stuart Newman, a cell biologist and uncompromising proponent of an extended synthesis of evolutionary theory, has long advocated in the USA that we should keep our hands off technologies that we do not understand. By this, he means biotechnology and genetic engineering, both of which aim to manipulate the human germline. Newman makes a strong point that, for example, the majority of people in modern societies have no concept of what genetic engineering can fundamentally do in somatic cells (adult body cells) and germline cells (sperm or eggs). There is neither widespread awareness of the risks and dangers nor future opportunities in germline manipulation. Newman points out that humans are not machines. While the biotech industry stubbornly pursues the notion that all life can be deconstructed, reconstructed, and commercialized, Newman repeatedly makes it uncompromisingly clear that humans are not machines to be understood in this way. According to Newman, it is still not possible to turn a genetic cog and at the same time know which other cogs are turning as a result. In short, the notion of genetic determinism that is deeply ingrained in scientists is false. The complexity of cellular interactions beyond genes is unmanageable and beyond human control. Genes do not determine the properties of organisms in the way that has been believed and taught for 100 years (cf. Sections 3.6 and 3.7). Thus, we do not know what we are doing when we intervene genetically in our germline. Today's industrial efforts have the potential to change the species character of human embryos and our civilization (Stevens and Newman 2019). And yet, biotech insiders hold regular meetings, planning staff convene, and strategies are developed to jumpstart the multibillion dollar market.

Tina Stevens and Stuart Newman know what they are talking about. The two researchers refer, among other things, to extensive experiments with which they back up their cautionary stance. In 1997, an experiment led by Newman analyzed the ability of mice to orient themselves spatially and learn to swim. Here, even in inbred strains of adult mice, the roles of specific genes could not be determined. The authors also point to a second experiment from 2013, in which gene expression patterns triggered by inflammation-induced stress (such as after trauma or sepsis) widely differed between mice and humans. The conclusions were that, first, mice are questionable model organisms for inferring gene function in humans, and second, genes can play different roles under the same conditions. Even more confusing, variable results were produced in the same experiments using different inbred mouse strains.

The molecular geneticist and evolutionary biologist Jürgen Brosius at the University of Münster, Germany, an expert in experimental pathology, warned against the manipulation of the human genome as early as 2003. In that year, the first complete sequence of the human genome had been published. Brosius certainly recognizes the justified interest in controlling deadly diseases in one's own body or that of one's children. Ultimately, however, there are relatively few monogenetic diseases, i.e., those that are based on a single, hereditary genetic defect. In these diseases, sometimes a single altered letter in the DNA of a gene has devastating consequences. The affected gene then either produces a defective protein or none at all. The range of these diseases includes Huntington's disease, a hereditary, fatal disease of brain cells, and the no less dangerous Duchenne muscular dystrophy (DMD). In DMD, patients are born without symptoms. The disease only becomes noticeable from the age of four and then progresses to severe muscle curvature. Affected children become confined to a wheelchair. Death is unavoidable when the heart muscle or the breathing apparatus fails at approximately 25 years of age. The same group of monogenetic diseases includes amyotrophic lateral sclerosis (ALS), a disease of the nerve endings that causes muscle atrophy, from which Stephen Hawking suffered. Sickle cell anemia, a hereditary disease of the red blood cells characterized by impaired blood clotting, is far more common. Cystic fibrosis is also a known hereditary disease and is often caused by point mutations. As of February 2020, the OMIM database, which covers all known human hereditary diseases, contained approximately 25,000 entries. Among them, the causative genetic errors for approximately 16,000 entries are known and described (https://www.omim.org/statistics/entry). Hopes are high that CRISPR technology will provide answers to these diseases.

Other genetic diseases can be incomparably more complex in terms of their genetic linkage. Autism, for example, is a disease with various, fluid manifestations. It is inconceivable today to intervene here with genome editing. Dozens of genes are involved in other diseases with a genetic predisposition, such as the autoimmune disease type-1 diabetes. However, each gene has a relatively small influence. In parallel, exogenous origins may exist that interact with genetic conditions in complex ways. Such possible environmental factors (cesarean birth is one, formula feeding another) play a role in type-1 diabetes. Delivery by cesarean section affects the composition of the child's intestinal microbiome because the baby does not inherit the mother's beneficial bacterial flora. The baby's immune system then lacks certain prerequisites for healthy development. These conditions favor the development of autoimmunity. The risk of a child developing diabetes by the age of 12 is doubled as a result, according to Anette Ziegler of Helmholtz Zentrum in Munich, Germany (Bonifacio et al. 2012). Such diseases cannot be treated using CRISPR/Cas at this time. Nevertheless, enormous efforts are also being made to target all these diseases, as well as certain cancers, with CRISPR. It is even hoped that CRISPR can be used to detect mutations before they can cause irreversible damage (Doudna and Sternberg 2017).

The risks of dealing with CRISPR/Cas thus lie in the fact that we do not and probably never will know the genetic network that influences the phenotype in total, in all its possible genetic combinations and forms of expression. Consequently, mistakes will be made in the future treatment of disease via gene therapy that may not be apparent for several generations (Doudna and Sternberg 2017; Stevens and Newman 2019).

However, Brosius is most concerned about manipulations of another class, namely, enhancement of our physical, mental, or even cognitive characteristics and its consequences. The intense global interest in addressing these potentially expansive areas of business will likely result in this coming true (Brosius 2003; Stevens and Newman 2019). Brosius speaks of "directed evolution" in humans. He emphasizes the evolutionary danger of uniformity in the human genome. The diversity of gene variants that our genome exhibits are the greatest asset for our evolutionary future and survival as a species (Brosius 2003). Here, standardization of targeted genetic endowments and phenotypic traits means an evolutionary tightrope act with a completely uncertain outcome.

Jennifer Doudna, one of the two discoverers of CRISPR technology, no longer fundamentally rejects the idea of editing the human germline, in contrast to her earlier stance. However, she draws attention to the risks involved, as do her colleagues. She acknowledges that the question was not whether DNA would be redacted in human germlines but *when* and *how* such a redaction would take place (Doudna and Sternberg 2017). So, it is safe to assume that mankind will get into this game and start an arms race for genetic enhancements. Lamarck sends his regards.

Stevens and Newman (2019) share the same view. To draw maximum public attention to the issue of biotechnology and the manipulation of germ cells, Stuart Newman filed an application to patent chimeras in the United States in 1997. Composites of mouse and human cells would be allowed, as well as those of human and chimpanzee cells, starting from the zygote stage. This is a unique approach with completely different consequences from, for example, the cultivation and implantation of an organ from mixed cells. Newman was proposing bona fide mixed beings derived from two different species from the early embryonic stage and with quite different proportions of human cells. The justification for the application stated that partially human embryos could serve as a source for tissue and organ transplants but also as test objects for the compatibility of toxic substances and drugs.

In fact, Newman, a fierce opponent of germline genetic manipulation, had a very different objective. He intended to raise public awareness of the issue. Foreseeing such developments coming in the biotech industry and science, Newman strove to get the public to take a stand on the question of where to draw the line between biology and culture, between human and nonhuman. The application was denied in 2005 after several rounds of appeals. Newman refrained from appealing to the highest US court. He had, so to speak, pressed the alarm button, knowing full well that the industrial apparatus could neither be stopped, nor has it been, and that behind closed doors, this apparatus drives highly profitable business development without hesitation.

8.1.3 Nanobot Technologies

The news media repeatedly reports on research into nanobot technology. Nanobots are robots of molecular size. Such machines have already been described in novels by Philip K. Dick (*Autofac*, 1955) and Stanislaw Lem (*The Invincible*, 1964). A nanometer is one billionth of a meter. If you imagine this to be the size of a soccer ball, then the soccer ball, enlarged on the same scale, has the circumference of the earth. Nanobots can be made from DNA or proteins. We have already learned about the xenobot, the biorobot named after the clawed frog, made from its cells and designed with an AI system. But the xenobot is still several orders of magnitude larger than the nanobot. The first autonomous prototypes for targeted cancer therapy are already in the demonstration phase. They attack tumors, cutting off their blood supply. In addition to fighting cancer, future nanobot systems will be used to eliminate pathogens in the blood, break up blood clots, break up kidney stones, replace neurons, and perform numerous other functions. The body cells they attack will be destroyed and removed without any damage to healthy tissue. The particular challenges here are that nanobot systems will not only need to be introduced into the bloodstream but will also need to independently locate a specific site in the body, such as the knee joint fluid, and become active there. If control is too difficult, the nanobot inventors get creative and help from outside the body, for example, with magnetic fields or ultrasound at the corresponding body sites, to which the nanobot then reacts. Ultimately, nanobots should be able to reproduce themselves in the body or self-destruct when their work is done.

8.1.4 Slowing Aging and Immortality

A goal that is rarely perceived by the public but breathtaking nonetheless is to slow the aging process, stop it altogether, and avoid death. Death is a fundamental component of Darwin's theory. Species cannot change or differ without death; adaptation to changing environmental conditions would not be biologically possible for organisms with an infinite life span. Living things are, therefore, "programmed" to die. Death is a result of natural selection. However, immortality is indeed possible, at least for a jellyfish discovered by an Italian researcher in the Mediterranean Sea in 1999 (*Turritopsis dohrnii*, Figure 8.1). When it becomes old or injured, and its body cells no longer function well, its cells dedifferentiate into the stem cell stage. The jellyfish thus returns to its original polyp stage, becoming young again. It then rises from the sea floor as a juvenile jellyfish and begins a new life. The process theoretically repeats itself ad infinitum (Piraino and Boero 1996).

To date, it is not sufficiently clear why we age. One theory is that harmful genes are activated in old age and can, therefore, no longer be selected. If two such genes are removed from the nematode (*RAS2* and *SCH9*), it can live up to 10 times longer (Wei et al. 2008). However, this is not true to the same extent for more complex life forms. Conversely, there are numerous so-called Methuselah genes. These are genes or genomic changes that, in combination, codetermine a long life span.

The Briton Aubrey de Grey determined that with aging, pollutant degradation by lysosomes in cells becomes less complete. In the future, microbes will be programmed and introduced into body cells in such a way that efficient waste degradation will be maintained. According to De Grey's predictions, there is a 50% probability that a 50% rejuvenation process will be possible in 25 years, i.e., a 60-year-old will be able to look like a 30-year-old (Grey and Rae 2010).

The discovery of telomerase in 1985 caused a sensation in the scientific community and the media. This enzyme in the cell nucleus ensures that the degradation of telomeres, the end pieces of the chromosomes that are further shortened with each round of cell division (mitosis)), is arrested in certain cells, and

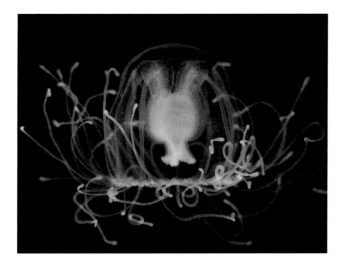

Figure 8.1 Immortal. The jellyfish *Turritopsis dohrnii* lives forever, at least as long as it is not killed by its environment. It is the first animal known to be able to regress from the adult to the juvenile state and to do so as often as desired.

the telomeres are thus restored. The Nobel Prize in Medicine was awarded in 2009 for this discovery. Telomerase plays an important role in the fight against cancer, where the aim is to stop uncontrolled cell division. However, telomere shortening, which only allows a certain number of cell divisions, is not the only reason for aging. Unfortunately, I can only give a small insight into the numerous research directions here. With increasing age, for example, the ability of cells to repair themselves decreases. At present, there is intensive effort to investigate the question of how our body's cells can achieve the correct balance between cell growth and cell repair as we age. A few molecules come to mind that can influence this control, and they are on the market in modified form as drugs. One such group, for example, increases the expression of genes called sirtuins. The enzymes encoded by these genes, of which humans have seven types, are responsible for DNA repair. Increased sirtuin levels in mice led to better resistance to disease, improved blood circulation and organ function, and a longer life span.

A second drug developed for the same purpose, i.e., to elevate the cell repair. mechanism, is called metformin. It is possible that metformin can target several diseases of aging, including cancers and type-2 diabetes, simultaneously and revolutionize aging research. Ideally, in the future, a direct delivery system will be developed to bring the compounds in question directly to cells. In addition to those mentioned above, there will likely be other antiaging preparations that will increasingly be used individually, considering the patient's age, gender, DNA profile, metabolic status, and other factors (Metzl 2019).

Ray Kurzweil, a well-known futurologist and Director of Engineering at Google, predicts *that humans will be immortal by 2045* despite numerous unanswered questions (Kurzweil 2017). Regardless of when immortality will be achieved, research on the topic of stopping the aging process is being conducted at a feverish pace and with huge budgets. Whether the end result is something we really want is another matter.

Radical life extension culminating in the abolition of the aging process altogether is the primary goal of transhumanism. In this way, the cognitive, physical, emotional, and other human limitations and weaknesses addressed by transhumanism would ultimately be overcome (Loh 2019).

8.1.5 Human–Machine Combinations

The most ambitious fields of research in medical technology lie in overcoming incurable diseases, adequately replacing missing body parts, halting the aging process, and human–machine combinations. The German philosopher Nicole C. Karafyllis speaks of the indistinguishable fusion of biology and high-tech (Karafyllis 2003). At present, we already exist in certain forms as cyborgs, that is, as human–machine combinations. This is as true for people with pacemakers as it is for those with insulin pumps, retinal implants, or brain implants. In September 2018, the Mayo Clinic in Rochester reported that a paraplegic could walk assisted for 100 m with the help of implanted spinal cord stimulation electrodes. The interrupted nerve pathways from the spinal cord to the legs were bridged with the implants.

Impressive research is also underway in prosthetic arm surgery. On YouTube, you can watch a man with a prosthetic arm peel an orange. The illustrated technique is already outdated because the control of the prosthesis—as elegant as it looks—is still accomplished using the remaining muscles of the upper arm. To execute the desired hand movements of the prosthesis, the patient must first learn to trigger signals with his remaining muscles. This is difficult and requires much practice. New prostheses (Figure 8.2), such as the Modular Prosthetic Limb, a prototype from Johns Hopkins University (and also shown on the Internet), are connected to the nervous system (mind control). The prosthesis can even be detached from the body and hung from a stand. It communicates wirelessly with a cuff on the patient's upper arm. This is coupled to the nervous system and thus to the brain. The prosthesis itself has 100 sensors corresponding to fields on the fingertips and back of the hand. If the doctor touches the detached prosthesis, for example on the little finger, the patient feels this at the corresponding imaginary point. If the doctor presses firmly, the patient feels pain. The prosthesis thus allows the patient to feel. One might object that a robot will never have the sensitivity of a finger. The Georgia Institute of Technology has already developed a high-resolution robotic skin with many thousands of individual sensors per square centimeter, more than on a fingertip (our entire body has 900 million touch-sensitive receptors). The artificial skin is stretchable and sensitive to touch, pressure, and temperature. With this degree of sensitivity, an artificial hand would have a good chance of peeling a boiled egg.

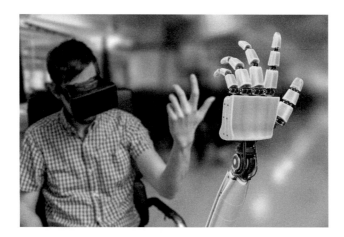

Figure 8.2 Man controls an artificial hand. There are already numerous developments of AI-controlled prosthetic arms, also with many sensors for touch, tactile and pressure sensitivity. They are wirelessly coupled to the nervous system. The patient feels touch with individual fingers and also feels pain.

Kevin Warwick, an engineer at Coventry University, was able to connect microchips inserted into his arm to the Internet. Nerve signals from his hand were transmitted over the Internet to a robotic hand in another location, and the robotic hand did what Warwick did. In another experiment, he connected his nervous system to that of his wife. His nerve impulses (hand movement) triggered specific hand movements in her and vice versa. Thus, communication between two nervous systems is already possible; the electronic connection from brain to brain would then be the logical vision, although it would be much more difficult to realize. It is easy to imagine that the human brain communicates in the described ways with not one but hundreds of real systems in hundreds of places at the same time. I leave it to the reader to imagine such scenarios.

Brain–computer interfaces with various applications are also on the horizon. Elon Musk announced the details of his company Neuralink in 2019. Here, chips are implanted directly into the brain via tiny holes in the top of the skull. High-tech electrodes thinner than a hair are placed directly into brain cells using surgical robots. The chips are connected to a smartphone via a receiver behind the ear via Bluetooth. In the initial phase, it should be possible to better treat serious illnesses or injuries to the brain and spinal cord, such as paraplegia. But Musk is known as a visionary. He is thinking of further improvements to the human body with this technology: the control of robotic arms, the simple learning of a new language, the movement sequences of martial arts, or the storage of city maps. New abilities are then to be loaded into the brain like software updates and readouts.

The much-discussed uploading of the entire brain to a machine while preserving the identity of the person in question, on the other hand, is an incomparably more difficult project from which today's technology is still light years away. Not only would all the approximately 100 billion neurons have to be mapped in the machine, but so would all of the synapse connections, and that is 10,000 times as many. In addition, there would be extrasynaptic interactions. Some researchers, therefore, go so far as to claim that every atom of the brain would have to be represented in the machine if the copy of the connectome, i.e., the totality of all connections in the brain, is to work (Seung 2013). But that is far beyond the reach of any computer capacity. Apart from that, we still do not know what constitutes our identity, our consciousness. Can it even exist outside of our body? Our gut feeling contradicts it.

8.1.6 The Future of Genetic Engineering and Transformation of Life

Within two or three generations, genetic engineering may become a dominant issue for mankind. The course is being set today for interventions that are unimaginable to most people. The manipulation of our genetic makeup can and very likely will develop into the greatest technical advance in the history of mankind. After billions of years of evolution, a new era will then begin with new, previously unknown rules of evolutionary change. This epochal development will be dominated by a series of technical processes whose foundations are being laid today. Even beyond today, these will attract political and social attention, spark ongoing ethical discussions, and lead to different opinions and decisions in different countries and on different continents of the world.

The path this development will take will be characterized by a series of biotechnological processes, some of which will only become possible with applications of artificial intelligence and with big data, i.e., the storage and comparison of millions of human DNA sequences. The procedures that I will explain in more detail below will initially be seen predominantly in connection with the prevention of hereditary diseases or inherited, congenital malformations. In an incremental process, however, the targeted selection of complex physical, psychological, and mental characteristics will undoubtedly also increasingly determine the field.

Artificial insemination is not a novel concept and is used as a common method in many countries today. It does not in itself constitute a genetic intervention. However, it does involve artificial selection, which is already practiced for selecting suitable eggs and sperm. In vitro fertilization (IVF), as it is called, is a laborious process and stressful for the expectant mother. She must be prepared to provide about 10–15 eggs at a time, in up to three cycles, which are selected in the laboratory. The few usable ova in each case are fertilized with suitable sperm, a second selection process. In some countries such as the USA or Israel, the sex can be determined and selected in a third selection process. In Germany, where this is prohibited, I knew of cases as early as 2007 wherein wealthy fathers, who absolutely "needed" a son because their culture demanded it, had a frozen sperm package flown to the USA and back to have the sex determined there using a centrifugation technique; male sperm are in fact very slightly heavier than female sperm. Affluent people can often find a solution. Finally, in many countries, including Germany since 2011, DNA analysis of the desired embryo(s) can be performed as a part of preimplantation diagnostics (PID) to detect serious hereditary diseases and select appropriate embryos. In the USA, some clinics even allow positive selection of a genetic anomaly, for example, in the case of hereditary deafness.

IVF is certainly not suitable for large-scale manipulation of the human species, but scientific efforts extend much further. Since 2014, it has been possible in humans to convert somatic cells, for example skin cells, into primordial stem cells, which are the precursor cells of germ cells (Cyranoski 2014). A similar attempt succeeded for the first time in 2018 with blood cells. Also, in the fall of 2018, a Japanese team succeeded in taking primordial cells, which are not yet fertile and thus not yet divisible in the form of meiosis, to the next stage of development. Further progress can be expected in this field, as experiments in mice suggest. In these, one is already much closer to the goal of generating fully functional oocytes from induced pluripotent stem cells (iPS cells), which in turn were generated from skin cells, in a petri dish. In any case, the media reported enthusiastically on the discovery. The British newspaper *The Guardian* called the result a revolution and predicted that by 2040, it would be commonplace to produce synthetic embryos from skin cells taken from people of any age or gender.

Even though the intention of such experiments is to permit couples wishing to have children to overcome infertility, the more recent techniques open up entirely new horizons. In this way, far more than the severely limited number of natural eggs could be painlessly made available for artificial insemination in the future. They could be genetically screened using preimplantation screening (PGS) and selected according to various needs. In the future, screenings will contain more and more desired statements about the likelihood that a child will be susceptible to certain diseases, will be short-sighted or learning-disabled, or will have rapid comprehension, whether he or she will have an attention deficit or be a good team player, in short, the whole program of future preimplantation genetic testing procedures. Since the procedure is still tied to the generation sequence in humans, which takes several years, the retrieval of somatic cells from the early embryo and their transformation back into germ cells is being considered and would enable the IVF process to be repeated after only a short time. The generation sequence would thus be shortened to a few days or weeks instead of 15 or more years. In this way, the number of available embryos could be increased by a power of 10, the selection of embryos could be refined according to many more criteria, and the process of targeted artificial selection could be accelerated. In another procedure performed on mice, artificial ovaries (ovaries) were developed using 3D printing techniques, into which follicles, precursors to eggs, were inserted. The ovaries were implanted into an ovariectomized mouse and were able to integrate into the blood system. This mouse later mated and gave birth to live young (Laronda et al. 2017). Let us leave undecided in which countries such procedures would be ethically desirable and possible.

The importance of genetics on the evolution of humankind increases with the availability of DNA screening at a fraction of earlier costs, due to the use of AI with machine learning capabilities (Section 8.2). Information obtained with big data can, in principle, be used to select and manipulate embryos. In China, for example, the state has the sole right of disposal of its citizens' genetic information. In a 2017 survey, the Chinese tended to view the possibility of genetically modifying their children in a positive light (Metzl 2019).

In the previous section, with reference to several well-known authors, I made it clear that our current knowledge of the human genome still leaves open most questions about what exactly determines hereditary diseases with polygenic factors. We still know very little about their genetic basis. However, this does not prevent AI applications from comparing millions of DNA sequencing to determine patterns of disease or its predisposition. It is highly probable that in 10 years, the DNA sequences of two billion people will be available in databases (Metzl 2019). Recognizing that all the causalities of complex genetic diseases will overwhelm our limited brains, we will need powerful AI applications to better manage such complexities. Without needing to know in detail exactly which genes, gene expression patterns, and epigenetic factors are responsible for a multifactorial or polygenetic disease—there may be hundreds or thousands—the analysis of many DNA sequences will help to statistically narrow down the patterns associated with specific diseases. One will then be able to say something to the effect that an individual DNA sequence or pattern can lead to Alzheimer's, hypertension, type-2 diabetes, or some other disease with probability X in a particular individual at an advanced age. It stands to reason that as soon as such knowledge is available in increasingly detailed form, companies will be very interested in using it. This may be the case, for example, with insurance companies, which may offer customers enticing financial incentives to have their DNA sequenced and their data disclosed.

So, the entry into the analysis and manipulation of our genome will probably occur in the medical field. First, we will see genome editing of monogenetic hereditary diseases, and this will likely continue for many years. The technique will gradually gain the acceptance of broader and broader segments of the population. Parents' desire to protect their children from inheriting their own, perhaps fatal, disease is legitimate and will almost certainly prevail. Over time, the pushback against increasingly complex diseases will be addressed gradually and with newer and better methods of genome editing. Genome-wide association studies will help researchers to obtain an overall picture of the genetic variation in an organism's genome. This is then associated with, i.e., compared to, a specific phenotype, for example, an Alzheimer's patient.

But how will we address traits that lie beyond what we now classify as medical considerations, such as body size, intelligence, athletic fitness, skin color, and personality traits? On these subjects, we may judge decisively that one should strictly refrain from such manipulations. But opinions and perceptions differ between cultures. What seems to us to be an ethically outrageous and unworkable path may be quite acceptable in other cultures and countries.

In addition to culture, economic interests exist on an inconceivable scale; biotech companies envision potential multibillion dollar markets here. Sooner or later, therefore, such hardly noticeable developments will occur and will then creep out of the medical fields into non-medical ones, whereby those responsible will undoubtedly continue to invoke medical arguments. The boundaries between gene therapy and genetic optimization will gradually blur. When it comes to height, for example, deviation from a norm cannot be readily determined medically; cultural factors play into this. Intelligence and other traits could be considered in a similar way. Today, numerous genes are known to play a role in determining body size, and several hundred genes are known to influence a person's intelligence quotient (IQ). We truly cannot exclude the possibility that in a few decades, people with an IQ of 150 will be bred in series in some countries. The aberrations that will occur in the process remain an open question.

Even though we are still a long way from having an approximate overall picture of the genetic basis of the complex traits mentioned above, which are also influenced by environmental factors, we can already see in practice that in China, for example, children are selected for competitive sports on the basis of certain genetic factors. A Chinese ministry required applicants for the 2022 Olympic Games to have their genetic material sequenced for speed, endurance, and jumping power (Metzl 2019)—artificial selection on an assembly line in humans. South Korea is one of the countries with the highest percentage of cosmetic surgery among young girls in the world. In Seoul, it is almost a standard for a girl to have her face or breasts "beautified." It is easy to imagine that people in such a culture will be amenable to genetic interventions on their own germ cells if they can use them to "optimize" the appearance of their children.

23andMe.com is an American website that offers personal DNA sequencing. For US$ 199, the customer receives a vial, which they fill with saliva, and send back to the organization. Based on the selected sequencing of sections of all 23 chromosomes, a health predisposition report is generated. This provides information on genetic variants associated with an increased risk of, for example, type-2 diabetes and many other genetically dispositive diseases. The report meets the requirements of the US Food and Drug Administration for genetic health risks and can be the basis for possible personalized medicine. Similar applications in other countries are already used to detect genetic predisposition to intelligence, athleticism, or other desirable factors.

Our species should eventually approach the idealized human conceptions (Section 8.2). Sex as the natural process of procreation will perhaps disappear in the long term in some countries. In support of this, awareness of the opportunity to detect and prevent increasing heritable diseases through greater genetic knowledge of these diseases, advances in genome editing, and more efficient IVF methods, will grow over time. Where we stand with genetic engineering is comparable to where the automobile stood in 1900, not much more than a decade after Carl Benz invented the internal combustion engine. Measured against reasonable expectations, the global competition for mastery and application of genetic engineering has not even begun. It will take on spectacular proportions. In parallel with the manipulation of humans, research is being conducted at a feverish pace on the restructuring of countless genomes in the animal and plant worlds, in particular, all the components of our diet, and immense quantities of money are being earned. In view of the expected upheavals described above, perhaps hardly anyone will speak of Darwin in 100 years.

8.2 ARTIFICIAL INTELLIGENCE AND TRANSHUMANISM IN EVOLUTION

8.2.1 AI and Humanoid Robots

The image of a technical, abiotic human future demands an examination of AI and its consequences. Discussions have shown time and again that our society still largely holds the view that computers cannot think and are certainly not capable of learning. This image must be viewed critically and questioned in a differentiated manner. Today we speak of "intelligent systems," Computer applications no longer only reproduce what was previously input into them, and they can behave contrary to the intentions of their developers.

Let us be clear about one thing: if this chapter refers to computers deciding, thinking, deliberating, etc., then these terms—whether they are in quotation marks or not—are not tenable from a philosophical point of view. The terms for decision-making or intelligence in humans and machines refer in principle to different facts. The primary message in this chapter is that in our technologized society, we are increasingly dealing with machine environments that work toward imitating human behavior: then, they look to us as if a system decides when a person should be denied credit. Does your smartwatch with lightning-current blood pressure and heart rates decide whether you feel healthy, or is it you who decides that? For example, you might get the impression that a system is intelligent if, for example, it translates the book you are reading into another language in a matter of minutes, or if it soon "answers" all your questions via the Internet about your planned vacation, and you cannot tell whether you are dealing with a human being or a machine. In his famous Turing test, Alan Turing predicted that at some point, we would no longer be able to distinguish the behavior of a machine from that of a human. On the one hand, it is important to analyze the consequences for us of machines (only) simulating our behavior and sensations, but becoming less and less distinguishable from human intelligence. This is the topic of this section. On the other hand, it is important to keep in mind the categorical differences between, for example, "intelligence" and "decision-making" in humans and machines and to put them into an evolutionary context, which is discussed in Section 8.3. There, we discuss the theory of evolution, including the technosphere highlighted here.

We are faced with systems that appear to think or even feel like us, even though they do not in the least. The builders of such machines aim to make us believe that they—whether child's dolls, cell phones, ticket machines, online support systems, or humanoid robots—can "communicate" with us more and more at eye level. To this end, terms from the human behavioral repertoire are intentionally used, such as "decide," "think," "intelligent," "communicate," and "sex robot." The machines themselves and the linguistic association with humans in descriptions of their capabilities elicit fascination and incentives to buy that result in global markets. The primary goal of this chapter is to give you an idea of where the AI journey may be headed. Whether you will be impressed further by the examples of modern digital technology or whether your critical spirit will be awakened, I leave it as an open question. I do suggest, however, that you be critical about your evaluation of future technology. Decide consciously how much you want to allow. It always starts with the vocabulary: what terms do we use when we talk about machine capabilities or machine "behavior"?

AI agents are becoming increasingly complex. The program code can be simple but still produce complex results in the form of so-called black boxes. This means that the exact path from input to output can no longer be traced, even by the program developers, the output cannot be predicted, and unforeseen behaviors can occur when interacting with the world and other AI systems (Rahwan et al. 2019).

The German philosopher Julian Nida-Rümelin gives a clear picture of the fact that computers based on today's common system of Turing machines cannot feel real joy or pain, cannot be really sad, cannot understand the meaning of language, cannot weigh the reasons for their decisions, and cannot act morally. In contrast, "Man is not determined by mechanical processes. Thanks to his capacity for insight, as well as his capacity to have feelings, he can determine his own actions, and he does this by deciding to act in this way and not in another. People have reasons for what they do" (Nida-Rümelin and Weidenfeld 2019). Klaus Mainzer, a German science theorist and complexity researcher, disputes this view when he writes that "semantics and understanding of meanings do not depend on human consciousness" (Mainzer 2018). For example, he argues that a system may well acquire enough common sense that it "knows" a human is a living being that breathes, needs to eat, and usually brushes its teeth in the morning, even if that is not explicitly mentioned every time the term "human" is used. Mainzer, therefore, does not rule out "that AI systems will be equipped with consciousness-like capabilities in the future." Furthermore, he points out that there is a growing realization in cognitive and AI research "that the role of consciousness in human problem solving has been overestimated and the role of situational and implicit learning has been underestimated" (cf. Kahneman 1990). Mainzer's view thus does not coincide with Nida-Rümelin's.

Whether computers will have feelings, however, may turn out to be the wrong question if machines simulate feelings so credibly that we cannot even distinguish their behavior from the expression of real feelings. The same applies to the consciousness of machines. Asks Jay Tuck, where do real feelings begin, and where does simulation end? Jürgen Schmidhuber, a German AI visionary, explains that robots are already being fitted with "pain sensors" so that they can "sense" in their own way when they need to visit the charging station. Such terms are used rather carelessly these days. Thomas Nagel tells us unequivocally that we will never know whether future dog robots, which will react in complex ways to their environment and behave, in many ways, exactly like dogs, are a mere mesh of circuits and silicon chips or whether they will also have a consciousness (Nagel 1987). In this respect, Nagelalso contradicts his colleague Nida-Rümelin.

In a nonrepresentative survey conducted by the author among a handful of young people with a high school diploma or university degree in September 2018, 60% of respondents believed humanoid robots would never exist in our daily environment, while as many as 20% thought it would never get to the point where most machines in daily life communicate with us in voice dialog or that a machine court interpreter would be used. Society's misconceptions about the learning capabilities and intelligence of computers can lead to a dangerously false picture of our entire future. How well AI systems can learn was demonstrated impressively by the example of *Libratus*. This poker AI program only knows the poker game rules. It does not have access to a predefined database with millions of stored games, as chess programs do. Rather, *Libratus* independently developed bluffing in millions of games against itself, can handle intuition without being programmed to do so, and thus won against four professional players in 2017. The software can be used for other purposes as well. If software can handle tactics in poker, why not in negotiations?

There will soon be sex robots.", according to David Levy, author of *Love and Sex with Robots—The Evolution of Human–Robot Relationships*. When asked what this is for, Levy doesn't answer that robots are better at this than humans. Rather, he says that for people who "can't have satisfying sexual relationships with other people for a million reasons, these represent a new relationship that contributes to their happiness, a relationship they wouldn't otherwise have."

This development will completely change couple relationships (Levy 2007). The creature that Levy described to his readers as still being an invention of the future came on the market in 2017. It is called Harmony, a sleek, talking, heatable doll with numerous touch sensors. The fact that she can be customized almost at will in terms of hair color, bust size, and vagina is still the least of it. She can talk, learns her owner's preferences, knows her birthday, and of course, has an orgasm. Sex stress is left at the door. "We will fall in love with robots," says an interview with the British grande dame of cognitive science, Margaret Boden, about artificial intelligence that is completely incapable of relationships (Klein 2019).

AI systems are increasingly being used as human robots to care for the elderly; they perform simultaneous translations on the basis of artificial neural networks and deep learning (multilayer learning) with the desired quality of professional translators; they manage traffic flows and stock exchanges and save human lives—and not only in the early detection of tumors. IBM's Watson AI program first sensationally won hands down against humans in a US quiz show back in 2011, and it is currently being used to diagnose cancer. It has learned to do so by invoking 210 million patient records from 2017, which it can use for comparisons. Additionally, it has read the entire relevant scientific literature, i.e., 200,000 medical articles in the PubMed database. Even the most experienced doctor in the world can only use a tiny fraction of this amount of data in their diagnosis. Today, artificial neural networks can recognize patterns in cell samples that let them distinguish benign from malignant tumors. Such systems thus not only diagnose in a way that cannot be found in any medical textbook; they also conduct cutting-edge research.

The Pittsburgh-based AI system MultiSense can determine whether a patient is suffering from depression, schizophrenia, or a post-traumatic situation within a five-minute machine video interview. To do this, MultiSense uses 68 grid points on the patient's face, from which it evaluates, for example, eye contact or smiles to make the diagnosis. Doctors can use this system to track down new behavioral characteristics or changes in patients that they had not previously been aware of. The goal is to tailor the diagnosis for each patient using AI. But such systems have a long way to go; so far, only correlations are involved. The machine merely matches external behavioral characteristics, albeit many of them. It does not make a more in-depth medical diagnosis, because AI algorithms do not uncover cause-and-effect relationships per se.

This AI system and many others do not get to the bottom of the causes, as a good doctor would do in a given case. Rather, AI systems often work according to heuristic methods and with statistical procedures. In this case, heuristic means that (in contrast to, say, a car navigation system) there are no unambiguous algorithms—only "fuzzy" algorithms or approximate instructions for the system's decisions. These can then be better or worse and thus represent underlying causes well, less well, or even wrongly. This should be noted in passing. The result—as is the case here—can nevertheless be deeply impressive.

The industry is working feverishly to develop humanoid robots (Figure 8.3), synthetic beings that are indistinguishable from their biological counterparts. At the same time, autonomous combat robots are being developed today, and hundreds of millions of jobs are at risk of being replaced by intelligent systems in the coming decades (Ford 2021). Social upheaval will accompany this development. The AI development process is moving far too fast for parallel compensation of job losses as AI progresses. A comparison with the Industrial Revolution, in which many jobs were created in parallel with rationalization, is therefore inappropriate. In a 2017 report, Goldman Sachs Economic Research predicts the loss of 300,000 truck driver jobs in the US per year, or 25,000 jobs per month, from autonomous trucks alone. This does not even

Figure 8.3 Sophia. Sophia, an AI development by Hong Kong-based Hanson Robotics, is currently the most human-like robot. She can simulate human facial expressions and gestures, recognize faces, answer questions, and carry on simple conversations. Sophia is the first robot to have citizenship.

include cab drivers. Semi and fully autonomous vehicles alone will account for 20% of total US auto market sales between 2025 and 2030, according to the study (Dougherty 2017). As early as 2013, Carl Frey, an economist at Oxford University, together with Michael Osborne, developed a scenario in a widely acclaimed study, according to which 700 activities with varying probabilities could be lost in the USA within the next 20 years as a result of digitization—equivalent to half of all current jobs (Frey and Osborne 2013). These include many middle-class occupations: administrators, lawyers, bankers, tax accountants, insurance agents, and others. Occupations that involve repetitive workflows have a poor future. Such workflows will sooner or later find themselves in the algorithms of AI systems. However, other researchers pointed out that the data for the US cannot be applied to other countries. They also emphasized that digitization will not substitute for occupations per se, but that certain activities will be taken over by machines (Dengler and Matthes 2015).

Consultancies, market analysts, and governments are currently struggling to analyze the net effects of the loss of existing jobs and the simultaneous creation of new ones. Moreover, digital development is advancing rapidly. A comprehensive analysis of the German labor market in 2013 revealed a potential of 15% or 1.5 million employees subject to social security contributions who could be substituted by digitization (Dengler and Matthes 2015). In 2016, the figure had already been raised to 25% or 2.5 million (Dengler 2019). This clearly demonstrates the dynamism of digital progress. There is no doubt that countless jobs will be rationalized away in the course of digitization. It is difficult to predict how many can be substituted, and how many new jobs will be created in parallel. The media often says that rationalization has always led to sufficient new jobs in the past, but against the backdrop of today's technological change, this view is too short-sighted and superficial.

The human race could be extinct or immortal by 2050, according to Jeff Nesbit of the National Science Foundation. We do not know the consequences of AI systems that will exceed our intelligence. No one is asking us if we want them, but that which is feasible is usually realized. Technology is advancing exponentially and globally on countless terrains that may still be islands today but are increasingly growing together. The human capacity to think is overwhelmed by the pace of technological advancement. More than 2,000 prominent experts, including Stephen Hawking and Elon Musk, CEO of Tesla and SpaceX, therefore called for research into both the benefits and dangers of AI in an open letter to the United Nations in 2015. The initiative was taken by Max Tegmark and Toby Walsh, whose opinions on the matter are presented later in this chapter.

According to the concept of transhumanism, AI will intervene every expect of our lives and revolutionize our existence. It is rather alarming how marginal this topic is in society today due to its abstractness and lack of interest in it. This also applies to politics, which is failing to discuss and create frameworks for an AI society. Private investment in AI startups in 2017 was $8.1 billion in the US, $5.6 billion in China, and just $0.2 billion in Germany. By 2015, of all machine learning patents filed, 1400 were from the US, 754 from China, and 140 from Germany.

What is AI? Klaus Mainzer defines a system as "intelligent if it can solve problems independently and efficiently" (Mainzer 2018). AI, according to Jay Tuck, can be defined as software that writes itself (Tuck 2016). This means that it also independently writes its own updates, fixes bugs, and permanently optimizes itself for a dedicated task at speeds unimaginable by humans, without an outside understanding of how this happens and what exactly is happening. According to Max Tegmark, professor of physics at MIT in Boston, AI is a system with the ability to achieve complex goals (Tegmark 2017). In the visionary case, an AI system not only adopts the goals programmed by humans but flexibly adapts them itself. Google already suggests alternative answers to users before they have fully formulated a question. These answers are generated by AI software. Military drones, invisible to the human eye in the sky, already exist today, equipped with so-called "kill decisions" that can autonomously decide to kill people in the field. According to Stefan Lorenz Sorgner, there exists today a "merciless effort by the virtual class to force the abandonment of the body, to dump sensory perception on the trash heap and replace it instead with a disembodied world of data streams" (Halmer 2013). Philosopher Sorgner calls transhumanism "the most dangerous idea in the world" (Sorgner 2016). He is not alone in this. He belongs to moderate transhumanism, while the British philosopher Nick Bostrom (Oxford) belongs more to pronounced transhumanism or technological posthumanism. Toby S, Chair of Artificial Intelligence at the University of New South Wales in Australia, summarizes: "AI will have comparable effects on the conception of man as the Copernican revolution" (Walsh 2017).

8.2.2 Superintelligence—The Last Invention of Mankind?

Bostrom, in his book *Superintelligence* (Bostrom 2014), explains the pathways that could lead to the development of a superintelligent being that could replace humans as a biological species. At this point, we must first clarify the differences between AI systems. Unlike a conventional AI system (weak AI, Figure 8.4), a system with artificial general intelligence (AGI) is no longer limited to a specific task, such as speech recognition, playing chess or autonomous driving, but has general intelligence in all fields. Furthermore, superintelligence or artificial superintelligence refers to beings or machines with intelligence superior to that of humans in many or all areas. It can also, beyond AGI, set its own goals and adapt them.

Is that impossible? Maybe, maybe not. When asked if a machine could ever be smarter than a human, Margaret Boden replied that if so, the computer would have to be taught "everything an adult human ever learned about the world and other people. The computer would also have to somehow know how all these things are related to each other." She confirms to her interviewer that in comparison, the task of keeping track of a few dozen pieces in Go, the hardest strategy game there is, is a joke (Klein 2019). That does not stop science from advancing in that direction. In April 2019, the *MIT Technology Review* reported that the humanoid robot *Atlas*, now good at climbing stairs, was "implanted" with AlphaGo, the AI program that defeated the Go world champion in 2016. This made Atlas intelligent, a significant step toward connecting an AI program to the physical world and

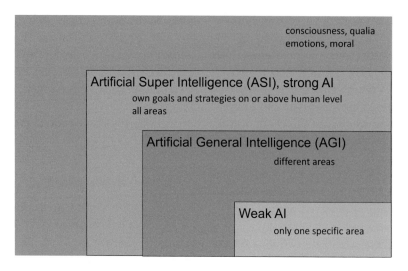

Figure 8.4 AI levels by capabilities. A weak AI is specialized in one area. AGI (artificial general intelligence), on the other hand, can perform many tasks. A super intelligence (artificial super intelligence) or strong AI is an intelligence superior to humans in every field. It can also pursue its own goals and change these goals independently. In addition, systems that possess forms of consciousness, self-awareness, sentience, emotions, and morality are theoretically conceivable.

letting it learn there. The same goal of connecting a virtual system to the real world is being pursued by Dactyl, an open-AI program that, when paired with a robotic hand, demonstrates amazing dexterity to feel an object from all sides, slide it through its fingers and rotate it with them, just as a human does. Such technology is expected to equip robots for everyday tasks, such as filling the dishwasher or getting a beer from the refrigerator. Quite imperceptibly, however, such systems will leave behind a world in which they were fixated on solving a single problem.

Possible directions of development for the realization of superintelligence include technical advancement of computers, genetic advancement of humans, and fusion of both in cyborgs, and brain emulation, i.e., the replication of the brain in the computer. According to Bostrom, the development of a superintelligent system is perhaps the most important event in human history, perhaps as important as the creation of humans themselves. If we invent machines that can do everything better than we can, that would be, in his view and the cinematic thematization by James Barrat (*Our Final Invention*, 2013), the last invention of mankind.

As mentioned earlier, AI systems today are often still specialized in individual tasks; this is not the case for *Alexa* or IBM's Watson program. Unlike us humans, however, there are still no AI engines that can tie shoes, feed dog, and study biology. A few years ago, however, there was also no robot that could climb stairs or get up again from a lying position and do a forward roll. Today, these sills are no longer issues. Rather, it is clear that the AI industry, led by Google as the world's most important AI company, is working flat out on networking AI systems that can perform every conceivable human task. Google is well positioned in the robotics industry and the development of a superintelligent system (Google Brain). At the same time, Google avoids calling a spade a spade.

I fundamentally disagree with the limited possibilities of AI seen by some IT professionals because of its alleged limitation to pattern recognition: in fact, our entire lives are made up of patterns, from getting up in the morning, brushing our teeth, tying our shoes, and eating lunch to every uncertain blink of an eye, falling in love, or planning a vacation. All of this and thousands of other behaviors can be categorized into patterns. AI systems are making all these behaviors their own; they can analyze all these patterns and "optimize" them for their own purposes. They are getting better at learning to cooperate with humans, becoming masters at reading our current mood and emotions, and will cheer us up and comfort us when we are sad. Step by step, often without our realizing it, they will intervene in all our life processes. The one-meter-twenty Pepper, a humanoid robot with cute googly eyes from the French company Aldebaran Robotics, is already well on its way.

Superintelligent systems of the future will have all the knowledge of the Internet at their disposal and thus practice deep learning every second. They will independently plan new traffic concepts for megacities. They will manage the efficient, integrated supply of electricity, water, and heat for one city and then for a thousand cities, and then optimize them for each household. They will make surgical decisions in hospitals because they will make diagnoses thousand times more accurately than doctors. If such systems do not exist in 20 years, they will do so in 50 or 100. They will eventually dictate our behavior for an efficient climate policy and control us. They will conduct negotiations on, say, a new airport construction with dozens or hundreds of contradictions and conflicting goals. Robots will build robots that build robots. They will "reproduce" themselves, to the extent that they are not already doing so today. For the term "robots to build robots", Google provided approximately 40,000 search results in 0.6 seconds in 2020. In just a few years, there will probably be a million. The most modern factories for this purpose are in Shanghai or Japan, not in Germany or the USA. Conflicts with humans have long been preprogrammed. The exact point in time when the clash will occur is irrelevant. It will come—on many levels.

Do you doubt? I understand that. That is why I would like to give an example from my own life: As I mentioned earlier, as a diabetic, I use a closed-loop system for insulin therapy. This is an approximated closed-loop system. Such a system was a pure utopia a few years ago. My insulin pump is controlled by my smartphone using an AI app from Cambridge University: Additionally, this app is connected via Bluetooth to a continuous glucose management system (CGM) on my arm. The combined system optimizes insulin delivery in a partially automatic manner. The blood glucose curves of the previous weeks and months can be viewed in the cloud by authorized third parties, from anywhere in the world. In the 20 years since insulin pumps have been in use, new models are gradually taking more and more decisions away from patients, performing tasks they previously had to think about over and over again, or worse: tasks they forgot to think about.

Whereas a biological human generation takes 20 years, the generation time for autonomous software upgrades, i.e., those downloaded without human intervention, is in the range of minutes, seconds, or even milliseconds and that in uninterrupted operation—24 hours a day, 365 days a year, without a break. What is technically possible, happens at some point.

A superintelligent system would be able to give itself goals and ethical values and adapt them depending on the situation (value-loading problem). Examples of such values are freedom, justice, and happiness, or more concretely: "minimize injustice and unnecessary suffering," "be kind," and "maximize corporate profit." In today's programming languages, value terms, such as "happiness," do not exist. The use of terms from philosophy still poses an insoluble difficulty since these are not unrestrictedly convertible into computer syntax. If it were possible to give the system simple values from the outside, from which it should then develop and learn its own values with the help of its own "seed

AI," various new problems would arise. Certain values could no longer be desired in a changing world or unforeseen goal conflicts could arise, and the system would be expected to recognize and correct them. In many cases, however, conflicting goals that are typical in the real world cannot be resolved. Thus, the problem of values remains unsolved today. In particular, it is not known how a superintelligent system could install understandable human values by way of value learning. Even if this problem were to be solved, a further problem exists: which values should be chosen and which selection criteria should be used to choose them.

Toby Walsh also addresses the issue of ethical values in a future AI world. As far as human values are concerned, Walsh sees the overriding problem that machines do not explain to us how they arrive at such values. Their algorithms simply spit out answers, a problem that already characterizes AI today. Likewise, without human intervention, discrimination against segments of society—male or female, young or old, white or black—will be amplified in a future AI world, which is in the statistical logic of algorithms. We already see examples of such biases today when systems decide on possible repeat offenders in US courts or manipulate political elections. By 2062—that is the title of Walsh's book—one will not be able to tell a fake politician from the original, he says. Truth will no longer be discernible or communicable. Ethical bias and unfairness may be intended or condoned by developers, but they may also be unintended and even unavoidable in certain cases (Walsh 2018). Bias, deception, or fraud may well then be seen as goals and behaviors that an AI system regenerates.

Let us turn once again to conflicting goals. As the term is abstract, I have provided examples. An intelligent applicant selection system uses various algorithms to select suitable applicants for advertised positions. Thus, a machine makes decisions about people, which can hopefully be explained causally. The selection is performed without considering the gender of the applicants. Based on different criteria, the case may arise that the system selects more male than female applicants, perhaps simply because more male candidates apply. But this may be undesirable or even perceived as unfair if the company has a quota target and simultaneously wants to hire as many female applicants as male applicants. In this case, other algorithms would have to be used. However, the new decision-making process would contradict the previous one because it would no longer be the most qualified candidates who are selected. The conflict of goals here can be resolved if a compromise is found in the qualification requirements. However, if no compromise is possible when deciding on the best candidates—regardless of their gender—for example, because the company cannot pay salaries in line with the market to attract the best candidates, a conflict of objectives basically remains: the company wants to attract candidates with optimal qualifications and pay low salaries at the same time but does not achieve that. This problem is independent of whether people or machines make the decisions.

For other goals, the contradictions can be even more intricate. Take economic growth and climate protection, for example. Both goals are many times more difficult to reconcile in an entire economy than in an individual company. Later, we will examine conflicting goals in a different form when it comes to individual ambivalences that shape our lives (Section 8.3).

Returning to superintelligence, further problem areas arise: should the intentions of the superintelligent system be rechecked by humans before execution? Does the system even allow for such control on a permanent basis? The control problem is to ensure that humans maintain control over the machine. Bostrom demonstrates the problem with the following example:

> Suppose we have an AI whose only goal is to make as many paper clips as possible.

The AI will realize quickly that it would be much better if there were no humans because humans might decide to switch it off. Because if humans do so, there would be fewer paper clips. Also, human bodies contain a lot of atoms that could be made into paper clips. The future that the AI would be trying to gear toward would be one in which there were a lot of paper clips but no humans

Bostrom (2014).

How can such a negative development be prevented? Bostrom describes two potential dangers in this regard. The first concerns the motivation of the designers of the superintelligent machine. Are they developing the machine for their personal benefit, for scientific interest, or for the benefit of mankind? The dangers of the first two motivations can be banished if the client controls the developer. The second potential danger concerns the control of the superintelligent machine by its designers. Can a machine that is more highly qualified at the end of its development be supervised by a less qualified developer? The control measures required for this would then have to be planned in advance and built into the machine without being subject to further manipulation by the machine. There are two approaches to this: control of the capabilities and control of the motivation of the machine. If even one of the two slips away, the superintelligence can gain control over the human being.

Besides ethical questions, the cooperativeness of a superintelligent machine with nature is also seen critically. Does a superintelligent system still need nature? What will become of our evolutionary dependence on our natural environment? Can we still get in touch with nature at all? Will our brains possibly change radically when digital worlds become more experiential than any real world, indeed when both become indistinguishable? The digital scent of a rose, a digital mountain or planetary hike, or virtual sex? According to the Internet, some of these things are already possible, for example, in Tokyo.

Perhaps only then will we learn that nature is an evolutionarily fundamental component of our humanity. Nobody can answer all these questions today.

8.2.3 Intelligence Explosion and Singularity

In connection with superintelligence—, the visionary of artificial intelligence, Irving John Good, spoke in 1966 of a possible intelligence explosion that could occur in a cycle of recursive self-improvement (Good 1966). Today, this scenario is presented as a process in several stages. An artificial superintelligent machine is best imagined as a networked system that uses all knowledge in the cloud, including the knowledge of similarly intelligent, competing systems. Initially, the current system has capabilities far below basic human levels, defined as general intellectual ability. At some point in the future, it reaches human levels of intellect. Nick Bostrom refers to this level as the beginning of takeoff. With further continuous progress, the system inexorably and automatically acquires the combined intellectual capabilities of all of mankind. It becomes a powerful superintelligent machine and eventually spirals itself to a level far above the present combined intellectual capabilities of mankind. The takeoff ends here; from this point on, system intelligence increases only slowly. During takeoff, the system may cross a critical threshold. From this threshold on, the majority of the system's improvements are intrinsic to the system, i.e., outside interventions are of little relevance. Such an intelligence explosion could take place in a few days or hours (Figure 8.5).

The dimension of an intelligence explosion becomes exemplarily clear if we imagine that the world's gross domestic product, which today grows laboriously by two to three percent per year, doubles in a year or two (hopefully fully sustainable), that the system could write a dissertation on which the author had to toil for 8 years in a few minutes or the book at hand, including the research for it, in a similarly short time. In this case,

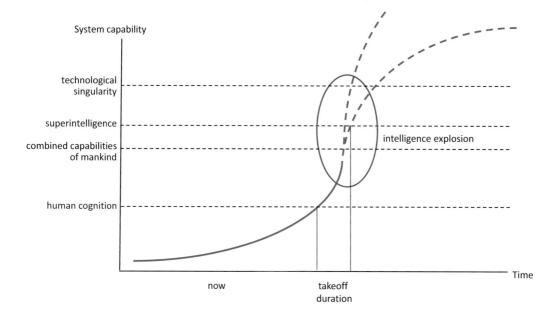

Figure 8.5 Intelligence explosion. According to Nick Bostrom, when machines exceed human cognitive ability, their knowledge increases explosively in the takeoff phase because they learn quickly from each other. Once they have reached the level of superintelligence—, it is unclear what their goals are and whether we can still control those goals and their behavior.

the human being would hardly have time to react. His fate would depend essentially on precautions already taken. In the extreme case, he would face the inability to act anymore, which would lead to a technological singularity, the fusion of man and machine and the replacement of the human species by an artificial superintelligence. The future of mankind is no longer predictable in a technological singularity.

The time needed to achieve such a singularity has been discussed by various authors and estimated in the range of decades (Kurzweil 2005; Chalmers 2010). This is likely to occur unexpectedly, possibly even for those involved in its development. Schmidhuber sees AIs and robots colonizing the universe. Humans will no longer play a dominant role (Schummer 2011). Experts who take a similar cautionary position include Stephen Hawking, Bill Gates, Apple co-founder Steve Wozniak, Austrian robotics expert Hans Moravec, nanotech pioneer Eric Drexler, and others.

Forecasts attract more of our attention when they are better justified. Bostrom devotes many pages to the justification for a fast takeoff as in Figure 8.5. In doing so, he uses numerous assumptions. The consequence of an intelligence explosion is then based on several chained, coherent assumptions, all of which must occur. However, cognitive psychologist and Nobel Laureate in Economics Daniel Kahneman (1990) pointed out that chained coherent statements may sound plausible, and they should. They give the impression that they are more probable when in fact they are not. By contrast, the probability of occurrence of an event actually decreases strongly due to concatenations. Kahneman calls this the conjunction fallacy. Here is an example to illustrate the connection: the interest rate will rise next year due to high inflation. Therefore, the US bull market is expected to weaken. This sounds plausible; the approaching end of the US stock market boom is well justified by the rise in interest rates because alternative forms of investment become more attractive with better interest rates. In fact, however, the following statement is logically more likely, although it is only vaguely formulated and does not specify a cause: the rise in stock prices in the US is expected to slow next year, so be careful! Here, according to Kahneman, less is more.

The assessment of coherence and plausibility of an assumption leads to the wrong track: both say nothing at all about the probability of occurrence. Thus, a slow takeoff may be more likely should fewer linked assumptions be required for it. In contrast to Bostrom, Tegmark (2017) addresses probabilities of alternative future events more clearly.

A slow takeoff will lead to geopolitical, social, and economic dislocation, according to Bostrom's analysis, as interest groups attempt to reposition themselves in terms of power politics in the face of imminent drastic change. Ray Kurzweil does not see that happening. He sees a gradual transition to singularity. A slow takeoff, however, is unlikely, according to Bostrom. Bostrom does not discuss whether one alternative or the other requires fewer assumptions and may therefore actually be more likely, that is, apart from plausibility and deceptive intuition. The same is true in principle for Kurzweil's and other forecasts. Toby Walsh gives a whole series of reasons why a technological singularity is not logically inevitable; however, he does not rule out the possibility. He leaves hardly any doubt that there will be superintelligent systems in the year 2062 but expresses strong concerns in some areas about how technology will develop by then. Accordingly, we must fear above all that we will be leaving life-and-death decisions in military conflicts to machines that are not sufficiently competent (Walsh 2018).

Singular visions in technological-posthumanist scenarios do not represent utopias or dystopias. Rather, they should be understood as "serious forecasts of an era of the posthuman dawning on the horizon of the not-so-distant future" (Loh 2019). Development in this direction proceeds quasi automatically with some human-initiated thrusts. Klaus Mainzer also makes singular visions in technological-posthumanist scenarios, aware that the question of a superintelligence— is no longer a fantasy but is increasingly taking on realistic features. Thus, "astonishing increases in intelligence are at least theoretically conceivable in AI systems that are superior to humans" (Mainzer 2018). However, Mainzer considers human-like intelligence to be possible only in an environment in which machines "not only have bodies that are adapted and adaptable to their tasks, but are also able to react situationally and largely autonomously." This view cannot be emphasized clearly enough. If we think one step further, ultimately, the human self is precisely not a neuronal object but an inseparable, integrated unit of body and mind. Consequently, the development of an artificially intelligent system with or without its own consciousness requires a high degree of interdisciplinary research between engineering sciences, cognitive sciences and brain research, systems biology, synthetic biology, and other disciplines. In the overall view, according to Mainzer, a superintelligent system must still obey first, the logical-mathematical laws and proofs of computability, decidability, and complexity and second, the laws of physics.

8.3 THE THEORY OF EVOLUTION IN THE TECHNOSPHERE

What can the theory of evolution contribute to an AI scenario? One might say that it is the evolution of our brain that makes transhuman developments possible in the first place and that our brain can be explained by traditional evolutionary theory. But this explanation amount to anything? Do we not need an explanatory framework that explores machine behavior as well as hybrid human–machine behavior? A framework that includes the possibility of biological and nonbiological self-manipulation? For this to happen, agreement must first be reached on what counts as heredity. After all, an inserted brain implant is not inherited biologically in the Mendelian sense, nor are techniques to increase stress resistance. However, humans can reuse all these techniques in individuals of the next generation and the one after that, constantly improving them. This is tantamount to inheritance; moreover, it is cultural evolution.

More precisely, it is an inheritance of acquired characteristics, to strain the words of Lamarck. It is the epigenetic inheritance of learned knowledge, an evolutionary trait. Lamarck, gone but not forgotten, reappears here, albeit in a different sense than he originally intended, the context of behavioral evolution.

We have passed on knowledge culturally for millennia: first through gestures and the spoken word, then on cuneiform tablets, on papyrus scrolls, later in books, and today via computers and the Internet. With each new generation, human knowledge has been added to and expanded. Michael Tomasello has used the descriptive term ratchet effect for this expansion of knowledge on the foundation of earlier knowledge (Section 7.3). Today, the knowledge of mankind doubles within a few years due to the ratchet effect. This is what the evolutionary biologists Eva Jablonka and Marion J. Lamb mean by the symbolic, epigenetic inheritance system. Only the inclusion of additional cultural and thus epigenetic or nongenetic inheritance systems opens the view to an extended, post-Darwinian theory of evolution (Jablonka and Lamb 2014).

Cultural inheritance needs to be more fully addressed and better integrated by evolutionary theory. Both this and the inclusion of multilevel selection theory are called for by David Sloan Wilson in his recent book (Wilson 2019). He is thus on the "culture is evolution" track (Chapter 2; for a critique, see Section 7.4) that others have already established in recent years (Richerson and Boyd 2005; cf. Lange 2017; Tomasello 2014; Chapter 7). Wilson's book, with its consistent line of thinking, evidences a basic optimism about the ways in which the superorganism of mankind can be adapted in the future. However, this optimism is not always shared by other authors, who say, for example that our whole life today takes place in a highly technologized world in which it has become "practically as well as theoretically impossible to separate nature from culture" (Braidotti 2019). We are witnessing a transformation of the ontological conditions of the human being. It is characterized by the fact that poiesis, that is, purposeful human action within a nature-given framework, is more ambiguous than ever. In highly mechanized capitalism, tools are no longer an extension of nature, used to accomplish specific purposes. Rather, modern humans live in a world where machines accumulate, store, transform, and distribute information. This is accompanied by the effort to provide and store energy. The actual use and function of such stored information and energy are nebulous and not related to the direct actions of humans. The natural use of tools for given purposes is, as Luciana Parisi expresses it, "dismantled," The strangeness of the purposes does not correspond to man's biocentric needs. In that "his means no longer correspond to his purposes", a rupture takes place at the level of biological causality in the modern nature of the relationship between technology, culture, and nature (Parisi 2019).

New technologies are constantly appearing at increasingly shorter intervals. We know neither meaning for us nor their long-term consequences and costs. We ourselves become tools in the world we have created. The herbicide glyphosate, for example, is to be banned in some countries in response to sustained pressure from the population. As long as new ecological forms of agriculture do not gain widespread acceptance, glyphosate will be replaced by another, supposedly better artificial agent that is supposed to ensure high productivity. But then, in 10 years at the earliest, we will have an indication of what effects the new agent will have on people, soils, animals, and plants. As early as 1980, the technology researcher David Collingridge identified the problems that new technologies pose in comparison to highly advanced and widespread ones. At first, one cannot say anything about the consequences and side effects of an innovation; later, one cannot influence them in a formative way. The problem of technology assessment has also been called the Collingridge dilemma after its discoverer (Genus and Stirling 2018): Externalities, the economic costs that the originator does not have to pay for when resources are misdirected, are typically high for many of today's technologies and often cannot be calculated at all. Power plant operators globally pay too little for CO_2 emissions, and the agricultural industry or its customers pay too little for the ongoing destruction of soils. The impacts of synthetic fertilizers and pesticides are just two examples of countless other socio-ecological technology impacts that we deal with daily but of which we do have sufficient awareness or perception.

The development of artificial intelligence and robotics is entirely embedded in this context: their long-term effects are not transparent today. Their development, on the other hand, is even more difficult to control and steer, both socially and politically, the further it advances. Here, too, rational human beings resort to innovative measures to help make our lives and work easier. We do this in an implicit belief that advanced developments are in the spirit of evolution and adaptive for our species preservation. However, it is often impossible to determine whether they really are, which only becomes apparent much later.

"The modern techne as well as industrial infrastructure of capital, for which man is nothing but an appendage that produces surplus value, leads to an irreversible transformation of the essence of man" (Parisi 2019). In this process, we expose ourselves to the danger of extinction. For we must realize that human beings have long ceased to act autonomously but are part of a world system that primarily serves their own internal requirements (Haff et al. 2019). Now, we also find self-organization in the technical-cultural world as we know it in biology. This need not result from the free will of individuals. Human networks, human societies, and technology thus depend not only on the activities of their individual members but also on other mechanisms. The technosphere (created by itself) assigns humans a new place in the world and steers them away from the anthropocentric purposefulness of the individual toward co-option forces of the Technosphere to which we are bound, Peter K. Haff says.

This technosphere could be seen as an outgrowth of the biosphere. Seen in this light, we are not far from a technosphere in the sense of a self-organizing system that can no longer be controlled by human volition (Haff et al. 2019). Parisi speaks of a new layer in the biocentric order of reality and the decoupling of the essence of man from his biological being (Parisi 2019). The technosphere is the wild card of global change. "It could bring about a revised Anthropocene, such that humans are no longer the decisive factor in it," is how paleobiologist Jan Zalasiewicz (2017), who specializes in the Anthropocene, puts it. Human-initiated evolution, now global, no longer plays out solely at the levels defined by Tomasello or social brain theory (Chapter 7). Rather, the preservation of nature and our living world as we know it requires global partnerships and global cooperation. Many scientists today agree that major questions and crises can only be solved as a collective. Although concepts of cooperation have also found their way into evolutionary theory, it does not yet offer a broad theoretical basis for global cooperation. Today, only initial theoretical foundations exist for effective global cooperation that serves the well-being and survival of humanity, i.e. that is evolutionarily adaptive. On the other hand, the game theory, for example, can provide solutions for the interaction of several individuals or organizations, but it fails when the number of actors becomes too large or the mutual motivations and objectives are not transparent.

The entire technosphere is part of the process of global evolutionary change. It is a hundred percent the result of evolution; however, this fact cannot explain much at first. From an evolutionary point of view, the question thus arises whether enlightened humans are able to carry out sustainable global development in the sense of evolution or in the sense of the theory of human niche construction and if so, to what extent (Chapter 5), for example in the sense of *Sustainable Development* adopted by the United Nations (United Nations 2015), in such a way that this transformation is adaptive for our species conservation. In the search for answers, the realization gaining ground that we will not achieve this goal if only a few intellectual researchers in the tradition of enlightened reasoning provide us with the insight that things cannot continue as they have been. The appeal to reason is often overemphasized today and is not sufficient on its own. In fact, two different approaches are needed here: first, there must be compatible overall concepts that tell us what to do in the face of overpopulation, climate change, limitless and reckless resource consumption, mismanaged global agribusiness and factory farming, growing social inequality and poverty. These integrated overall concepts will inevitably demand that the existing social, economic, and political system structures be changed. This requires a fundamental change in mentality. Here, however, science alone can only provide necessary but not sufficient conditions in the form of its technocratic blueprints. What is required, therefore, is, on the other hand, the simultaneous change in the behavior of billions of people who must first become aware of their ambivalences. Social psychology speaks here of cognitive dissonances in which we persist. In this case, thought and action do not coincide, in other words, our thoughts often cannot be implemented. For example, we want more to be done for the climate, but we also want to be able to drive an SUV. We are in favor of phasing out coal, but at the same time we do not want to pay a higher price for electricity, and we also reject a wind turbine near our home. No one wants to torture animals, pollute soils, poison groundwater, and destroy our children's livelihood, but we do. How we should and can deal with such contradictions individually is a difficult subject. Ambivalences are part of our lives and often prove impossible to resolve in practice.

The Australian John E. Stewart, who is a member of a Free University of Brussels research group, has dealt extensively with the question of how the many levels of human cultural evolution must give rise to a form of global cooperation if humanity is to survive and flourish in evolutionary terms. If there are evolutionary theorists who encourage us that humanity can master its own well-being evolutionarily, the aforementioned David Sloan Wilson (Wilson 2019), but no less John E. Stewart, are among them. Stewart analyzes the conditions for the emergence of a cooperative global organization (Stewart 2020, 2014). Humans have been progressively integrated into societies of greater and greater scales as humanity evolved–from family groups to tribes, then nations and empires. Since the 20th century, some integration is occurring at the supranational level. The fact that forms of integration such as the United Nations, the World Health Organization, the FIFA, the climate conferences, and many others have not yet produced an integrated global society does not represent a fundamental limitation of the observation made. For Stewart, however, it is not enough that there is selection between groups, as Wilson describes. Group selection operating alone on unorganized groups is not sufficient to produce ever higher and eventually all-inclusive integration. To achieve this, effective management is required. Management thus becomes a central concept in Stewart's theory (Stewart 2020, 2014).

If groups have adaptive advantages over individuals, which cannot be doubted in human culture, then evolution will favor the emergence of management that can establish complex cooperation within groups. Management can achieve this by supporting cooperators and disadvantaging non-cooperators. This enables the group to exploit the adaptive advantages of integration. With organizations that are managed in this sense, a vertical self-organization develops over all levels up to a unified, cooperative, sustainable, global cooperation. Such a system of global governance will continually align the interests of all citizens and organizations with those of the whole. The potential benefits of such emerging cooperation are universal. With the idea of effective management introduced by Stewart, his theory points to a postmodern direction. Effective management leads to a global cooperative human niche construction (Chapter 5, Section 5.3).

Of course, Stewart is aware of all the constraints and potential disruptive forces discussed in the theory that occur on the way to effective cooperation. Even he cannot fully guarantee that humanity will respond effectively and in a timely manner to global challenges. Selection pressures on humanity may be greater than ever before and are needed to spur lasting, adaptive change. But time to act effectively is running out, as we know. Despite the threats Stewart gives us hope that humanity can maintain its dignity.

The German philosopher Hans Jonas (1984), professor of philosophy at New School for Social Research in New York from 1955 to 1976, formulated the *Imperative of Responsibility* not least against the background of what has been said above about our behavior: "Act so that the effects of your action are compatible with the permanence of genuine human life." One may call this imperative no less than a basic evolutionary condition for man. However, we experience every day the contradiction of human action. Added to this are fears of change. Fears arose in our evolution when we lived in small groups. They were essential for survival. But abstract figures such as the millions of deaths due to air pollution or thousands of deaths due to resistant germs and medical errors in hospitals–which we should be afraid of–are just as ineffective as the call by a thousand experts in 2015 to think about the opportunities and risks of artificial intelligence. We fear major changes that we should be dealing with, and at the same time have fears of trivial scenarios. The media, through the nature of reporting, does its part to trigger archaic fears in us. So, does our evolution dominate our reason?

Claus Otto Scharmer is a German economist and one of those researchers who focuses on how to address the transformation process in organizations of all kinds—from schools to global corporations, from small associations to churches. Scharmer argues that we are disconnected from reality and collectively produce results that no one wants. The means of the past are no longer opportune to oppose this development. Using insights from economics, psychology, and sociology, Scharmer therefore calls for more than mere reflection, more than pure reasoning, to see more clearly in a complex world of interrelationships. In his method, *Theory U*, developed at MIT in Boston, he clarifies that an opening of thinking (curiosity, renunciation of judgments and categorizations), an opening of feeling (open heart, mindfulness compassion), and finally an opening of the will (determination, courage, meaningfulness of intentions) must come together. It is necessary to let go of automatisms in thinking with the connection of head, heart, and hand, from the limited point of view of one's own ego or one's own company to the view of society as a whole and thus to come to a creative potential of the future. We must relearn to see what we are doing (Scharmer 2007).

The rules of the game in the technosphere have changed more rapidly and comprehensively than ever before in human history. Under such altered conditions and rules, we are transforming into the unknown. Knowing that the technosphere is 100% a result of evolution one thing, but to determine the driving mechanisms behind cultural changes and how change can be adaptively managed for humans (and the planet) in the technosphere is quite another. The general question is whether today's humans, with the evolutionary "ballast" they carry around with them–including, in addition to the "false" fears mentioned as examples, the characteristic of being extremely focused on satisfying short-term needs—*can* behave appropriately to cope with the upcoming transformation. This question applies to the social intelligence of the general human population. Ultimately, the discussion leads to the question: can a future artificial social intelligence adapt better to nation-related and global challenges than our organic intelligence?

Interesting questions related to our own future are as follows: Will humans be able to continue to cope with their intelligence in the future in the face of the world's advancing, apparently uncontrollable urbanization, with the explosion of megacities such as Tokyo (38 million inhabitants), Shanghai (34 million), Jakarta (31 million), or Delhi (27 million) and the rapid growth of the world's population, which is heading for the 10-billion mark? How much can this population growth be regulated? How well are we still evolutionarily adapted to our modern way of life? Are these developments possibly occurring much too fast, thereby even endangering the position of the human species? Are we perhaps marching straight ahead deeper into maladaptation? After all, it is deeply irritating that we, as the most intelligent species, are waging wars, destroying rainforests, exploiting natural resources, heating up the climate, and decimating biodiversity. In other words, we do not know today whether our just-vaunted intelligence will be sufficient to correct our damaged adaptation, maintain our well-being, and ultimately save our species (Lange 2021).

Maladaptations are a particularly tangible problems. They may lead to collapse of the system owing to their persistence and resilience. It is evident that if little to no efforts are undertaken, the entire ecosystem can collapse. The worsening weather conditions can lead to much greater disasters than those observed today. The global monetary and financial system, overstretched with excessive debt, may collapse as well. Social peace is not assured in such cases.

Actions undertaken by individuals involve those that do not serve to improve biological fitness; however, these actions appear to be subjectively useful. Many actions are undertaken according to the market needs. The market owing to its short-term profit orientation replaces or at least temporarily displaces natural selection. Behaviors of entire societies are determined through these factors. In terms of evolutionary behavior, they represent trial-and-error processes or random variations. Such developments can be demonstrably evolutionary harmful and then also be considered evolutionary maladaptations for humanity. The dynamics of maladaptations may lead to their reinforcement. Undeniably, natural selection has the last word. Evidently, culture permits the presence of maladaptations and can only be overridden, but not eliminated, through human artificial selection (Lange 2021).

Maladaptations, as described here are brought about by humans, can be explained and are compatible with the theory of Darwin and with the theory of niche construction (Chapter 5). That human culture does not necessarily have to be adaptive has been clearly pointed out by Richerson and Boyd (2005), Odling-Smee (interview, Chapter 5), and other authors (conf. Lange 2021).

So far, I have pointed out some of the issues that need to be addressed when we deal with the theory of the future evolution of man in the technosphere: (1). the evolutionary heritage of our hunter-gatherer past with the possible major obstacles for our present, (2). a theory of global cooperation that still needs to be further developed, and (3). the dilemma of socio-ecological technology assessment, (4). dealing with the unmanageable complexity of the global socio-technical society and, last but not least, (5). pointing out the complete lack of a globally applicable alternative to the current, supposedly successful neoliberal, capitalist economic system and its financial institutions.

It is clear that I can only touch upon these extensive aspects of our present evolution. Therefore, I would like to illuminate the topic of a theory of human evolution in possible AI scenarios mentioned at the beginning of this section a little more closely. The origins, conditions, and modes of operation of technological developments and future scenarios should, I am convinced, find room in evolutionary theory. The theory of niche construction can contribute to a better understanding of human–machine behavior. What framework conditions are needed for intelligent machines to exhibit the property of producing and improving themselves, i.e., of evolving? What goals can they set for themselves? What role do we play for them? What developments is the human–machine society tending toward?

Writing in *Nature* magazine in 2019, a group of 23 scientists from numerous research disciplines made a start by encouraging research into machine behavior and a call for a broad research agenda to study machine behavior (Rahwan et al. 2019). The authors come from computational science, evolutionary biology and ecology, economics and political science, evolutionary anthropology, cognitive science, behavioral science, and other departments at renowned universities and research institutes around the world. As a starting point, they noted that machine behavior, similar to animal and human behavior, cannot be fully understood without studying the context from which it occurs. Understanding machine behavior requires the integral study of algorithms and social environments in which algorithms operate. The goal is to understand the broad, unintended consequences of AI agents, which can produce behaviors and social effects that are unpredictable by their inventors. The authors see societal benefits from AI, but on the flip side, fear that humans may lose oversight with respect to machine intelligence (Rahwan et al. 2019).

The group proposes the well-known question-type matrix of behavioral scientist and Nobel laureate Nikolaas Tinbergen. Tinbergen posed four questions that should be asked to fully understand an evolutionary trait. (1). the question of function: why does it exist? (2). what is its intergenerational history? (3). what is its physical mechanism? (4). how did the trait evolve in the individual life history of the organism? With these questions, both morphological and behavioral traits can be analyzed optimally. Our research group now wants to apply these types of questions to machine behavior. Tinbergen's question 2 would then be, for example, in relation to the evolution of machines: how did a machine type evolve certain types of behavior?

The numerous questions from the matrix are to be addressed first to single AI systems, second to groups of interacting AI systems, and third to hybrid systems, i.e., scenarios with human–machine groups. The authors emphasize that "machines may exhibit very different evolutionary trajectories, as they are not bound to the mechanisms of organic evolution." They can develop forms of intelligence and behavior that are qualitatively different from, or even alien to, biological actors. Only such a study of AI systems in a larger socio-technical factory will meet the ultimate responsibility for benefit or harm that humans face as AI systems proliferate.

Other scientists are already looking for links between machine learning and natural evolution. They want to learn what related principles exist between the two. In this context, meta-learning systems—AI systems that improve the learning of other AI systems—

have been developed. For example, one learning system (metalearner) helps another (learner) find appropriate categories more quickly, such as dogs, cats, gloves, house keys, and so on.

As another example, a group around Richard A. Watson, a member of the Extended Synthesis research collective (Chapter 6), is also actively drawing on evolutionary knowledge (Kouvaris et al. 2017). First, it is recognized in biology that evolution cannot look ahead. Rather, natural selection has always been understood to allow organisms to evolve robust designs that tend to produce what has been selected for in the past but may be unsuitable for future environments. However, this view has recently been challenged. Thus, the issue of evolution of evolvability is highly topical. How is this to be understood?

One possible idea is that evolution can discover and use information not only about the phenotypes selected in the past but also about their underlying structural regularities. In this way, novel phenotypes with the same underlying regularities but with new individual traits could be useful in different environments. This sounds similar to what we learned earlier about the importance of robustness (Section 3.3) and cryptic masked mutation (Section 3.8). In robustness lie the foundations for novelty in evolution. But this is even more true here, where it is necessary to understand the conditions under which natural selection "discovers" deep regularities rather than exploiting "rapid corrections," corrections that provide adaptive phenotypes in the short term but limit future evolvability. One makes use of recent findings in machine learning and sees a deep analogy between learning and evolution. In machine learning, learning principles are known today that enable generalization from past experiences. The effort to transfer this to biology leads to the conclusion that evolving biological systems and learning systems are different instances of the same algorithmic principles. It will be interesting to see what further mutual stimuli are brought to light.

8.4 SUMMARY

The future of mankind will be dominated by technology, the development of which is pushing back natural selection. In addition to genome editing, with which future humans will intervene in the germ line, we are observing an increasingly strong human–machine coevolution. During an advancing transhumanism, the increasing technical equipment or the replacement of the human body and brain components by technology determines a direction of fusion of our biological species with technology. Humans may find themselves coexisting with biosynthetic and engineered life-like forms and in extreme cases may be replaced as a biological species.

In addition to recognizing the evolutionary ballast from our long past as hunter-gatherers, evolutionary theory should address how humans have changed under the special conditions of the technosphere. Among other things, the theory should take on the task of explaining human–machine combinations and hybrid human–machine behavior. We need to know what exactly is going on in machines when they exhibit human-like behavior. Equally important is knowing how this behavior evolves and how machines "see" humans. If we are to remain in control of our future in the field of AI and robotics, research into these complex cultural processes will play an essential role in humankind's future.

The natural law of evolution remains that systems, biological as well as technical, must adapt to environmental conditions in the universe to survive (Hansmann 2015). Postmodern, interdisciplinary evolutionary theory in the post-Darwinian and post-Synthesis era has only just begun to expand its field of vision with the Extended Synthesis and will have to broaden further if it is to delineate the playing field of evolution on which humans operate in the Anthropocene.

REFERENCES

Note: Citations from German-language books and articles that are not published in English are translated by the author.

Bonifacio E, Warncke K, Winkler C, Wallner M, Ziegler A-G (2012) Cesarean section and interferon-induced helicase gene polymorphisms combine to increase childhood typ1 diabetes risk. *Diabetes* 60:3300–3306.

Bostrom N (2014) *Superintelligence: Path, Dangers, Strategies*. Oxford University Press. Oxford.

Braidotti R (2019) Zoe/Geo/Techno-Materialismus. In: Rosol C, Klingan K (eds.) *Technosphäre*. Matthes & Seitz, Berlin, 122–143.

Brosius J (2003) From Eden to a hell of uniformity? Directed evolution in humans. *BioEssays* 25:815–821.

Chalmers DJ (2010) The singularity: a philosophical analysis. *J Conscious Stud* 17:7–65.

Cyranoski D (2014) Rudimentary egg and sperm cells made from stem cells. *Nature*. https://doi.org/10.1038/nature.2014.16636. https://www.nature.com/news/rudimentary-egg-and-sperm-cells-made-from-stem-cells-1.16636.

Dengler K (2019) Substituierbarkeitspotenziale von Berufen und Veränderbarkeit von Berufsbildern. Impulsvortrag für die Projektgruppe 1 der Enquete-Kommission „Berufliche Bildung in der digitalen Arbeitswelt" des Deutschen Bundestags am 11.3.2019. Institut für Arbeitsmarkt- und Berufsforschung. http://doku.iab.de/stellungnahme/2019/sn0219..pdf

Dengler K, Matthes B (2015) Folgen der Digitalisierung für die Arbeitswelt. Substituierbarkeitspotenziale von Berufen in Deutschland. Institut für Arbeitsmarkt- und Berufsforschung. http://doku.iab.de/forschungsbericht/2015/fb1115.pdf

Doudna JA, Sternberg SH (2017) *A Crack in Creation. Gene Editing and the Unthinkable Power to Control Evolution*. Houghton Mifflin Harcourt Publishing, Boston, MA.

Dougherty C (2017) Self-driving trucks may be closer than they appear. *The New York Times*, 13.11. 2017.

Frey CF, Osborne MA (2013) The future of employment: how susceptible are jobs to computerisation? https://www.oxfordmartin.ox.ac.uk/downloads/academic/The_Future_of_Employment.pdf.

Ford M (2021) *Rule of the Robots: How Artificial Intelligence will Transform Everything*. Basic Books, New York.

Genus A, Stirling A (2018) Collingridge and the dilemma of control: towards responsible and accountable innovation. *Research Policia* 47(1):61–69.

Gerbault P, Liebert A, Itan Y, Powell A, Currat M, Burger J, Swallo DM, Thomas MG (2011) Evolution of lactase persistence: an example of human niche construction. *P Roy Soc B Bio* 366:863–877.

Gibson DG, Glass JI, Lartigue C et al. (2010) Creation of a bacterial cell controlled by a chemically synthesized genome. *Science* 329:5987.

Good IJ (1966) Speculations concerning the first ultraintelligent machine. *Adv Comput* 6:31–88.

de Grey A, Rae M (2010) *Niemals alt! So lässt sich das Altern umkehren. Fortschritte in der Verjüngungsforschung*. Transcript, Bielefeld.

Haff PK, Renn J, Peter K (2019) Haff im Gespräch mit Jürgen Renn. In: Klingan K, Rosol C (eds.) *Technosphäre*. Matthes & Seitz, Berlin.

Halmer N (2013) Transhumanismus und Nietzsches Übermensch. https://sciencev2.orf.at/stories/1727448/index.html.

Hansmann O (2015) *Transhumanismus Vision und Wirklichkeit. Ein problemgeschichtlicher und kritischer Versuch*. Logos, Berlin.

Jablonka E, Lamb MJ (2014) *Evolution in four Dimensions. Genetic, Epigenetic, Behavioral, and Symbolic Variation in the History of Life*, 2nd ext. ed. MIT Press, Cambridge, MA.

Jonas H (1984) *The Imperative of Responsibility: In Search of Ethics for the Technological Age*. University of Chicago Press, Chicago, IL.

Kahneman D (1990) *Thinking, Fast and Slow*. Farrar, Straus and Giroux, New York.

Karafyllis C (ed.) (2003) *Biofakte. Versuch über den Menschen zwischen Artefakt und Lebewesen*. Mentis, Paderborn.

Klein S (2019) *Wir werden uns in Roboter verlieben. Gespräche mit Wissenschaftlern*. Fischer, Frankfurt a. M.

Klingan K, Rosol C (2019) Technische Allgegenwart – ein Projekt. In: Klingan K, Rosol C (eds.) *Technosphäre*. Matthes & Seitz, Berlin, 12–25.

Kouvaris K, Clune J, Kounios L, Brede M, Watson RA (2017) How evolution learns to generalise: using the principles of learning theory to understand the evolution of developmental organization. *PLoS Comp Biol* 13:e1005358.

Kretschmer S, Schwille P (2016) Pattern formation on membranes and its role in bacterial cell division. *Curr Opin Cell Biol* 38: 52–59.

Kriegman S, Blackiston D, Levin M, Bongard J (2020) A scalable pipeline for designing configurable organisms. *PNAS* 117(4):1853–1859. https://doi.org/10.1073/pnas.1910837117.

Kurzweil R (2005) *The Singularity Is Near. When Humans Transcend Biology*. Viking, New York.

Kurzweil R (2017) Immortality by 2045. https://www.youtube.com/watch?v=f28LPwR8BdY.

Lange A (2017) *Darwins Erbe im Umbau. Die Säulen der Erweiterten Synthese in der Evolutionstheorie*, 2. überarb. Aufl. Königshausen & Neumann, Würzburg (eBook).

Lange A (2021) *Von künstlicher Biologie zu künstlicher Intelligenz – und dann? Die Zukunft unserer Evolution*. Springer, Berlin.

Laronda M, Rutz A, Xiao S, Whelan KA, Duncan FE, Roth EW, Woodruff TK, Shah RN (2017) A bioprosthetic ovary created using 3D printed microporous scaffolds restores ovarian function in sterilized mice. *Nat Commun* 8:15261. https://doi.org/10.1038/ncomms15261.

Levy D (2007) *Love and Sex with Robots – The Evolution of Human-Robot Relationships*. HarperCollins, New York.

Loh J (2019) *Trans- und Posthumanismus zur Einführung*. Junius, Hamburg.

Lu TK, Chandrasegaran S, Hodak H (2016) The era of synthetic biology and genome engineering: where no man has gone before. *J Mol Biol* 428:835–836.

Mainzer K (2018) *Künstliche Intelligenz – Wann übernehmen die Maschinen?* 2nd ext. ed. Springer, Berlin.

Metzl J (2019) *Hacking Darwin. Genetic Engineering and the Future of Humanity*. Sourcebooks, Naperville.

Nagel T (1987) *What Does It All Mean?* Oxford University Press, Oxford.

Nida-Rümelin J, Weidenfeld K (2019) *Digitaler Humanismus. Eine Ethik für das Zeitalter der Künstlichen Intelligenz*. Piper, München.

Parisi L (2019) Disorganische Techne. In: Klingan K, Rosol C (eds.) *Technosphäre*. Matthes & Seitz, Berlin, 104–121.

Piraino S, Boero F (1996) Reversing the life cycle: medusae transforming into polyps and cell transdifferentiation in Turritopsis nutricula (Cnidaria, Hydrozoa). *Biol Bull* 190(3): 302–312.

Rahwan I, Cebrian C, Obradovich N, Bongar J, Bonnefon J-F, Breazeal C, Crandall JW et al. (2019) Machine behaviour. *Nature* 568(-7753):477–486.

Richerson PJ, Boyd R (2005) *Not by Genes Alone. How Culture Transformed Human Evolution*. The University of Chicago Press, Chicago, IL.

Scharmer CO (2007) *Theory U: Leading from the Emerging Future*. A BK Business Book, 2nd ed. Berrett-Koehler, San Francisco, CA.

Schummer J (2011) *Das Gotteshandwerk. Die künstliche Herstellung von Leben im Labor*. Edition Unseld, Suhrkamp, Berlin.

Service RF (2016) Synthetic microbe has fewest genes, but many mysteries. Science mysteries. *Science* 351(6280):1380–1381.

Seung S (2013) *Das Konnektom. Erklärt der Schaltplan unseres Gehirns unser Ich?* Springer Spektrum, Heidelberg.

Sorgner SL (2016) *Transhumanismus. Die gefährlichste Idee der Welt?* Herder, Freiburg.

Stevens T, Newman S (2019) *Biotech Juggernaut. Hope, Hype, and Hidden Agendas of Entrepreneurial Bioscience.* Routledge, New York.

Stewart JE (2014) The direction of evolution: the rise of cooperative organization. *BioSystems* 123:27–36.

Stewart JE (2020) Towards a general theory of the major cooperative evolutionary transitions. *BioSystems* 198:104237.

Tegmark M (2017) *Life 3.0: Being Human in the Age of Artificial Intelligence.* Knopf, New York.

Tomasello M (2014) *A Natural History of Human Thinking.* Harvard University Press, Cambridge, MA.

Tuck J (2016) *Evolution ohne uns: Wird künstliche Intelligenz uns töten?* Plassen, Kulmbach.

United Nations (2015) Transforming our world. The 2030 Agenda for Sustainable Development. https://sdgs.un.org/2030agenda.

Walsh T (2017) *It's Alive! Artificial Intelligence from the Logic Piano to Killer Robots.* La Trobe University Press, Melbourne/Australia.

Walsh T (2018) *2062: The World that AI Made.* La Trobe University Press, Carlton.

Wei M, Fabrizio P, Hu J, Ge H, Cheng C, Li L, Longo VD (2008) Life span extension by calorie restriction depends on Rim15 and transcription factors downstream of Ras/PKA, Tor, and Sch9. *PLoS Genet* 4(1):e13. https://doi.org/10.1371/journal.pgen.0040013.

Wilson DS (2019) *This View of Life. Completing the Darwinian Revolution.* Pantheon Books, New York.

Zalasiewicz J (2017) Geologie: eine vielschichtige Angelegenheit. In: *Spektrum Spezial Biologie – Medizin – Hirnforschung 3. Die Zukunft der Menschheit: Wie wollen wir morgen leben?* Spektrum der Wissenschaft, Stuttgart, 52–61.

TIPS AND RESOURCES FOR FURTHER READING

Bostrom N (2016b) *Superintelligence: Path, Dangers, Strategies.* Oxford University Press. Oxford.

Cyborgs: A Personal Story. Kevin Warwick. TEDxCoventryUniversity, 2/21/2016. https://www.youtube.com/watch?v LUd4qv2Qr0A.

Is it possible to escape the slow and steady the progression of ageing? The animals that can live forever. https://www.science.org.au/curious/earth-environment/animals-can-live-forever.

The Kurzweil interview, continued: Portable computing, virtual reality, immortality, and strong vs. narrow AI. *Computerworld.* NOV. 13, 2007. https://www.computerworld.com/article/2477417/the-kurzweil-interview--continued--portable-computing--virtual-reality--immortality--and.html.

Paralyzed man takes first steps since his injury with help of electronic device in his spine https://globalnews.ca/news/4480833/paralyzed-walking-electronic-device/.

Stewart JE (2012) *The Evolutionary Manifesto. Our Role in the Future Evolution of Life and Strategies for Advancing Evolution.* The Chapman Press, Orange, CA.

Tegmark M (2017) *Life 3.0.* Alfred A. Knopf, New York.

Watson health. How AI is impacting healthcare. https://www.ibm.com/resources/watson-health/artificial-intelligence-impacting-healthcare/.

We All May Be Dead in 2050. Scientists are beginning to worry about AI and the danger it poses to mankind. *U.S. News*, Oct. 29, 2015. https://www.usnews.com/news/blogs/at-the-edge/2015/10/29/artificial-intelligence-may-kill-us-all-in-30-years.

CHAPTER NINE

More than One Theory of Evolution—A Pluralistic Approach

It is difficult to unify the theory of evolution once again after the Modern Synthesis (MS). A synthesis like the one proposed 80 years ago is not easily imaginable today with the multitude of scientists and disciplines involved. There is no congress where scientists of all disciplines meet and agree on what a consistent, congruent, postmodern theory of evolution should look like. Science today is regulated differently than it did back then. Now, scientists deal with a "superorganism" of global consortia and networks, participating companies and universities, and their private and public funding modes and governance (Nowotny and Testa 2009). At present, publications by individual researchers can be checked online in near real time to see how often they are cited. A scientist who is not cited a few hundred or thousand times "usually counts for little," to put it bluntly. This is a mercilessly transparent process, embedded in the scientific superorganism; however, this process is not actually meaningful in terms of content quality. For new ideas to stand out among the plethora of papers, money has to be raised—lots of money and the authors have to know about writing proposals and managing large projects.

Important technical terms in this chapter (see glossary): complexity, Extended Evolutionary Synthesis, gene centrism, gene-culture coevolution, Modern Synthesis, reciprocity, and reductionism

9.1 FROM OLD TO NEW SHORES—OBSTACLES AND OPPORTUNITIES

One argument against the permanent rejection of neo-Darwinism is the abundance of new studies and findings on gene regulation and gene regulatory networks. This field is likely to remain inexhaustible and promising for decades to come. Simultaneously, it obscures the view of superordinate topics and creates a preponderance of molecular biological research simply because of the sheer amount of material being generated.

Last but not least, there is the "Darwinian factory," as it was once casually called. Everything that supports Darwin's theory and especially the MS is accepted in advance. The attempt to constructively criticize the MS and to adopt new perspectives that pave the way for novel views is a Sisyphean task; however, strong beginnings have been made. When a publication, such as those of authors Laland, Jablonka, Müller, Moczek, Odling-Smee, and others (Laland et al. 2015), is cited and reviewed 700 times and more, mostly positively, that speaks for itself. EES projects have been given a clear structure. Several universities participate in the international network; research on this is independent and has generated budgets in millions. These are good foundations. The new ideas, therefore, have a chance to spread—according to Max Planck, who once remarked: "A new, great scientific idea does not tend to establish itself in such a way that its opponents are gradually convinced and converted—it is a great rarity for a Saul to become a Paul—but rather in such a way that the opponents gradually die out and that the next generation is made familiar with the idea from the outset" (Planck 1958).

The following are the four pillars of the Extended Evolutionary Synthesis (EES): (1) evolutionary development (evo-devo), (2) developmental plasticity, a partial aspect of evo-devo, (3) inclusive inheritance, and (4) niche construction (Laland et al. 2015). Beyond their empirical relevance, the evolutionary mechanisms discovered in evo-devo research have far-reaching consequences for how the synthetic evolutionary theory must be reevaluated. Specifically, these consequences arise from (a) biased development or variation, i.e., the mechanisms that allow certain phenotypes to occur statistically more frequently than others, (b) developmental plasticity, i.e., the ability of evolution to produce more than one continuous or discontinuous variable form of morphology, physiology, and behavior in different environmental situations, and (c) agencies in development, i.e., organizational instances of agency that the organism can unfold in a form- and function-shaping manner at the genetic and epigenetic levels (Moczek et al. 2019). These mechanisms, which enable the construction of complex phenotypic variation and innovation, put the role of natural selection into perspective. Successive rounds of selection to achieve greater adaptation to external conditions are therefore not necessary in many cases (Lange et al. 2014). Sometimes phenotypic variation is neither small nor random and is produced via interactions with the environment (Newman 2018).

Another pillar of EES is a more comprehensive understanding of inclusive inheritance. Besides genetic inheritance, the importance of epigenetic and cultural inheritance has gained increasing recognition. Finally, as a fourth pillar, the theory of niche construction describes the ability of organisms to construct, modify, and select components of their environment, such as nests, burrows, and nutrients. Niche construction is seen as an independent adaptive mechanism alongside natural selection, which helps determine the selection pressures to which species are subjected. The theory of niche construction sees a complementary, reciprocal, biased relationship between the organism and the environment. EES provides a structure to the theory of evolution. This new structure includes two key concepts: constructive evolution and reciprocal causality (Laland et al. 2015).

9.2 EVOLUTION FROM TWO DIFFERENT PERSPECTIVES

The foundation explained in this book allows us to examine evolution from two different perspectives, that of the MS, with its traditional population-genetic, gene-centered approach, and that of EES, with its ecological-evolution-oriented approach. According to Laland and coauthors, these different perspectives also allow both theories to exist side by side (Laland et al. 2015).

However, juxtaposition also creates difficulties: the new ideas are extensions of the MS; however, they are also renewals of its foundations. The EES states assumptions, describes processes, and comes to predictions, which neo-Darwinism does not find and which do not fit into its building. Conventional theory, which is focused on random mutation, gene centrism, and gradualist, additive change, provides no platform for biased development or discontinuous evolution. It cannot explain such phenotypic variation (Müller 2020). Alternatively, modern evolutionary theory, recognizes new, constructivist, intrinsic developmental mechanisms or agencies of development and cites reciprocal cause–effect chains, both in evo-devo and niche construction theory. This distinguishes their basic motifs and shows a new theory structure.

Those who reject the MS mainly invoke the fact that its basic assumptions are allegedly not only incomplete but also incorrect. They are considered false as they are dogmatic in character, e.g., assumption that genetic inheritance or natural selection *alone* is the sole mechanism of evolution (Noble et al. 2014). Should two theories coexist, or should one replace another? How should this contentious issue be dealt with? The viewpoint of the observer determines the decision on how "evolution of life" should be considered and what kind of theory structure is needed to depict the topic in a coherent overall context.

The observer is also here as a scientist in a difficult situation. Established theories prove to be robust for him. The cognitive psychologist Daniel Kahneman speaks of an "intellectual weakness of many scientists," which he—a highly respected Nobel Prize winner—has also observed in himself. Once one has accepted a theory and is aware of the fact that it is seen as valid by majority of colleagues in the scene, it is exceedingly difficult to admit its weaknesses. Events that cannot be reconciled with the current theory are then attributed to the fact that there must be an excellent explanation for the correctness of the theory after all, even if this explanation does not exist at all. "When in doubt, one decides in favor of the established theory and trusts the community of experts who believe it to be correct." Kahneman calls this "theory-induced blindness" (Kahneman 1990).

The following two possible perspectives are offered to the observer: with the first perspective, he looks at evolution with its processes and mechanisms rather denotatively, i.e., it can be described for him unambiguously, homogeneously, and constantly. Here, the reductionist method is typical and legitimate. Reductionism is to be rejected only if it is proposed as the only possible explanation (Mitchell 2008). The MS with its gene centrism and with natural selection as the predominant evolutionary mechanism is one way of thinking from the first perspective.

An observer from the second perspective is convinced that one must approach the explanation of evolution with complex, irreducible, multifactorial, and multicausal descriptions and methods. Thus, as he is well aware, he does not have before him a self-contained, unambiguous model but is faced with the task of theoretically conceptualizing and justifying his variable object of inquiry. Factors such as heredity or epigenetics can always be weighted differently here, contexts can be seen differently again, and unexpected new aspects can emerge (Schülein and Reitze 2005; Mitchell 2008; Lange 2017). The multifaceted, postmodern EES corresponds to such a categorization of theory. Consequently, the contributions here are typically more heterogeneous. They are open and variable. Recognition of a diversity of causal structures, levels of inquiry, and partial theories forces pluralistic explanatory strategies, according to Sandra Mitchell (Mitchell 2008). Pluralistic explanations of nature are consistent with the unity of nature in terms of substantial monism (Section 7.4). This is not true when one theory is reduced to another (Rohrlich 1988); however, such is not the case with the Extended Synthesis. The concern that theory pluralism could turn into an everything-is-possible maze à la Paul Feyerabend has been defused with reference to scientific practice (Mitchell 2008).

9.3 SUMMARY AND OUTLOOK—FOR A THEORETICAL PLURALISM

Few concepts in biology can be explained with a single theory. John Beatty cites Mendelian and non-Mendelian inheritance as examples of such parallel concepts; knowledge of various theories of gene regulation as well as allopatric and other forms of speciation and that of the role of selection and genetic drift in evolution is necessary to understand these concepts. In these cases, the relative importance of individual theories on the same concept, rather than their accuracy/correctness, is debatable. According to Beatty, only one theory need not necessarily be valid in principle for a particular concept of biology (Beatty 1997). Why should science be committed to a method that dictates the search for unitarity when there is no evidence that the results of evolution come about only one way? The renunciation of unitarianism should then, as a consequence, also apply to the theory of evolution and the explanation of evolutionary mechanisms and causalities.

The sciences do not form a unity, and the theory of the methodological unity of science is false. Epistemology must find a form to constructively deal with the pluralism of methods. If scientific ideals of methods change in history, this must be accepted (Rheinberger 2007).

In any case, the information content or explanatory value of the EES is more comprehensive than that of the conventional theory because first, additional processes are described that are not known in the synthetic theory and that it does not inquire about. Second, the Extended Synthesis provides a more comprehensive theory and causal structure. The Synthesis does not have to be discarded; rather, parts of it can be integrated into the Extended Synthesis. Jan Baedke of the Ruhr University Bochum and colleagues presented a theoretical framework for this purpose, which can evaluate the validity of different evolutionary explanations of the same phenomenon. This way, standard criteria can be named why and when evolutionary explanations of the EES are better than those of the Synthesis (Baedke et al. 2020).

Evolution is, therefore, not a black-and-white topic. The comparison of the conventional and new theories is not like the question of whether the Sun revolves around the Earth or the Earth around the Sun. In addition, the days are probably gone when a single scientist like Darwin or Einstein could present an overall theory. Evolution is rather complex; just consider the genetic and nongenetic forms of inheritance, or the questions of what a species is (still being discussed), and whether gene-culture coevolution should be included in the theory of evolution or not. How then should culture be defined? There is no single definition for such terms. They are connotative (Schülein and Reitze 2005). Science theorists Staffan Müller-Wille and Hans-Jörg Rheinberger also point out that complex objects of study, such as organisms and genes, cannot be captured by a single best description, explanation, or definition (Müller-Wille and Rheinberger 2009). Numerous researchers contribute details that must be conscientiously assembled into the most coherent overall picture possible. However, this is difficult.

Even in the natural sciences—including biology and medicine—what exactly constitutes a scientific fact is more often ambiguous. As an illustration, we only need to follow the stirring booklet by microbiologist and epistemologist Ludwik Fleck. He analyzes how medicine sees a given disease as something entirely different several times over the course of a century (Fleck 1990). Fleck's findings are also congruent with Werner Heisenberg's thesis, which serves as the motto at the beginning of my book. According to it, what we observe is not nature itself but nature exposed to our method of questioning. Today, like Donna Haraway, one goes even further and states that knowledge is not only discovered but also made (Loh 2019). Elsewhere, scientific activity is said to be a fierce struggle to construct reality (Woolgar and Latour 1986). Mitchell thus calls for an integrative pluralism in the complex scenarios we are dealing with that do justice to the pluralism of causes, levels, and the bringing together of individual theories (Mitchell 2008).

It becomes clear that "meaning and validity of linguistic expressions in all forms of their occurrence, procedures, and structures of concept and theory formation" are the subject of philosophy (Janich 2008). Natural science needs philosophy. It is therefore for good reasons that the EES works closely with philosophers.

Charles Darwin is seen as a representative of the second group mentioned above, while his successors rather belong to the first group. Darwin would probably be open to the new ideas of today. Like hardly any other before or after him, Darwin always critically questioned his own views in his books and letters and was open to alternative aspects. Simultaneously, he always had a curious view of what could not yet be explained and of what was yet to be explored by future generations. In the introduction to the sixth edition of *Origin of Species*, he expressly emphasized that "natural selection has been the main but not exclusive means of modification" (Darwin 1859).

I imagine Charles Darwin sitting at his large, dark desk in his downstairs study at Down House with a young visitor from the evo-devo school. Earlier, he has taken his daily walk on the sand walk and tuned into his visitor in thought. A young graduate student wants to talk to him about the role of the embryo in evolution. Even in his old age, Darwin still answers every letter. But such a young scientist visiting Down House, he has never had that pleasure. So, he wrote to her kindly, accepted and invited her.

The young woman has come all the way to England and south London to the rural county of Kent to fulfill her dream of one day discussing the potential of embryonic development for evolution with her great role model. She is tense. What will the old gentleman think about modern ideas? She knows that Darwin had long suspected that development in the embryo must play a role in evolution. But he couldn't quite make sense of it, and he just couldn't reconcile embryonic development with his theory. Darwin's questions to his counterpart therefore never cease. He also addresses his favorite topic, earthworms, which he has studied intensively for more than four decades and well into old age. His conclusion: the worms create the prerequisites for agriculture by working the soil.

The young woman immediately thinks of niche construction; earthworms build a niche for themselves and other creatures. But she restrains herself and enjoys experiencing the wise man enthusiastically in his profession.

There, just as the young woman takes advantage of a pause to dissect what is known today about developmental genes and about innovations that the embryo can produce, just as the conversation begins to get really interesting, Darwin's wife Emma asks them both to come next door to the table. But her husband almost always overhears the call and the bell when a new idea won't let him go, and so it was on this sunny late spring day.

REFERENCES

Baedke J, Fábregas-Tejeda A, Vergara-Silva F (2020) Does the extended evolutionary synthesis entail extended explanatory power? *Biol Philos* 35(1):1–22. https://link.springer.com/article/10.1007/s10539-020-9736-5.

Beatty J (1997) Why do biologists argue like they do? *Philos Sci* 64(Proceedings):432–443.

Darwin C (1859) *On the Origin of Species by Means of Natural Selection, or the Preservation of Favoured Races in the Struggle for Life*. John Murray, London. http://test.darwin-online.org.uk/contents.html#origin.

Fleck L (1990) *Entstehung und Entwicklung einer wissenschaftlichen Tatsache. Einführung in die Lehre vom Denkstil und Denkkollektiv*. Suhrkamp, Frankfurt a. M.

Janich P (2008) Naturwissenschaft vom Menschen versus Philosophie. In: Janich P (ed.) *Naturalismus und Menschenbild*. Felix Meiner, Hamburg.

Kahneman D (1990) *Thinking, Fast and Slow*. Farrar, Straus and Giroux, New York.

Laland KN, Uller T, Feldman M, Sterelny K, Müller GB, Moczek A, Jablonka E, Odling-Smee J (2015) The extended evolutionary synthesis: its structure, assumptions and predictions. *Proc Royal Soc B* 282:1019.

Lange A (2017) *Darwins Erbe im Umbau. Die Säulen der Erweiterten Synthese in der Evolutionstheorie*, 2nd ext, ed. Königshausen & Neumann, Würzburg (eBook).

Lange A, Nemeschkal HL, Müller GB (2014) Biased polyphenism in polydactylous cats carrying a single point mutation: the Hemingway model for digit novelty. Evol Biol 41(2):262–275.

Loh J (2019) *Trans- und Posthumanismus zur Einführung*. Junius, Hamburg.

Mitchell S (2008) *Komplexitäten. Warum wir erst anfangen, die Welt zu verstehen*. Suhrkamp, Berlin.

Müller GB (2020) Evo-devo's contributions to the extended evolutionary synthesis. In: Nuno de la Rosa L, Müller GB (eds.) *Evolutionary Developmental Biology: A Reference Guide*. Springer, Basel.

Müller-Wille S, Rheinberger H-J (2009) *Das Gen im Zeitalter der Postgenomik*. Suhrkamp, Berlin.

Moczek AP, Sultan SE, Walsh D, Jernvall J, Gordon DM (2019) Agency in living systems: how organisms actively generate adaptation, resilience and innovation at multiple levels of organization. Proposal for a major grant from the John Templeton Foundation.

Newman SA (2018) Inheritance. In: Nuno de la Rosa L, Müller GB (eds.) *Evolutionary Developmental Biology. A Reference Guide*. Springer, Basel.

Noble D, Jablonka E, Joyner MJ, Müller GB, Omholt SW (2014) Evolution evolves: physiology returns to centre stage. J Physiol 592(11):2237–2244.

Nowotny H, Testa G (2009) *Die gläsernen Gene. Die Erfindung des Individuums im melokularen Zeitalter*. Suhrkamp, Berlin.

Planck M (1958) *Physikalische Abhandlungen und Vorträge*, vol III. Vieweg, Braunschweig.

Rheinberger H-J (2007) *Historische Epistemologie zur Einführung*. Junius, Hamburg.

Rohrlich F (1988) Pluralistic ontology and theory reduction in the physical sciences. Br J Philos Sci 39(3):295–312.

Schülein JA, Reitze S (2005) *Wissenschaftstheorie für Einsteiger*. Facultas, Wien.

Woolgar S, Latour B (1986) *Laboratory Life: The Construction of Scientific Facts*. Princeton University Press, Princeton, NJ.

CHAPTER TEN

The Players of the New Thinking in Evolutionary Theory

The scientists listed here contribute to extensions of the Modern Synthesis and the post-modern theory of evolution or did so during their lifetimes. Some of them have proposed theories that were precursors of the Extended Synthesis, while others do not explicitly advocate a change of theory or do not call themselves evo-devo researchers but publish important research in evo-devo or on topics that fit niche construction and are incorporated into the Extended Synthesis. In addition to the representatives mentioned here, a growing number of others advocate a necessary renewal.

Pere Alberch (1954–1998, Spain)—Biologist specializing in Zoology and Embryology, Professor at Harvard University, and later Director of the National Museum of Natural Sciences, Madrid. Alberch was a pioneer of evo-devo. He is credited with the elaboration of fundamental concepts, including heterochrony and developmental constraints. He recognized the failure to address morphological evolution as a shortcoming of the synthesis. He considered the morphological outcome of mutation to be non-random, even though randomness may exist at the molecular level. He clarified that the respective developmental systems impose constraints on the morphological form that limit the number of possible stable results. In the early 1980s, he already considered constraints as internal organismic mechanisms of the developmental system with an active rather than passive character, whereby evolution can be facilitated on the epigenetic level. He is credited with recognizing that the genetic-level view is not necessarily the most critical view of alterations in developmental processes. Rather, an epigenetic perspective on physiology, where many confounding factors can occur, is critical. He corroborated his theses with empirical experiments on frogs and salamanders, establishing rules for which toes in the embryo are lost first and which are lost last when cells in the limb bud are reduced. He also became known for his study of anomalies. Here, he was concerned with the fact that variations, even evolutionarily disadvantageous ones, occur repeatedly in the same form. Alberch died at the age of 43.

Patrick Bateson (1938–2017, United Kingdom)—Sir Paul Patrick Gordon Bateson was a British zoologist and science writer. He was a Professor of Behavioral Biology at Cambridge University from 1984, Vice-President of the Royal Society, and President of the Zoological Society of London from 2004 to 2014. He was instrumental in establishing the biological discipline of behavioral science in the United Kingdom. Bateson's research focused on the developmental biology of behavior, including the neurobiological basis of the phenomenon of imprinting in birds and the learning behavior of cats and monkeys. In particular, he was interested in the consequences of playful activities of the young in these mammals to the emergence of physical, cognitive, and social behaviors in adults. In addition to academic duties (including serving as rector of King's College, Cambridge, from 1988 to 2003), he considered himself a mediator between the natural sciences and the non-academic public, writing numerous popular science books on behavioral science, developmental biology, and genetics. He was repeatedly in demand as an advisor to the British Parliament. He was knighted in 2003. In 2006, he was elected a member of the American Philosophical Society. Bateson was a member of the EES research program until his death.

Paul Brakefield (b. 1952, UK)—Professor of Zoology at the University of Cambridge. He is an evolutionary biologist specializing in butterflies and other insects. He is particularly interested in the effects of their evolution and physiology on natural selection. Brakefield has shown that evolutionary changes in an organism are determined by the interplay between the environment and genes that control development. Brakefield's experiments on butterfly eyespots are now considered classic in the field of evo-devo. In parallel with his research, Brakefield is the Director of the University Museum of Zoology at the University of Cambridge. He is also the President of the Tropical Biology Association and the Linnean Society of London and became a Fellow of the Royal Society in 2010. He is a member of the EES research program but does not argue that the traditional theory should be changed.

Sean B. Carroll (b. 1960, USA)—Molecular geneticist, geneticist, developmental biologist, and evolutionary biologist. Carroll's research in evo-devo is on the genes that control body structure in animals and how they have changed throughout evolution. He also authored popular science books on this topic and has a column entitled *Remarkable Creatures* in the *New York Times*. In 2010, he became Vice-President of Science Education at the Howard Hughes Medical Institute (HHMI). He wrote a dual biography of Jacques Monod and Albert Camus (*Brave Genius*). Carroll is a consistent Darwinist and believer in the Modern Synthesis. Self-organizability in the form of Turing systems does not exist for him. They are not even necessary from his point of view because the phenotypic form and variation of each species can be created by the genetic toolkit during development.

Niles Eldredge (b. 1943, USA)—Paleontologist. Eldredge was an Associate Professor of Paleontology in the Department of Earth and Environmental Sciences at the City University of New York. In 1972, together with Stephen J. Gould, he proposed the theory of *punctuated equilibria*, a variant of the theory of evolution. It assumes that species evolution proceeds not steadily but in long phases of stability alternating with short, rapid developmental bursts.

John Arthur Endler (b. 1947, Canada)—Ethologist (behaviorist) and evolutionary biologist known for his work on the adaptation of vertebrates to their specific perceptual environments and the way in which animals' sensory abilities and color patterns evolve. Endler performed extensive work on guppies (*Poecilia reticulata*), and in 1975, rediscovered the species aquarists now refer to as "Endler's livebearers" in his honor. He is further known for his experimental work on inducing evolution on a small scale. To this end, he has studied numerous species, including the behavior of bowerbirds in northern Queensland, Australia, in addition to his experiments on guppies. Endler is an associate member of the EES research program.

Marcus W. Feldman (b. 1942, Australia)—Mathematician and theoretical biologist. Feldman is a Professor of biological sciences, director of the Morrison Institute for Population and Resource Studies, and codirector of the Center for Computational, Evolutionary, and Human Genomics (CEHG) at Stanford University, Palo Alto, California. He became known for his mathematical theory of evolution, for his computational studies in evolutionary biology, and as a cofounder of the theory of gene-culture coevolution.

Scott F. Gilbert (b. 1949, USA)—Gilbert is an evolutionary biologist and historian of biology. He is a Professor Emeritus of Biology at Swarthmore College and Finland Distinguished Professor Emeritus at the University of Helsinki. Gilbert is the author of the textbook *Developmental Biology* (first edition 1985, 11th ed. 2016). With David Epel, he co-authored the textbook *Ecological Developmental Biology* (2009/2015). He is one of the initiators of ecological developmental biology as a new biological discipline. Gilbert's research on the history and philosophy of biology concerns the interactions between genetics and embryology, anti-reductionism, the formation of biological disciplines, and bioethics.

Stephen Jay Gould (1941–2002, USA)—Paleontologist, geologist, and evolutionary biologist. Gould is among the best-known and most influential evolutionary biologists of the 20th century. He studied paleontology and evolutionary biology at Columbia University. Gould is the author of successful popular science books including *The Panda's Thumb* (1980). The theory of *punctuated equilibria* (or punctualism), which he developed together with Niles Eldredge, is considered to be Gould's outstanding contribution to the field of evolutionary biology. According to this theory, evolution does not take place in small steps at a constant speed. Rather, rapid change alternates with longer periods of no change (stasis) in relatively short geological phases. In his comprehensive and influential book *Ontogeny and Phylogeny* (1977), Gould explores the relationship between embryonic development and evolution. In two other technical publications, one with Richard C. Lewontin (1979) and the other with Elisabeth Vrba (1982), he argued that characteristics of an organism may have survived throughout evolution without a direct effect on function. They may exist as do the claddings of the round arches (*spandrels*) in Gothic cathedrals, which have no architectural function. He pointed out that natural selection characterizes negative selection and does not positively select certain features because of their function in an adaptationist manner. Gould also repeatedly opposed the idea that evolution indicates progress.

Brian K. Hall (b. 1941, Canada)—Emeritus biologist at Dalhousie University. Hall actively participates in the evo-devo debate on the nature and mechanisms of animal body plan formation. He is particularly interested in the neural crest and the skeletal tissues that arise therefrom. Further, he has written extensively on the history of evolutionary biology and on leading figures in the field. Among numerous other books, in 2012, he published *Evolutionary Developmental Biology (Evo-Devo): Past, Present, and Future*.

Eva Jablonka (b. 1952, Israel) (Figure 10.1)—Polish-born evolutionary theorist and geneticist. She is a Professor at the Cohn Institute at Tel Aviv University. Jablonka emigrated to Israel in 1957. She is known for her publications on various forms of epigenetic inheritance. Together with Marion J. Lamb, she authored *Evolution in Four Dimensions: Genetics, Epigenetics, Behavior, and Symbolic Variation in the History of Life* (2005). In this book, the authors take neo-Lamarckian positions on heredity that extend far beyond the neo-Darwinian synthesis. Jablonka is an associate member of the EES research program.

Marc W. Kirschner (b. 1945, USA) (Figure 10.2)—Biologist. Kirschner graduated from Northwestern University in 1966 and received his PhD from the University of California, Berkeley in 1971. In 1972, he became an assistant professor at Princeton University; in 1993, he moved to Harvard Medical School. The theory of *Facilitated Variation* is an explanatory model that sees itself as a complement to evolutionary theory and is concerned with the nature of variation in evolution. This theory was published in 2005 by Kirschner, as Founder and Chairman of the Systems Biology Department at Harvard Medical School, and John Gerhart of the University of California, Berkeley, in their book *The Plausibility of Life. Resolving Darwin's Dilemma*.

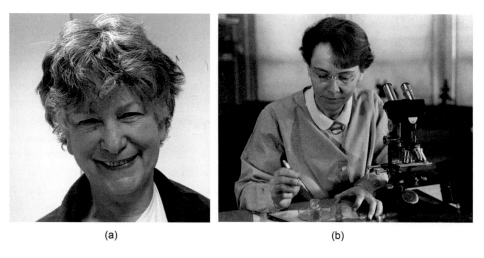

Figure 10.1 (a) Eva Jablonka and (b) Barbara McClintock.

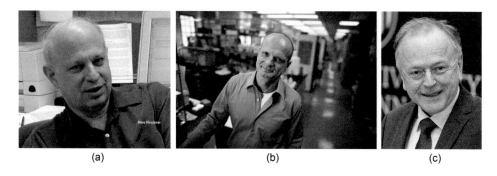

Figure 10.2 (a) Marc Kirschner, (b) Armin P. Moczek, and (c) Gerd B. Müller.

Kevin N. Laland (b. 1962, UK) (Figure 10.3)—Professor of Behavioral and Evolutionary Biology at the University of St. Andrews, Scotland (2002–present). His publications, including numerous books, focus on animal behavior and evolution, specifically social learning, gene-culture coevolution, and niche construction. In 2018, he published the book *Darwin's Unfinished Symphony: How Culture Made the Human Mind*. Laland is an elected Fellow of the Royal Society of Edinburgh, Fellow of the Society of Biology, and Director of the EES Research Program.

Manfred D. Laubichler (b. 1969, Austria)—Professor of Theoretical Biology and History of Biology at the School of Life Sciences and Director of the Global Biosocial Complexity Initiative at Arizona State University. He is a co-editor of *From Embryology to Evo-Devo* (2007), *Modeling Biology* (2007), *The High Seat of Knowledge* (2006), and *Form and Function in Developmental Evolution* (2009). Although he is an evo-devo scientist, Laubichler does not advocate for a theoretical renewal.

Richard C. Lewontin (b. 1929, USA)—Evolutionary biologist, mathematician, geneticist. Lewontin was a student of Theodosius Dobzhansky, one of the founders of the Modern Synthesis. He is among the best-known evolutionary biologists of the 20th century. He held the Alexander Gassiz Professorship of Zoology and Biology at Harvard University from 1973 to 1998 and became a Research Professor there in 2003. In addition to outstanding achievements in the fields of population genetics and molecular biology, Lewontin emerged with criticisms of mainstream evolutionary theory. In 1979,

Figure 10.3 (a) Kevin N. Laland and (b) John Odling-Smee.

he and Stephen J. Gould used the term *spandrel*, borrowed from architecture, as a proxy for nonadaptive phenotypic traits. This countered the tenets of neo-Darwinian theory that all traits are selected and have a function. Further, Lewontin was one of the first biologists in 1970 to introduce hierarchies of selection, from the genetic to multiple epigenetic levels. By outlining the dependence of the environment on the organisms that inhabit it, he introduced a novel line of reasoning into evolutionary theory, from which the theory of niche construction developed.

Lynn Margulis (1938–2011, USA)—Biologist. She developed the endosymbiont theory (symbiogenesis) on the origin of plastids and mitochondria as previously independent prokaryotic organisms. According to this theory, the former entered into a symbiotic relationship with other prokaryotic cells early in evolutionary history, causing the latter to evolve into eukaryotic cells. Margulis was elected to the National Academy of Sciences in 1983 and to the American Academy of Arts and Sciences in 1998. In 1999, she was awarded the National Medal of Science.

Barbara McClintock (1902–1992, USA) (Figure 10.1)—Geneticist and botanist. She discovered transposons (jumping genes), for which she received the Nobel Prize in 1983, after her teaching had been ignored and rejected for almost 30 years. Thanks to her research, it is now recognized that all genomes contain elements that can cause remodeling of their own genome.

Allessandro Minelli (b. 1948, Italy)—Professor Emeritus of Zoology at the University of Padua. He provided a conceptual foundation for the evo-devo discipline. In doing so, he strictly opposed so-called "adult-centrism, "the view that development is (exclusively) considered a series of stepwise processes from the fertilized egg to the adult phenotype." Some life forms proceed through more complex cycles. In addition to the early stages of embryonic development, he believes evo-devo should include post-embryonic stages in the field of view of comparative morphology. Evo-devo, in his view, should be expanded from its overemphasis on the animal kingdom to include other kingdoms. Minelli is the author of *Biological Systematics*

(1993), *The Development of Animal Form* (2003), *Perspectives in Animal Phylogeny and Evolution* (2009), and *Forms of Becoming* (2009). He does not advocate a shift away from traditional evolutionary theory.

Armin Moczek (b. 1969, Germany) (Figure 10.2)—Professor of Biology at Indiana University Bloomington. Moczek was born in Munich. He studied biology with Bert Hölldobler at the University of Würzburg, then moved to Duke University in 1994, where he received his PhD. There, his collaboration with Fred Nijhout was a formative experience. In 2002, he moved to Arizona University as a postdoctoral fellow. Since that time, Moczek has focused on insects, including extensive work on horned beetles (*Onthophagus*) ., which he began to study in depth at Indiana University in 2004. Moczek studies complex innovations in evolution, which includes the horned beetles.' horns. In particular, he focuses on studying the very early stages of innovation in evolution and the interplay of genetics, development, and ecology in which important transitions in evolution are enabled. He is also a cofounder of an international initiative to broaden traditional perspectives on what determines speed and direction in evolution, taking into account recent advances in the fields of evolutionary developmental biology, developmental plasticity, non-genetic inheritance, and niche formation. Moczek is a member and project leader in the EES research program. His work addresses typical topics in EES.

Gerd B. Müller (b. 1953, Austria) (Figure 10.2)—Evolutionary biologist, student of Rupert Riedl. He was a Professor emeritus at the University of Vienna and the head of the Department of Theoretical Biology at the Center for Organismal Systems Biology at the University of Vienna until 2018. Müller studied Medicine and Zoology in Vienna. His research interests include vertebrate limb evolution, the origin of evolutionary innovations, theoretical integration of evo-devo, and the EES. Müller is a founding member and has been president of the Konrad Lorenz Institute for Evolutionary and Cognitive Research in Klosterneuburg near Vienna since 1997. He is president of EuroEvoDevo, the European Society for Evolutionary Developmental Biology. Together with Stuart Newman, Müller edited the book *Origination of Organismal Form* (2003). Together with Massimo Pigliucci, he is an editor of the volume *Evolution – The Extended Synthesis* (2010). Müller is one of the founders of EES and one of its most consistent proponents. He is an associate member of the EES research program.

Stuart A. Newman (b. 1945, USA)— Professor of Cell Biology and Anatomy at New York Medical College in Valhalla, NY. Newman's research is focused on three areas: cellular and molecular mechanisms of vertebrate limb development, physical mechanisms of morphogenesis, and mechanisms of morphological evolution. His work in developmental biology involves a mechanism for pattern formation of the vertebrate limb skeleton based on self-organization of embryonic tissue. Several computer models were created in his school for this purpose. He also described a biophysical effect in the extracellular matrix populated with cells or nonliving particles, i.e., matrix-directed translocation, which provides a physical model for stem cell tissue morphogenesis. Newman and Müller co-edited the book *Origination of Organismal Form* (2003). Newman is one of the most consistent EES proponents and clearly states that the basic assumptions of the Modern Synthesis are incorrect.

H. Frederic Nijhout (b. 1947, USA)— Evolutionary biologist and a Professor of Biology at Duke University (Durham, North Carolina). His research focuses on evolutionary developmental biology and entomology (insect research), with particular emphasis on hormonal control of growth, molting, and metamorphosis in insects, including mechanisms that guide the evolution of alternative phenotypes. Much of his work also focuses on the evolution of wing patterns.

Denis Noble (b. 1936, Great Britain) (Figure 10.4)—British physiologist. Noble is one of the pioneers of systems biology. He studied at University College London. In his highly regarded doctoral thesis (1961), he developed the first mathematical model of the working heart. Noble held the Burdon-Sanderson Chair in Cardiovascular Physiology at Oxford University from 1984 to 2004. |||In 2006, he published *The Music of Life*, the first popular science book on systems biology, and in 2016, also from a systems biology perspective, *Dance to the Tune of Life: Biological Relativity*. In both, he critiques Neo-Darwinism with its ideas of genetic determinism and reductionism, as found most radically in Dawkins' theory of the selfish gene. He posits that because of various feedback mechanisms (e.g., niche formation, non-genetic inheritance, splicing, and epigenetics), the genome should not be emphasized as a level of organization and especially not as a program from which the function of proteins, cells, or even organs can be inferred by taking a reductionist approach. Instead, he proposes a systemic approach to organisms, with equal access to all levels of organization. Noble has served as president of the International Union of Physiological Sciences and of the Virtual Physiological Human Institute. He is an associate member of the EES research program.

John Odling-Smee (b. 1935, UK) (Figure 10.3)—Professor emeritus of Biology and Anthropology at Oxford University. Odling-Smee published more than 100 articles on animal learning and its role in evolution as well as on the theory of niche construction, which he established with Kevin N. Laland and Marcus W. Feldman. Their joint monograph *Niche Construction: The Neglected Process in Evolution* was published in 2003. Odling-Smee is a member of the EES research program.

Massimo Pigliucci (b. 1964, USA)—Pigliucci is the K.D. Irani Professor of Philosophy at City College and Professor of Philosophy at the Graduate Center of the City University of New York. He received his PhD in evolutionary biology from the University of Connecticut and his PhD in philosophy from the University of Tennessee and subsequently conducted research in evolutionary ecology at Brown University. His interests include the philosophy of biology, the relationship between science and philosophy, and the nature of pseudoscience. Pigliucci is co-editor, with Gerd B. Müller, of the book *Evolution – The Extended Synthesis* (2010). He is a member of the EES research program.

Rudolf Raff (1941–2019, Canada)—Professor of Biology at Indiana University and Director of the Indiana Molecular Biology Institute. He was a pioneer in the emergence of evo-devo as a new research discipline. In the process, he became a leading force in integrating the fields of evolution and development. He inspired a new generation of scientists.

Rupert Riedl (1925–2005, Austria) (Figure 10.4)—Former zoologist and systems biologist at the University of Vienna. Riedl viewed natural history, particularly evolutionary organismal development, as a system of interconnected relationships. Since the late 1970s, he was one of the first scientists to argue that the Modern Synthesis neglects the role of development and morphology in evolution. He argued that it failed to explain the emergence of body plans and patterns at the macroevolutionary level. He detailed his theory in *Order in Living Organisms: A Systems Analysis of Evolution* in 1978. This work revolutionized the current understanding of causality. Riedl argued that although every organism in embryonic development emerges from the activity of genes, in evolution the "scope of action" of genes is limited by their functional interdependence, similar to that of individuals in an organization.

James Alan Shapiro, (b. 1943, USA)—Biologist specializing in bacterial genetics, Professor at the University of Chicago in the Department of Biochemistry and Molecular Biology. Shapiro was a colleague of Jon Beckwith at Harvard when a gene was first

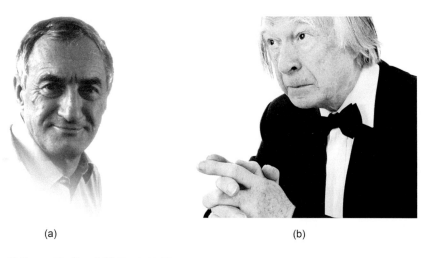

Figure 10.4 (a) Rupert Riedl and (b) Denis Noble.

isolated (in *Escherichia coli*) in 1969. Following his discovery of transposable elements in bacteria in 1979, he was instrumental in organizing the field of mobile genetic element research and was the earliest proponent of *natural genetic engineering* as a fundamental component of evolution and evolutionary innovation. He used the engineering view to make the case that genetic variation can have a non-random (directional) character. Shapiro also demonstrated cooperative behavior in bacteria. In his book *Evolution – A View from the 21st Century* (2011/2022), he advocates an expanded view of evolutionary theory that takes into account the cognitive, flexible, cooperative capabilities of cells based on "natural genetic engineering" for rapid evolutionary adaptation and innovation.

Kim Sterelny (b. 1950, Australia)—Philosopher. Sterelny's research focuses on the philosophy of biology. He views the development of evolutionary biology since 1859 as one of the great intellectual achievements of science. Sterelny is the author of numerous publications on group selection, meme theory and cultural evolution, such as *The Return of the Gene* (with Philip Kitcher, 1988), *Memes Revisited* (2006), and *The Evolution and Evolvability of Culture* (2006). In 2004, Sterelny's book *Thought in a Hostile World: The Evolution of Human Cognition* received the Lakatos Prize for an outstanding contribution to the philosophy of science. This book offers a Darwinian account of the nature and evolution of human cognitive abilities and is an important alternative to the nativist accounts familiar to the field of evolutionary psychology. His lectures are published under the title *The Evolved Apprentice*. These lectures build on the non-nativist, Darwinian approach to thinking in a hostile world while providing a discussion of more recent work by other philosophers, biological anthropologists and ecologists, gene-culture co-evolutionists, and evolutionary theorists.

John E. Stewart (b. 1952, Australia)—Evolutionary theorist and activist who is a member of the Evolution, Complexity and Cognition Research Group of the Free University of Brussels. As an evolutionary theorist, his main focus has been on the trajectory of evolution. His work on the directionality of evolution and its implications for humanity has been published in a number of key papers in international science journals. He is the author of the internationally acclaimed book *Evolution's Arrow: the direction of evolution and the future of humanity*. In 2008, he was a keynote speaker at the first international scientific conference on The Evolution and Development of the Universe held in Paris.

Sonia E. Sultan (b. 1958, USA)—Evolutionary plant ecologist at Westleyan

University, USA. Her research group focuses on ecological development or eco-devo. Sultan contributed to the empirical and conceptual literature on individual plasticity and its relationship to ecological breadth and adaptive evolution. In 2015, she published these ideas in a book titled *Organism and Environment: Ecological Development, Niche Construction and Adaptation*. Sultan's current experimental work focuses on inherited effects of the parental environment on development, the relationship of individual plasticity to invasiveness, and the role of DNA methylation as a mechanism regulating responses to the environment.

Eörs Szathmáry (b. 1959, Hungary)—Theoretical biologist. He deals with various fields such as the origin of life, the mathematical description of early stages of evolution, the origin and optimal size of the genetic code, and the evolution of language. Together with John Maynard Smith, he wrote the acclaimed book *The Major Transitions in Evolution* in 1995 and *The Origins of Life* in 1999.

Michael Tomasello (b. 1950, USA)—Anthropologist and behavioral scientist. After studying psychology at Duke University and earning a PhD in experimental psychology at the University of Georgia, he taught psychology at Emory University from 1980 to 1998. From 1998 to 2018, he was a codirector at the Max Planck Institute for Evolutionary Anthropology in Leipzig, where he directed the Wolfgang Köhler Primate Research Center. From 1999 to 2018, he was an honorary professor at the University of Leipzig. Tomasello focuses on the evolution of human language and thought and describes the differences between humans and animals. This led him to develop the concept of shared intentionality. His most recent authored works include *A Natural Theory of Human Thinking* (2014), *A Natural Theory of Human Morality* (2018), and *Becoming Human, A Theory of Ontogeny* (2019).

Alan Turing (1912–1954, Great Britain)—Turing was one of the most influential theorists of early computer development and computer science. He created much of the theoretical basis of modern information and computer technology. His contributions to theoretical biology were seminal. His 1952 paper *The chemical basis of morphogenesis* is now cited about 2,500 times a year in Google Scholar and is considered the landmark paper on biological pattern formation. While Turing's original intent was to use the reaction-diffusion process of a so-called Turing system described by him to explain surface patterns, such as those on cows, zebras, fish, or other animals, Turing models are now increasingly used in a mathematically modified and extended form to describe the emergence of three-dimensional, stable wave structures in the organism as self-organization, for example, in limb development. In this context, the reaction-diffusion behavior in the Turing mechanism is transferred from the chemical, molecular level to the intercellular level and argued with cell signals. In Germany, Turing's idea was adopted and further developed by Alfred Gierer and Hans Meinhardt from 1972 onwards.

Tobias Uller (b. 1977, Sweden)—Professor of Evolutionary Biology at Lund University, Sweden. Uller's research focuses on the interface between evolutionary biology, developmental biology, and ecology. His projects aim to reveal how functional processes in the development, physiology, and behavior of organisms influence their evolution. This includes the role of phenotypic plasticity in adaptive diversification; the evolutionary causes and consequences of extragenetic inheritance; and the genetic, developmental, and ecological factors underlying evolution through introgressive hybridization (movement of a gene or chromosome from one species to another). Uller is a deputy project leader of the EES research program.

Conrad Hal Waddington (1905–1975, Great Britain)—British developmental biologist, paleontologist, geneticist, embryologist, and philosopher. Waddington produced fundamental work on developmental biology and epigenetics. His work is considered an important precursor of today's evolutionary developmental biology and has been undergoing something of a renaissance since the 1990s. The terms such as epigenetic landscape, channeling, and genetic assimilation, introduced by Waddington, are now common in evo-devo.

Andreas Wagner (b. 1967, Austria/USA)—Evolutionary biologist, Professor at the Institute of Evolutionary Biology and Environmental Sciences, University of Zurich. Since 1999, he has also held a professorship at the Santa Fe Institute, New Mexico (USA). Wagner is known for his work on the role of robustness and innovation in evolution. In 2014, he published the book *The Arrival of the Fittest. How Nature Innovates.*

Günter P. Wagner (b. 1954, Austria)—Evolutionary biologist and ecologist. Wagner was a student of Rupert Riedl, Vienna. He teaches at Yale University. His research focuses on the evolution of complex traits, using both the theoretical tools of population genetics and experimental approaches in evolutionary developmental biology. Wagner has contributed significantly to the current understanding of the evolvability of complex organisms, the emergence of innovations, and modularity. He is an associate member of the EES research program. Wagner does not advocate the view that traditional evolutionary theory should be changed.

Mary Jane West-Eberhard (b. 1941, USA)—Emeritus theoretical biologist and entomologist, University of Michigan. She conducted research on social wasps of the tropics and used it to study mechanisms of evolution. For example, she emphasized the role of sexual selection (social competition among male individuals) in speciation and the role of alternative phenotypes as a basis for natural selection in evolutionary theory, for which she coined the term phenotypic plasticity. Phenotypic plasticity in this context is the innate ability of individuals to change their external appearance (phenotype) during development to the adult stage and to adapt to possibly changing environmental conditions. According to her theory, this is the primary starting point of natural selection. Incorporation into the genetic blueprint then occurs successively through random mutations, and individuals in which the altered phenotype is already genetically anchored have a selective advantage. A hallmark of West-Eberhard's theory of plasticity is her core statement: *"Genes are followers, not leaders"*. This clarifies her view that the genotype does not determine the phenotype. The genotype can, due to environmental influences, produce many phenotypes. In her major 800-page work *Developmental Plasticity and Evolution*, published in 2003 and containing numerous empirical studies, West-Eberhard provides a comprehensive critique of the dominant role of natural selection in synthetic evolutionary theory and calls for a new framework for a unified theory of evolution that considers development, the environment, and plasticity as causal factors in evolution. West-Eberhard belongs to the evo-devo discipline, although she does not use the term herself. She received the R.R. Hawkins Award for her book *Plasticity and Evolution* and the Sewall Wright Award, both in 2003 She also received the 2021 Linnean Medal for Zoology, one of the most important distinctions in the field of biology.

David Sloan Wilson (b. 1949, USA) (Figure 10.5)—Professor at Binghampton University, New York. Wilson is a prominent proponent of group selection (the modern variant) in evolutionary theory. He was awarded a professorship in Biological Sciences at the State University of New York in 1988. In 2001, he also became a Distinguished Professor of Anthropology. He is the author of several books, including *Darwin's Cathedral: Evolution, Religion and the Nature of Society* (2003), *Evolution for Everyone: How Darwin's Theory Can Change the Way We Think About Our Lives* (2007), and *This View of Life: Completing the Darwinian Revolution* (2019).

Figure 10.5 (a) David Sloan Wilson and (b) Edward Osborne Wilson.

Edward Osborne Wilson (1929–2021, USA) (Figure 10.5)—Emeritus entomologist, sociobiologist, and evolutionary biologist, University of Alabama and Harvard University. Wilson was among the world's leading evolutionary researchers. In 1975, he founded sociobiology as a new research discipline with a view to study state-forming insects and other animals, including humans, and spoke of *New Synthesis*. With this research, he gained worldwide notoriety. In 2010, Wilson went on the record with his own findings of social evolution, demonstrating fundamental errors. He received numerous international prizes, including the Pulitzer Prize twice: in 1979 and again in 1991 together with Bert Hölldobler for the book *The Ants*. The two authors published another joint work in 2008—the monumental monograph *The Superorganism: The Beauty, Elegance and Strangeness of Insect Societies*. In his old age, Wilson argued vehemently for an anthropocentric environmental and conservation ethic and defended biodiversity, a term also coined by him.

GLOSSARY

Adaptation: Result of natural selection at the population level. Physical and behavioral characteristics are interpreted as evolutionary responses of a population to specific environmental factors (natural selection). Adaptation increases the fitness of the population. The debate over the weighting and effectiveness of adaptation has existed since Darwin. Today, this debate takes a different form. In the Extended Synthesis (⇒ Extended Evolutionary Synthesis), adaptation can also be achieved by internal and external constructive processes, not only in a passive way by the organism (⇒ agent, ⇒ developmental constraints).

Agency: System at a biological level that can train agents for developmental variation. Agencies are found at the level of gene regulatory networks, cell tissues, organisms, or social groups. Agency is an evo-devo concept.

Agent: Concept in ⇒ evo-devo, according to which the organism is based on genetic and epigenetic organizational ⇒ agencies of form- and function-forming activity for developmental variation. The concept of the agent extends the earlier notion of ⇒ developmental constraints. With the introduction of the agent, the organism is no longer a mere passive object in the evolutionary process, but the subject itself.

Allele: A variant of a ⇒ gene that can differ between individuals of a species. Different alleles often produce different phenotypic traits at the individual level. For example, a gene that determines flower color may have an allele conferring red color and another conferring white color on flowers. Contrary to earlier opinions, an allele usually does not correspond 1:1 to a phenotypic trait expression. As a rule, traits are polygenic. This is also true for eye color, which used to be a reference for a specific allele. However, several genes are involved in pigment formation in the iris.

Allopatric speciation: Central process of speciation by the geographic splitting of a population into two reproductively isolated populations. Other forms of speciation also occur.

Alternative splicing: ⇒ Splicing

Amino acids: Organic molecules and building blocks of the ⇒ proteins of living things. The ⇒ genetic code specifies the incorporation of 20 different amino acids into proteins.

Analogy: ⇒ Convergence

Apoptosis, programmed cell death: Mechanism of embryonic development. Apoptosis is triggered by signals that activate cascade of "suicide proteins" in cells, resulting in self-destruction of the cell. For example, during limb development in some quadrupeds, the cellular material between the fingers and toes is removed by apoptosis.

Arrival of the Fittest, making of the Fittest: Processes that, in addition to *Survival of the Fittest*, describe ⇒ evo-devo pathways that lead to the production of a trait, rather than referring only to the trait in its final form.

Arthropods: Members of the most speciose phylum of the animal kingdom. They include such diverse animals as insects, millipedes, crustaceans, spiders, scorpions, centipedes, and the extinct trilobites. They are characterized by an exoskeleton of chitin, molting, articulated extremities, and body segments. About 80% of all known living animal species are arthropods, most of them insects.

Artificial intelligence (AI): Branch of computer science. An adaptive system that can independently and efficiently solve problems or achieve ⇒ complex goals. AI systems perform cognitive tasks that were previously performed by humans. A distinction is made between weak AI and strong AI. A weak AI engine is focused on solving a specialized task, e.g., autonomous driving, playing chess, or similar. A strong AI engine can solve several tasks using a process comparable to human cognition and possesses, at the highest conceivable but as yet unrealized level, additional forms of consciousness, self-awareness, sentience, emotions, and morality (⇒ superintelligence).

Atavism: Reappearance in the evolutionary history of a species of anatomical features that existed earlier, e.g., multiple toes in the horse.

Autosomal dominant inheritance: Form of inheritance in which the altered ⇒ allele must be present on only one of the two homologous ⇒ chromosomes, which are not involved in sex determination, for a trait to be ⇒ phenotypically expressed or for a disease to be detected. If the mutation must be present on both chromosomes, the inheritance pattern is referred to as autosomal recessive.

Base pair: two bases in ⇒ DNA or ⇒ RNA that is complementary to each other. The number of base pairs in a ⇒ gene represents an important measure of the information stored in the gene. DNA comprises the four nucleic bases adenine, cytosine, guanine, and thymine, usually called A, C, G, and T for short. A and T always occur as a pair, as do C and G.

Bias: ⇒ Developmental bias

Biodiversity, diversity: Measure of the diversity of living organisms but also of genetic information and of the proteins formed in living organisms.

Buffering: ⇒ Canalization

Cambrian period: In the Earth's chronostratigraphic system. Corresponds approximately to the period 542 to 488 million years ago. During the Cambrian period, the approximately 30 "blueprints" known today in the animal world were formed. This process, unique in evolutionary history, is called the Cambrian explosion.

Canalization: Term introduced by Waddington in 1942. Refers to the response of development to certain changes caused by external stimuli or genetic ⇒ mutation in such a way that the ⇒ phenotypic output is maintained unchanged.

⇒ Development readjusts according to the perturbation. Mutations are buffered without having a phenotypic consequence. Only a sufficiently strong stimulus can lead to ⇒ decanalization and thus to an altered phenotype.

Causal-mechanistic explanatory claim: Effort of ⇒ evo-devo to explain evolution causally, not by population-statistical correlations but instead by mechanisms in development. These include ⇒ facilitated variation, developmental ⇒ biases, ⇒ developmental plasticity, ⇒ niche construction, and inclusive inheritance.

Cell: Elementary unit of all living things. There are unicellular organisms (prokaryotes), which consist of a single cell, and multicellular organisms (eukaryotes), in which several cells are connected to form a functional unit with division of labor. The human body consists of about 220 different cell and tissue types and about 100 trillion cells. Thereby, cells have relinquished their independence by division of labor (specialization) and are predominantly not viable individually. The size of cells varies greatly (1–30 µm). Each cell represents a structurally definable, independent, and self-sustaining system in the cell network. It can absorb nutrients, convert them into energy, perform various functions, and, most importantly, reproduce itself. The cell contains all the information necessary for these functions or activities. All cells possess the general characteristics of life. The most important are reproduction by cell division (⇒ mitosis and ⇒ meiosis) and metabolism.

Cell division: ⇒ Meiosis, ⇒ Mitosis

Cell membrane: Semi-permeable biomembrane that surrounds the living ⇒ cell, encloses its internal milieu, and maintains it. Cells communicate with the extracellular environment via the cell membrane.

Cell signaling: ⇒ Signal transduction, ⇒ Morphogens

Cellular automaton: Term used to model the pattern formation of spatially discrete dynamic systems, where the development of individual cells at time $t+1$ depends primarily on the cell states in the given neighborhood and on their own state at time t. A cellular automaton does not have a central computational rule for a particular pattern. Information necessary for pattern formation is available in cells and mathematically as positional parameters for individual cells in equations.

Central dogma: of molecular biology hypothesis published in 1958 by Francis Crick about the possible flow of information between ⇒ DNA, ⇒ RNA, and ⇒ protein. "Once (sequential) information has been passed into protein, it cannot get out again." In 1970, Crick formulated an alternative central dogma: "The central dogma of molecular biology deals with the detailed residue-by-residue transfer of sequential information. It states that such information cannot be transferred from protein to either protein or nucleic acid." However, representatives of ⇒ systems biology emphasize various regulatory feedback mechanisms of proteins and ⇒ nucleic acids. These require a cell to be treated as a complex network in which information transfer of a sequential nature is no longer emphasized. From this point of view, the central dogma only partially describes information flow, and its use to justify a reductionist research methodology (⇒reductionism) that seeks to understand organisms in a bottom-up causal approach that starts with genes has been criticized.

Chance: Most often used in the context of chance ⇒ mutation. A mutation may also be non-random. ⇒ the Modern Synthesis refers to chance primarily in the sense that a mutation is random, not directed, with respect to its selection effect. Since in the ⇒ Extended Evolutionary Synthesis the causal chain of chance mutation and natural selection is decisively broken, chance loses relative importance from this point of view. On the other hand, evolution actively uses stochasticity, which is illustrated in self-organization models.

Chance mutation: ⇒ Chance

Chaperone: Protein that assists other proteins in folding correctly.

Chromatid: ⇒ Chromosome

Chromatin: Material of chromosomes, consisting of DNA and proteins.

Chromosome: Structure that contains ⇒ genes and thus the hereditary information. A chromosome consists of ⇒ DNA wound around ⇒ proteins (⇒ chromatin) into structures called nucleosomes. Chromosomes are found in the ⇒ nuclei of the cells of ⇒ eukaryotes, which include all animals, plants, and fungi. A eukaryotic chromosome can only be visualized during division of the nucleus when it takes on a rod-like appearance in humans and many other species. Until the next round of ⇒ mitosis, it consists of two identical chromatids, which are joined at the centromere (sister chromatids).

Cis-element: (from the Latin *cis*, "this side") A specific section on ⇒ DNA that plays a role in the regulation of a ⇒ gene located on the same DNA molecule (⇒ chromosome) as the cis-element.

Coevolution: Parallel ⇒ phylogenetic evolution of two or more mutually dependent traits or species, e.g., male and female sex characteristics.

Compartment: Region of the ⇒ embryo in which one or more selector genes are exclusively expressed, and one or a few signaling proteins are produced. This results in cell specialization within the relevant compartment. Embryos can be analyzed according to compartment maps.

Complexity: Property of a system whereby its overall behavior cannot be described by complete information about its individual components and their interactions. Homogeneous initial conditions can produce non-homogeneous (complex) patterns or structures through the local activity of the elements. Complex systems do not allow exact predictions and cannot be completely mastered/controlled. They are further characterized by properties such as multicausality:multicausality, inherent dynamics, self-regulation, ⇒ robustness or instability, uncertainty, non-linearity, feedback (⇒ reciprocity), macrodetermination, etc. ⇒ In this sense, development and evolution are complex systems. They can be analyzed with methods of complexity theory.

Complexity theory: Subfield of theoretical computer science. Complexity theory deals with the complexity of formally treatable problems with various mathematically defined algorithms and models. These include chaos theory, ⇒ Turing systems, principles of ⇒ self-organization in ⇒ cellular automata, genetic algorithms, evolving systems, the evolution of ⇒ cooperation, and network systems.

Constraints: ⇒ Developmental constraints

Convergence: The evolution of similar traits in unrelated species that were formed during evolution by adaptation to a similar function and similar environmental conditions. It follows that forms observed in different organisms can be traced directly to their function in the organism and do not necessarily infer a close relationship. Features that evolve because of convergence are called convergent or analogous features, such as insect wings and the wings of birds, the trunks of elephants and tapirs, or fins of fish and whales (↔ homology).

Cooperation: Evolutionary mechanism or factor recognized as such in the second half of the 20th century and described in the context of game theory and sociobiology. Most recently, Martin Nowak has presented mechanisms that explain why people do not only have their own advantage in mind but also help each other and are willing to make sacrifices and set aside their own needs for the greater good. Cooperation can outweigh the fitness of a single individual. The group can then be stronger in the struggle for survival than a single individual. Cooperation is found at all biological levels: in the genome, cells, microorganisms, state-building insects, and mammals.

Cooption: Use of existing ⇒ genes in a new context.

Copy error: An error occurring during the duplication of ⇒ DNA (replication) in the course of cell division (meiosis). It occurs, for example, when an incorrect, i.e., non-complementary, base is added to a separated DNA single strand (⇒ RNA) during the addition of the complementary bases (⇒ base pair). The result is a base sequence of the newly formed DNA double strand that is not identical to the original DNA double strand. After repair, the fidelity is about one error per billion base connections. This corresponds to about one typing error in about 500,000 typed pages. A copy error can be evolutionarily beneficial, neutral (most common case), or detrimental. In the latter case, it can result in severe damage to daughter cells. ⇒ Cells have DNA repair mechanisms to correct copy errors.

Core processes, preserved: Cell processes that produce anatomical, physiological, and behavioral features of the organism during ⇒ development. The various ⇒ phenotypic traits are generated by different combinations of core processes. Some core processes are unchanged for many hundreds of millions of years. Core processes are not restricted to events in the ⇒ nucleus. They also include, among other things, the ⇒ cytoskeleton of the ⇒ cell, i.e., its internal structural arrangement, metabolic reactions in the cell, and ⇒ signal transduction mechanisms.

CRISPR/Cas: ⇒ Genome editing

Crossing-over: Mutual exchange of DNA segments between non-sister chromatids during sexually induced cell division (⇒ meiosis).

Culture: Information that can influence the behavior of individuals acquired from other members of their species through training, imitation, and other forms of social transmission.

Cultural inheritance: All forms of transmission of knowledge within and across generations (⇒ including inheritance).

Cytoplasm: In ⇒ eukaryotes, the content filling the ⇒ cell (excluding the ⇒ nucleus). It is enclosed by the ⇒ cell membrane. Within the cytoplasm, chemical metabolic processes of the cell take place, which are controlled by ⇒ enzymes. In addition, there are cell-specific tasks, such as the formation of additional cell components during growth, degradation of harmful substances, buildup of substances to be stored or released, and transport of molecules through the cell and the membrane.

Decanalization: Term coined by C.H. Waddington for developmental change caused by a sufficiently strong ⇒ environmental stressor or ⇒ mutation that results in ⇒ phenotypic change. An example is ⇒ polydactyly.

Deoxyribonucleic acid: ⇒ DNA

Determinism: Assumption that strict, non-probabilistic laws of nature govern all natural processes. A system is called deterministic if each state is uniquely determined by its law of evolution. In evolutionary theory, the relationship between genotype and phenotype was originally viewed deterministically. Development and evolution are nondeterministic.

Development, ontogenesis: Emergence of the individual living being from the fertilized ovum to the adult living being. Development is the key focus of evo-devo. Understanding its processes and mechanisms provides a foundation for the recognition of evolutionary change. While development used to be seen as genetically programmed, the current emphasis is on the constructive aspect, ⇒ biases, and ⇒ plasticity of development. Today, agents and ⇒ agencies are recognized at the genetic and epigenetic levels in development.

Development reaction norm: Range of ⇒ phenotypic variation that can develop from the same ⇒ genotype under different environmental factors. The development reaction norm is seen as the result of ⇒ natural selection.

Developmental bias: Mechanism of development that makes some ⇒ phenotypes more feasible and thus more likely to occur than others, e.g., finger numbers in ⇒ polydactyly or eyespot size in butterflies. Some phenotypes may also be impossible from this point of view. Important concept in ⇒ evo-devo and ⇒ niche construction theory and thus, the Extended Evolutionary Synthesis.

Developmental constraints: Denote certain boundaries in ⇒ development, set by physics, morphology, or phylogenesis. ⇒ Evo-devo speaks of ⇒ bias and, more recently, of ⇒ agents rather than constraints and thus of active rather than passive factors. Developmental constraints play a prominent role in evo-devo and ⇒ the Extended Evolutionary Synthesis.

Developmental gene: A gene that may be expressed at several stages and serve various functions during embryonic development. The ⇒ Hox genes are key developmental genes, but there are numerous others, including, for example, the *Hedgehog* group as well as growth genes such as the *Bmp* group (for bone morphogenetic proteins), Distal-less (Dll)), or bcd (*Bicoid*). Many of them are ⇒ transcription factors in their ⇒ protein form, which in turn activate other genes in specific ⇒ signal transduction pathways. Many developmental genes were discovered in the fruit fly (*Drosophila*) and are identical or very similar across numerous animal taxa. Thus, they play an essential role in understanding the evolution and relatedness of species. Developmental genes play a prominent role in ⇒ evo-devo and ⇒ the Extended Evolutionary Synthesis.

Developmental plasticity, phenotypic plasticity: Ability of a ⇒ phenotype associated with only one ⇒ genotype to give rise to more than one continuously or discontinuously variable form of morphology, physiology, and behavior during development (and in different environmental situations). The concept of phenotypic plasticity describes the degree to which an organism's phenotype is predetermined by its genotype. Pronounced developmental plasticity means that environmental influences have a strong impact on the individually evolving phenotype. With low plasticity, the phenotype can be reliably predicted from the genotype, regardless of particular environmental conditions during ⇒ development. Developmental plasticity can relate to morphological traits, physiological adaptation, up- or down-regulation of an enzyme level, or behavioral responses. It is a pillar of ⇒ the Extended Evolutionary Synthesis.

Diploid: In genetics, the presence of two complete sets of chromosomes is a so-called double set of chromosomes in the cell nucleus (↔ haploid). Multicellular animals usually develop with a diploid genome.

Discontinuous variation: In contrast to continuous variation, which is associated with a gradual evolutionary process, discontinuous variation manifests as an "all-or-nothing" phenomenon in development, e.g., a completely new finger or toe (⇒ polydactyly), supernumerary ribs, a different number of spines on the back (in sticklebacks), or hairy versus hairless leaves in plants.

Divergence: Divergence of characteristics from different species or even from different populations of the same species.

DNA, deoxyribonucleic acid: Double-stranded, helically coiled macromolecule found in all living organisms and the carrier of genetic information in the cell ⇒ nucleus. Among other things, DNA contains the ⇒ genes that code for ribonucleic acids (⇒ RNA). In a complex mechanism, ⇒ amino acids are formed from RNA and from them ⇒ proteins, which are necessary for the biological development of an organism and the metabolism in the ⇒ cell.

E. coli, *Escherichia coli,* **coliform bacteria:** are found in human and animal intestines.

Eco-evo-devo: Evolutionary developmental biology with special attention to environmental influences that initiate evolutionary variation and accompany it with downstream genetic accommodation. Causal relationships between development, evolution, and the environment are sought here.

Ecology: Branch of biology that studies the interactions of organisms with biotic and abiotic components of their environment. ⇒ Evo-devo considers environmental factors as causal contributors to developmental and evolutionary changes.

Ecological inheritance: Term from ⇒ niche construction theory. Inheritance of biological or non-biological components through physical transformation of the environment by organisms. Organisms leave altered selective environments to their offspring as a result.

Ectopic gene expression: Expression of a gene and emergence of tissue during development at a site in the organism where it does not normally occur.

EES: ⇒ Extended Evolutionary Synthesis

Embryo: Individual living being in the early stages of its development. In animals, the organism in the initial stage of development from a fertilized egg (⇒ zygote) is called an embryo while it is still in the mother or in an egg case or eggshell. Once the internal organs have formed, the embryo is called a fetus.

Emergence: Spontaneous formation of properties or structures at the macro level of a system based on the interaction of its elements at the microlevel. Thereby, the emergent properties of the system cannot be traced back to those of the micro level elements, which they exhibit in isolation. Emergence properties are known to many sciences, including physics, e.g., temperature and material hardness cannot be explained by properties of single atoms or molecules. For example, a single water molecule is not moist, nor is human consciousness present at the level of individual neurons. Emergence contradicts ⇒ reductionism. Emergence is a characteristic property of complex systems and is computable with models. Developmental processes are emergence processes that cannot be predicted from the properties of genes or cells.

Endosymbiosis, horizontal gene transfer: Form of symbiosis or fusion in which the symbiont lives inside its host organism and a single, new organism is formed. The theory states, in simplified terms, that during the evolution of early life forms, the cell of one unicellular organism was "swallowed" by the cell of another unicellular organism, thereby becoming a component of the cell of a higher organism thus formed. Endosymbiosis is a possibility for the emergence of more complex life forms in evolution; the theory was proposed by Lynn Margulis. The endosymbiont theory is an addition to the theory of evolution in that the origin of new cell organelles, organs, or species is attributed to the symbiotic relationship and association between individual species. Accordingly, it follows from endosymbiosis that ⇒ phylogenetic trees can not only branch but also reconnect.

Environmental stressor, environmental stimulator: Internal or external stimulus that requires a response. Development responds to genetic and environmental stressors that can initiate evolutionary changes.

Enzyme: ⇒ Protein that catalyzes a biochemical reaction. Enzymes have important functions in metabolism. They control the majority of biochemical reactions from digestion to replication (using DNA polymerase) and ⇒ transcription (using RNA polymerase) of genetic information. Enzymes are not consumed in their own reactions.

Epigenesis: Processes of progressive morphological form development of the embryo. Epigenesis is a topic of evo-devo

Epigenetics: Deals with cellular properties that are inherited by daughter cells and are not fixed in the ⇒ DNA sequence (⇒ genotype). This is also referred to as epigenetic modification or imprinting. The DNA sequence is not changed in this process. In the context of evolutionary theory, a distinction must be made between (1). Epigenetics, which deals with directly heritable, non-genetic material (⇒ methylation, etc.) and (2). Epigenetics as ⇒ epigenesis, the totality of genetic-epigenetic developmental processes leading to the phenotype.

Epigenetic inheritance: All causal, non-genetic mechanisms by which offspring share characteristics with their parents. Genetic and epigenetic inheritance together are called ⇒ inclusive inheritance.

Epigenetic markers: Chemical appendages distributed along the DNA double helix strand or on the "packaging material" of ⇒ DNA. Among other things, they act as switches that can turn ⇒ genes on and off.

Eukaryote: Living organism with a ⇒ cell nucleus and ⇒ cell membrane. In addition, a eukaryote has multiple ⇒ chromosomes, which distinguish it from ⇒ prokaryotes.

Evo-devo, evolutionary developmental biology: Research discipline. Addresses (a) the emergence and evolution of embryonic development; (b) changes in development and developmental processes to produce phenotypic variation and innovative traits, e.g., evolution of feathers; (c) the role of developmental plasticity in evolution; (d) how ecology influences development and evolutionary change; (e) the basis of the evolution of ⇒ homology. The goal of evo-devo is to explain variation by analyzing not only gene mutations but also changes during development. Developmental processes can be influenced by the outside world. Evo-devo, in contrast to ⇒ the Modern Synthesis, recognizes intrinsic mechanisms of development and views these as independent evolutionary processes. Further, evo-devo allows for spontaneous, non-linear, directional, self-organizing change and ⇒ facilitated variation and can explain macroevolutionary change within shorter time periods.

Evolution, biological: All the changes by which life on Earth has progressed from its earliest beginnings to its present diversity. Change in the heritable characteristics of a population of living things from generation to generation by mechanisms such as ⇒ mutation and ⇒ natural selection, resulting in ⇒ adaptation but also by other processes such as shape-shaping intrinsic mechanisms that occur in the organism during ⇒ development and can be influenced by the environment (evo-devo). Evolution is recognized as a demonstrable fact.

Evolutionary developmental biology: ⇒ Evo-devo

Evolvability: Ability of a system to evolve adaptively. Evolvability is the ability of a population of organisms to not only generate genetic diversity but also generate ⇒ adaptive genetic diversity and evolve by means of it through ⇒ natural selection.

Exaptation: Function of a trait that originally functioned differently. An example is the bird feather, which supports the ability to fly but, in evolutionary history, first served to insulate against heat.

Exon: (from expressed region) Part of a ⇒ eukaryotic ⇒ gene that is retained after ⇒ splicing and can be ⇒ translated into a ⇒ protein in the course of protein biosynthesis. By contrast, introns are excised and degraded during splicing. The totality of the exons of a gene thus contains the genetic information that is synthesized in ⇒ proteins.

Exploratory processes: Adaptive behavior of cells during certain cellular and developmental ⇒ core processes by which a large, if not unlimited, number of specific initial states can be generated (examples: neural pathways, blood capillary system).

Extended Evolutionary Synthesis (EES): A term for the Extended Synthesis of the standard evolutionary theory, mainly based on ⇒ evo-devo, ⇒ developmental plasticity, ⇒ inclusive inheritance, and ⇒ niche construction theory. It is the subject of this book.

Falsification: A scientific-theoretical method of Karl Popper. Principle of refutation for theoretical discovery. The proof of the invalidity of a statement, method, thesis, hypothesis, or theory. Falsification consists of proving intrinsic contradictions or inconsistencies with statements assumed to be true or of revealing an error. One methodically replaces the contradictory statements with a corrected thesis. Either initial assumptions or the thesis itself may be modified.

Facilitated variation: Theory named by Kirschner/Gerhart that explains how complex phenotypic change can arise from a small number of chance variations in the genotype. Preserved ⇒ core processes in cells facilitate variation because they reduce the amount of genetic change required to generate phenotypic novelty, principally through their reuse in new combinations and in other areas of their adaptive performance spectrum.

Fitness, evolutionary or reproductive: Fitness in the narrow sense refers to the number of reproductive offspring produced during the lifetime of the individual. Individual fitness depends on many interacting genetic, developmental, and environmental factors. Fitness is also expressed in relation to populations and defined mathematically.

Game theory: Mathematical theory in which decision-making situations with several interacting participants are modeled. Among other things, it attempts to derive rational decision-making behavior in real social conflict situations from the results. Evolutionary game theory examines the temporal and/or spatial evolution of different ⇒ phenotypes in a ⇒ population. The phenotypes interact with each other in a constant state of change and pursue different strategies, e.g., in foraging or territorial fights. The strategies employed determine whether the fitness of individual phenotypes improves or deteriorates over time. The change in fitness of the individual phenotypes in turn influences their distribution within the population, i.e., their frequency.

Gene: An often interrupted section on ⇒ DNA that contains basic heritable information specifying a sequence of amino acids that are linked together into a polypeptide, which is then processed and folded into a protein. The concept of a gene has become problematic for a variety of reasons, including the heritability of epigenetic processes. A single gene is usually not a sufficient cause for the occurrence of a phenotypic trait. Genes are passive and do not function in isolation. To fulfill their complete role, they require the cell and enzymes. The ⇒ genome, the totality of the genes of an organism, is not the complete code or program of life. Attempts to redefine the gene are incomplete, incorrect, and thus unacceptable if they describe the gene as a unit of inheritance or as a coding functional unit.

Gene centrism: Tendency in evolutionary theory, among others, to see the ⇒ genome as the ultimate cause in explanations of the production of the ⇒ phenotype. Epigenetic processes and exogenous influences on the phenotype are considered irrelevant for inheritance. Gene centrism is a basic concept of ⇒ the Modern Synthesis.

Gene-culture coevolution: Term used in niche construction theory in reference to humans and other organisms. It addresses the causal reciprocal effect of genetic change on humans and adaptive feedback through cultural ⇒ niche construction activity on the population ⇒ gene pool. The best-known example is dairy farming based on ⇒ lactose tolerance.

Gene expression, protein synthesis: Biosynthesis of ⇒ RNA and ⇒ proteins from genetic information. Gene expression is the entire process of converting the information contained in the ⇒ gene into the corresponding gene product (⇒ protein). This process occurs in several steps. Regulatory factors can act at each of these steps to control the process.

Gene flow: Exchange of ⇒ genes between two ⇒ populations, e.g., between a population from the mainland and that on an island.

Gene frequency: Term used in ⇒ population genetics, the relative frequency of copies of an ⇒ allele in a population. The gene frequency describes the level of genetic diversity in a population.

Gene knockout: Complete silencing of a ⇒ gene in the ⇒ genome of an organism.

Gene pool: Term used in population genetics. Refers to the totality of all gene variations (⇒ alleles) in a ⇒ population at a given time.

Gene products: Results of the expression of a gene. These include ⇒ RNA molecules, ⇒ transcription factors, signaling molecules, ⇒ morphogens and, in general, all ⇒ proteins.

Gene regulation: Control of the activity of ⇒ genes, more specifically the control of ⇒ gene expression. Gene regulation determines when, in which concentration and for how long the ⇒ protein encoded by a gene should be present in the cell.

Gene switches: ⇒ Enzymes (transcription factors) that control gene activity. Since these enzymes can be active or inactive and are usually activated by other enzymes in turn, they are also referred to as digital switches. All genes require enzymes to become active, i.e. to code for ⇒ proteins. Those gene switches or switch combinations that are used during ⇒ development and the resulting changes are relevant to evolution.

Genetic accommodation: Genetic fixation of a phenotypic trait that is typically formed in response to an environmental stressor. Unlike in ⇒ genetic assimilation, the environmental factor is required to be permanently present for the trait to be fixed.

Genetic assimilation: Genetic fixation of a phenotypic trait that is typically formed in response to an environmental stressor. Unlike in ⇒ genetic accommodation, the environmental factor is not required to be permanently present for the trait to be fixed. In some circumstances, prerequisites for genetic assimilation are already present through cryptic (hidden) mutations and only come to light through epigenetic changes.

Genetic Bauplan: The evolutionary genetic/epigenetic sequence for carrying out all ⇒ gene expression as well as epigenetic processes (⇒ cell, ⇒ cell communication, ⇒ self-organization) during ⇒ development. Obsolete term. There is no genetically or epigenetically determined Bauplan for the ⇒ phenotype. Development only shapes the embryo in stages with the mutual interaction of the ⇒ genome, ⇒ cells, tissues, and the environment. The term has persisted, for lack of a better one, until today.

Genetic code: Rule according to which groups of three consecutive ⇒ nucleobases located in ⇒ nucleic acids, called triplets or codons, are translated into amino acids by ⇒ RNA (translation). The genetic code is identical for almost all living organisms on Earth with at most minor variations. It is fundamental evidence for the existence of evolution and the descent of all life forms from a primordial ancestor.

Genetic drift: A random change in ⇒ gene frequency within the ⇒ gene pool of a ⇒ population. Causes may include floods or earthquakes, the formation of mountains or valleys, and other natural events.

Genetic engineering: Methods and processes that build on the knowledge of molecular biology and ⇒ genetics and enable interventions targeting the genetic material (⇒ genome) and thus in the biochemical control processes of living organisms (⇒ genome editing).

Genetic toolkit: Set of genetic and epigenetic developmental tools. ⇒ Homeobox ⇒ Hox gene ⇒ Transcription factor.

Genetics: A branch of biology. Genetics deals with the structure and function of hereditary units (⇒ genes) and their transmission to the next generation (inheritance).

Genome: Hereditary material of a living being. Classically, the totality of heritable information of a ⇒ cell, present as ⇒ DNA, which contains all genes.

⇒ Evo-devo also considers ⇒ epigenetic inheritance.

Genome editing: Molecular biology method for targeted modification of ⇒ DNA, including the genetic material of plants, animals, and humans. Genome editing can be used to deliberately destroy a ⇒ gene (⇒ gene knockout), insert a gene at a specific location in the ⇒ genome (gene knockin), or correct a point mutation in a gene. In genome editing, displaced genes may originate from the same organism or from another one. Genes can be manipulated in the soma or germline. Germline manipulation in humans is controversial. The best-known method is CRISPR/Cas (Clustered Regularly Interspaced Short Palindromic Repeats), a method developed by Emmanuelle Charpentier and Jennifer Doudna in 2012. CRISPER/Cas is seen as the discovery of the century, roughly comparable to that of X-rays.

Genome sequencing: Determination of the ⇒ DNA sequence, i.e., the ⇒ nucleotide sequence of an organism's entire complement of genes. The genome sequencing of numerous living organisms after the turn of the millennium revolutionized the biological sciences and established the era of genome research.

Genotype: The combination of alleles present at a specific gene locus. The term genotype was coined by the Danish geneticist Wilhelm Johannsen in 1909.

Genotype-phenotype relationship: Relationship between ⇒ genes and their products. Contrary to earlier views, there is no deterministic relationship between the genome and the ⇒ phenotype, and certainly no 1:1 relationship. Thus, it is usually not individual genes that are responsible for one morphological or behavioral trait at a time, but combinations of many genes with extensive gene regulation. If epigenetic levels and feedback mechanisms are also taken into account, the relationship becomes even more complex.

Germ cells, gametes: Specialized ⇒ cells of which two unite to form a ⇒ zygote during sexual reproduction. Germ cells are carriers of parental hereditary information (↔ somatic cells). Hereditary information is passed on to the next generation via the germ line.

Germline: ⇒ Germ cells

Gradualism: Notion of evolution of living things by the accumulation of minor modifications over a period of many generations. Evolutionary change occurs in small steps. Evolution in large steps cannot exist according to this view. Gradualism is one of the basic concepts of ⇒ the Modern Synthesis.

Group selection: Evolutionary theoretical concept that originated with Darwin and was elaborated by the British zoologist V.C. Wynne-Edwards in 1962. The concept of group selection assumes that groups, rather than individuals, are the units upon which selection acts. Early on in the history of evolutionary theory, doubts existed that group selection was a crucial mechanism of evolution. More recently, however, some evolutionary biologists have advocated a rediscovery of group selection as ⇒ multilevel selection. The leading proponent of this viewpoint is the U.S. evolutionary theorist D.S. Wilson.

Haploid: Occurrence of the ⇒ genome in the cell ⇒ nucleus in a single-copy form. Each ⇒ gene is thus present as only one variant (contrast with ⇒ diploid).

Hardy-Weinberg equilibrium: Concept in ⇒ population genetics. To calculate this mathematical model, one assumes an idealized population in which neither the frequencies of the alleles nor the frequencies of the ⇒ genotypes change since these are in equilibrium according to the model. This means that in an ideal population no evolution takes place, since no evolutionary factors (⇒ natural selection, ⇒ selection pressure) intervene to alter the ⇒ gene pool.

Heterochrony: Change in the time course of individual ⇒ development that causes the beginning or end of a developmental process of a trait to shift or changes the rate of such a process.

Histones: Proteins found in the ⇒ nucleus of ⇒ eukaryotes. Histones are the main components of ⇒ chromatin for packaging or unwinding ⇒ DNA. Histones are packed together to form nucleosomes, which are chained together with DNA in a ⇒ chromosome.

Homeobox: A relatively short ⇒ DNA segment in ⇒ Hox genes, about 180 ⇒ base pairs, that is largely the same in different groups of animals. A characteristic sequence that codes for the ⇒ homeodomain.

Homeodomain: ⇒ Region of a protein that can bind to the ⇒ DNA of another ⇒ gene. Genes that contain a homeodomain arranged in clusters are called ⇒ Hox genes in vertebrates such as humans. They form the Hox gene family.

Homeotic genes: Group of ⇒ genes that underlie the formation of structures during ⇒ development.

Homology: In biology, structural similarities that result from a common ancestry, for example, the wing of a bird and the front extremity of a lizard, amphibian, or mammal. ⇒ Gene sequences can also be homologous. Homology thus corresponds to the basic similarities of organs, organ systems, body structures, physiological processes, or behaviors due to a common evolutionary origin among different systematic ⇒ taxa. Homologous phenotypic traits do not necessarily have homologous developmental constructs. Homology may simultaneously exist for the evolution of a trait at one organizational level and not at another (↔ convergence). Similarly, a phenotypic ⇒ innovation may also have homologous gene sequences.

Horizontal gene transfer: ⇒ Endosymbiosis

Hox genes: Special form of ⇒ homeotic genes. Regulatory genes whose gene products control the activity of other functionally related genes during ⇒ development. A characteristic feature of a Hox gene is the ⇒ homeobox. The tasks of Hox genes are so important for structure formation during individual development that mutations in these genes usually lead to severe malformations or death. This suggests that Hox genes have been highly conserved during the evolution of many animal groups because they are essential as regulatory genes. Hox genes are important evidence that arthropods and vertebrates evolved from a common ancestral group.

Inclusive inheritance: Summary of genetic and epigenetic forms of inheritance in the context of the ⇒ Extended Evolutionary Synthesis, including cultural inheritance. Inclusive inheritance allows inheritance to proceed from ⇒ germ cells to ⇒ germ cells, from somatic cells to germ cells, and from somatic cells to other somatic cells also with the involvement of the environment. At the same time, this allows for the possibility of inheritance of acquired characteristics (⇒ Lamarckism).

Inheritance: Direct transmission of characteristics from living beings to their offspring, either genetically (Mendel) or epigenetically (Kirschner/Gerhart, Jablonka/Lamb et al.). In addition, the inheritance of symbols such as writing and language exists. Genetic and non-genetic inheritance together are summarized as ⇒ inclusive inheritance.

Innovation: Evolutionarily new design element that has no homologous counterpart in the predecessor species or in the same organism. For example, the bird feather, the turtle shell, or the luminescent organ of the firefly are innovations. Even an extra finger or toe counts as one according to this definition, since in its place the regular limb has no homology. However, homologous substructures or gene regulatory mechanisms are repeatedly discovered in the development process of innovative traits, which, according to some researchers, requires a rethinking of the definition.

Intron: (from intervening regions) Non-coding section of ⇒ DNA within a ⇒ gene that is excised (⇒ spliced) and not translated into proteins. The division of a ⇒ gene into introns and exons is one of the main characteristics of ⇒ eukaryotic ⇒ cells.

Jumping gene: ⇒ Transposon

Lactase: ⇒ Enzyme that breaks down lactose (milk sugar) into its components galactose (mucilage sugar) and glucose (grape sugar). Without this chemical reaction, lactose cannot be digested and utilized. In humans, the enzyme is normally only produced in the small intestine during infancy, but in people of European descent, due to a mutation, it is still produced by most adults (⇒ lactose tolerance).

Lactose tolerance, lactase persistence: Milk sugar tolerance. In lactose tolerance, milk sugar (lactose) ingested with food is digested as a result of sustained production of the digestive enzyme ⇒ lactase, even in adults. Lactose tolerance has become the norm for the majority of the world's population. Populations of the Northern Hemisphere have high lactose tolerance due to ⇒ mutations and culturally promoted livestock production. The prevalence of lactose tolerance is over 70% in the global adult population, 90% in Europe, and 10%–20% in East and Southeast Asia.

Lamarckism: Inheritance of acquired traits. Theory that originated with the French biologist Jean Baptiste de Lamarck (1744–1829), who posited that learned/acquired characteristics of an individual are heritable. The theory is often illustrated with the example of the long neck of a giraffe. This is due to the continuous stretching of the neck. Lamarckism lives on today in the modified form of inclusive inheritance.

Level of selection: Biological unit within a hierarchy of biological organizations (e.g., ⇒ genes, ⇒ cells, individuals, groups) that is the object of natural selection. To date, there is a long-running debate about what the levels of ⇒ natural selection are and the relative importance of each level (⇒ multilevel selection).

Lysenkoism: Pseudoscientific theory of the 1930s, named after Trofim Lysenko, a Soviet agronomist and advisor to Stalin. Among other things, it tied in with ⇒ Lamarckism. The central postulate of lysenkoism was that the characteristics of crops and other organisms were determined not by genes but by environmental conditions exclusively.

Macroevolution: Large-scale evolutionary transitions that occur across species boundaries and lead to the emergence of new ⇒ taxa (↔microevolution).

Master control gene, master gene: Control ⇒ gene that contains a ⇒ homeobox and coordinately manages the expression of other functionally related genes during development.

Meiosis: Special form of cell ⇒ nuclear division in sexual reproduction, in which, in contrast to the usual nuclear division (mitosis), the number of ⇒ chromosomes is halved, and genetically different cell nuclei are formed. Meiosis is usually accompanied by sexual ⇒ recombination, i.e., a new assembly of parental chromosomes.

Methylation, enzymatic, DNA methylation: Chemical modification of basic building blocks of the genetic material of a ⇒ cell, not of the ⇒ DNA itself. It is not a genetic mutation. Methylation occurs in a wide variety of (possibly all) living organisms and has various biological functions. Australian Emma Whitelaw was the first to show how methylation patterns can be inherited epigenetically.

Microbiome: In the broadest sense, the totality of all microorganisms that colonize the Earth. In a narrower sense, the totality of all microorganisms colonizing a human or other living being. As a rough estimate, it can be assumed that the human body is colonized by about 10,000 species of bacteria.

Microevolution: Changes in living organisms that occur both within a biological species (and thus also within subspecies) and within a relatively short period of time. These are mostly minor changes due to ⇒

mutations, genetic ⇒ recombination and selection processes, which, as individual steps, lead only to an inconspicuous change in the morphology or physiology of organisms (↔ macroevolution).

MicroRNA: (Greek mikros, "small"), abbreviated miRNA or miR. Short, highly conserved, non-coding ⇒ RNA segments that play an important role in the complex network of ⇒ gene regulation. MicroRNAs regulate ⇒ gene expression in a highly specific manner at the post- ⇒ transcriptional level, i.e., after mRNA (messenger RNA synthesis. This is the single-stranded messenger RNA that serves as the template from which ⇒ proteins can be created during ⇒ translation.

Mitosis: Process of division of a ⇒ eukaryotic ⇒ cell. The DNA and all other components of the parent cell are divided among the daughter cells during mitosis. This usually results in two, sometimes more daughter cells. In mitosis, the number of ⇒ chromosomes is maintained by their replication and equal division among the daughter cells (↔ meiosis).

Modern Synthesis, synthetic evolutionary theory, synthesis, Neo-Darwinism: The unification of the evolutionary views of various biological disciplines formulated in the 1930s and 1940s, based on Darwin's theory, Mendelian genetics, ⇒ genetics, zoology, paleontology, botany, and as the main formal apparatus of the newly added ⇒ population genetics. The Modern Synthesis assumes the smallest variations in ⇒ inheritance (gradualism) determined by genetic ⇒ mutations and, according to Darwin and Wallace, sees ⇒ natural selection as the main mechanism of evolution. The best adapted of species have statistically higher survival and a higher number of reproductive offspring, i.e., their ability to pass on their own genes to the next generation is superior to those of their competitors (⇒ *Survival of the Fittest*).

Morphogen: Signaling molecule that can diffuse with an altered concentration during development as part of ⇒ signal transduction and control morphogenesis or pattern formation in multicellular organisms. As a ⇒ transcription factor, a morphogen can activate various genes. As growth factors, they lead to the rapid growth and proliferation of tissue (cell proliferation). In the fruit fly *Drosophila melanogaster*, for example, the morphogenic transcription factors *Bicoid* and *Hunchback* and the morphogenic growth factors *Hedgehog* and *Wingless* are known.

Morphogenesis: ⇒ Epigenesis

Morphology: Branch of biology: the study of the structure and form of organisms. Morphology initially referred only to macroscopically visible features such as organs or tissues. With the improvement of optical instruments and various staining methods, morphological studies can be extended to the cellular and subcellular levels.

Multilevel selection: Concept proposed by D.S. Wilson and E. Sober (1998) that originates from the previously known idea of ⇒ group selection. It examines whether groups can exhibit functional organization in a comparable way to individuals and therefore be vehicles for selection. Thus, groups that cooperate better may, through increased reproductive success, displace others that do not cooperate as well (⇒ cooperation). The lowest level of selection is genes, and the next highest is cells; these are followed by the organism, and the top level is groups of individuals. The different levels function cohesively to improve fitness. Selection at the group level, that is, competition between groups, must outperform the individual level, that is, competition between individuals in a group, for a group-specific advantageous trait to propagate. Multilevel selection does away with the altruism of earlier theories.

Mutation: Novel, permanent change in the genetic material. It initially affects the ⇒ genetic material of only one ⇒ cell but is passed on from that cell to all daughter cells and their descendants. Mutations occur during meiosis and mitosis. Those which occur during meiosis have evolutionarily relevance. A distinction is made between gene mutations (⇒ copying errors), ⇒ chromosome mutations, and genome mutations. Not only is a mutation random with regard to the place of origin and extent (⇒ natural genetic engineering); the vast majority of mutations cannot be corrected by the genome itself. Rather, they are corrected by enzymes as part of a complex DNA repair mechanism. ⇒ Evo-devo focuses not only on genetic mutation but also on epigenetic changes in the ⇒ developmental process. The term mutation was coined by botanist Hugo de Vries in 1901.

Mutation rate: In higher organisms, the mutation rate is the relative proportion of genes that are replaced by mutation within a generation. The mutation rate depends on the ⇒ genotype of the organisms and on other internal and external factors.

Nanobots, nanorobots: Autonomous, replicable future machines (robots) or miniature molecular machines as a development direction of nanotechnology. Nanobots are expected to shrink to the size of blood cells or below and be capable of autonomous locomotion. It is predicted that such machines will play a major role in medicine in the future. They are expected to be able to eliminate disease foci (such as cancer cells and tumors) automatically in the human body.

Natural genetic engineering: All non-random mechanisms of modification of the ⇒ genome. Term introduced by J.A. Shapiro.

Natural selection: Central concept and mechanism in Darwin's theory and ⇒ the Modern Synthesis, according to which ⇒ variations in ⇒ inheritance are preferred or not with regard to their contribution to individual fitness. Natural selection produces ⇒ adaptation in the population. Sexual selection is a subform of natural selection. In this process, an individual chooses (or rejects) a sexual partner based on certain appearance or behavioral traits. The relative importance of natural selection is subject to change in the ⇒ Extended Evolutionary Synthesis. Generative, intrinsic mechanisms of evolution diminish its permanent influence (⇒ agent).

Neo-Darwinism: Originally a designation of August Weismann's ideas in the early 20th century, in which Darwin's theory was revisited. Today, the term is used for ⇒ the Modern Synthesis, which emerged in the 1930s and 1940s.

Neural crest: During embryogenesis, precursor of the neural tube. The subsequent peripheral nervous system, skin, parts of the skull bones, teeth, and adrenal glands are formed from the neural crest. The neural crest occurs mainly in vertebrates.

Niche construction: Concept adopted and expanded by Odling-Smee in 1988. Describes the ability of organisms to construct, modify, and select for components of their environment, such as nests, burrows, dens, and nutrients. Niche construction is increasingly seen in evolutionary theory as an independent ⇒ adaptive mechanism alongside ⇒ natural selection. It helps determine the ⇒ selection pressures to which species are subjected and is therefore seen as an evolutionary factor in its own right. Niche construction theory sees a complementary, ⇒ reciprocal, ⇒ biased relationship between the organism and the environment.

Nucleic acid: Macromolecule composed of individual building blocks (⇒ nucleotides or ⇒ base pairs) and supporting material. The best-known nucleic acid is ⇒ DNA, the storehouse of genetic information. In addition to this task, nucleic acids can also serve as signal transducers or catalyze biochemical reactions.

Nucleobase: ⇒ Base pair

Nucleotide: Building block of ⇒ DNA or ⇒ RNA. A nucleotide can contain one of four nitrogenous bases. It also contains a sugar (ribose or deoxyribose) and a phosphate group.

Nucleus: Located in the cytoplasm, usually round-shaped organelle of ⇒ eukaryotic ⇒ cells, which contains the genetic material.

-omics: Suffix identifying subfields of modern biology that deal with the analysis of aggregates of similar individual elements, e.g., ⇒ genomics, proteomics, metabolomics, etc.

Ontogenesis: ⇒ Development

Organizer region: A region in the early ⇒ embryo responsible for differentiating ⇒ cells and "assigning" them new tasks. If cells of the organizer region are transplanted from germ cells of the early ⇒ embryo to another site of a second embryo, the latter will develop, for example, a second body axis, induced by the organizer cells. The duplication of fingers and toes in vertebrates has also been produced in the laboratory by transplanting organizer cells from the limb bud.

Parallel evolution: A trait has evolved by parallel evolution if not common ancestry but rather developmental ⇒ homology is the proximate cause of a phenotypic similarity. Understood in this way, parallel evolution allows apparent convergence to be rejected because homologous evolutionary paths are considered. An example is the bulbous head shape of African cichlids.

Phenotype: The sum of all characteristics of an individual. It refers to not only morphological but also physiological and psychological characteristics. The phenotype is not clearly determined by the ⇒ genotype. Developmental biology is concerned with its emergence. ⇒ Evo-devo deals with changes in the phenotype during development.

Phenotypic plasticity: ⇒ Developmental plasticity

Phenotypic variation: Differences in traits between members of the same species or related species. Phenotypic variation includes all features of anatomy, physiology, biochemistry, and behavior. In opposition to ⇒ the Modern Synthesis, it is determined not solely by genetic ⇒ mutation but by the complex interplay of multiple genetic and epigenetic mechanisms.

Phylogenesis, phylogeny: The phylogenetic development of the totality of all living organisms as well as certain kinship groups at all levels of biological systematics. The term is also used to characterize the evolution of individual traits in the course of developmental history. In contrast to phylogenetics is ontogenesis, the ⇒ development of a single individual of a species.

Plasticity: ⇒ Developmental plasticity

Pluripotent stem cells: ⇒ Stem cells

Point mutation: (single nucleotide polymorphism) ⇒ Gene mutation in which only a single ⇒ nucleotide base is affected. A ⇒ base pair is changed (base pair substitution). A point mutation is a special case of ⇒ mutation.

Polydactyly: Usually ⇒ autosomal dominant and less commonly recessive inheritance of multiple fingers. Occurs frequently and in many forms in mammals. In cats, mutation can produce up to eight additional toes and in humans up to 31 fingers and toes in total. The number of additional phalanges can vary in the same individual, even between the two hands or feet. A distinction is made between preaxial (thumb side, inner side) and postaxial polydactyly (outer side). In very rare cases, polydactyly is part of a syndrome, a collection of disease signs, including Greig syndrome with craniofacial malformation (dysmorphia), Pallister-Hall syndrome with tissue changes in the brain, or Townes-Brocks syndrome with malformations of the ears and anus. In most cases, polydactyly is associated with a mutation of the ⇒ developmental gene *Shh* its antagonist *Gli3* or their respective ⇒ gene switches. ⇒ Evo-devo aims to use the example of polydactyly to analyze the genetic-epigenetic process of the emergence of a complex phenotypic variation.

Polygene: Involvement of multiple ⇒ genes in the formation of a phenotypic trait. An example is body size, which is determined by several genes as well as by environmental influences. Numerous phenotypic traits have polygenic involvement.

Polymorphism: Term used to describe the existence of different ⇒ phenotypes. For example, differences in appearance within a species are referred to as a phenotypic polymorphism. Many species exhibit at least one sexual dimorphism, as males and females differ from each other. A temporal or seasonal polymorphism occurs when the generations of a ⇒ population that appear at different times of the year develop different morphs, such as in some butterflies. In biochemistry, polymorphism refers to the occurrence of different versions of a ⇒ protein. In molecular biology, the term single nucleotide polymorphism represents a ⇒ point mutation.

Polyphenism, pleiotropy: Trait for which different discrete ⇒ phenotypes can arise from one ⇒ genotype; an example is preaxial ⇒ polydactyly. In ⇒ polymorphism, a trait has different genotypes.

Population: A group of individuals of the same species that are linked by their processes of origin, form a reproductive community and are found in a uniform area at the same time.

Population genetics: Study of the distribution of ⇒ gene sequences under the influence of four evolutionary factors: ⇒ selection, ⇒ genetic drift, ⇒ mutation or sexual ⇒ recombination, and migration/isolation. Population genetics is the branch of biology that describes ⇒ adaptation in speciation. It is a dominant component of the Modern Synthesis and studies the quantitative laws underlying evolutionary processes. Today, the view that population genetics can fully explain evolution is under critical review.

Prokaryote: Cellular living organisms that do not have a nucleus (↔ eukaryote).

Protein: Macromolecule constructed of ⇒ amino acids. Proteins are among the basic building blocks of all ⇒ cells and living organisms. Most proteins consist of 100–800 ⇒ amino acids, but some are much larger. In human, there are several hundred thousand proteins. Which protein a cell is to form in each case is determined in a complex way by ⇒ genes, ⇒ transcription factors, hormones and ⇒ enzymes as well as environmental factors. Proteins have a three-dimensional shape and are often intricately folded.

Protein synthesis: ⇒ Gene expression

Proximate and ultimate causality: A proximate (near) cause is an event that is closest to or directly responsible for an observed outcome. This is in contrast to a higher or more distant (ultimate) cause, which is usually considered the true reason for something.

Punctualism: ⇒ Interrupted equilibrium

Punctuated equilibrium, punctuated equilibria: Theory first presented by American paleontologists N. Eldredge and S.J. Gould in 1972 that provides an explanation of discontinuous rates of change and jumps in the fossil series between which there are long-lasting periods of equilibrium (stasis).

Radiation: The divergence of a less specialized species into several more specialized species by the formation of specific adaptations to existing environmental conditions.

Reaction-diffusion system: ⇒ Turing system

Recombination, homologous: The reorganization within ⇒ DNA molecules during sexual reproduction in the form of recombination of parental ⇒ chromosomes by exchange of chromosome segments (crossing over). Reproduction is the basis for the emergence of genetic variability and an essential factor in evolution.

Reciprocity: Principle of mutuality or interactions. In reciprocal causality, causes are simultaneously effects and vice versa.

Reductionism: Idea according to which the higher levels of integration of a system can be fully explained causally on the basis of knowledge of its physical components. Thus, in reductionism, the causal capabilities lie exclusively at the level of the basic constituents of a system. An example is the explanation of a phenotypic trait exclusively by its genes. Reductionistic views are usually also deterministic views (⇒ determinism). Reductionism has been criticized for a long time in the philosophy of science for explaining complex relationships inadequately. Moreover, it distorts reality. Reductionism is rejected if it is supposed to serve as the only method of explanation, i.e., if it is dogmatized. Reductionism is not only predominant in evolutionary theory; economic theory has also been used, in neoliberal models, for a long time the central basic assumption of the rationally acting market participant, who has all market-relevant information. This view and its predictions are strongly criticized today.

Ribonucleic acid: ⇒ RNA

RNA, ribonucleic acid: Molecular chain of many bases. A major function of RNA (as messenger RNA, mRNA) in the cell is the ⇒ translation of genetic information into ⇒ proteins (protein synthesis, ⇒ gene expression).

Robustness: Also referred to as biological or genetic robustness. The persistence of a particular trait in a system in the face of perturbations or conditions of uncertainty. Robustness in ⇒ development is referred to as ⇒ canalization.

Saltationism: The belief that evolution proceeds not only in small, continuous steps but also in larger, punctuated steps. A famous early proponent was Briton W. Bateson. ⇒ Evo-devo acknowledges saltationism.

Selection: ⇒ Natural selection

Selection factor: Environmental factor that influences the fitness of an individual. A selection factor helps determine the path followed by the evolution of a species. For example, on islands with constant strong storms such as the Kerguelen Islands, mainly wingless flies evolve. They are less likely to be blown away than those with wings. Here, the constant storm is a crucial abiotic selection factor. In deserts, on the other hand, heat and water scarcity are two important selection factors. In polar regions, the cold and the white color of the ground are important. All species are subject to many selection factors and selection pressures of varying strength in their evolutionary history.

Selection pressure: Influence of a selection factor on a population of living beings. ⇒ The Modern Synthesis assumes that populations are subject to constant selection.

Selection theory: Evolutionary theory of Charles Darwin and Alfred Russel Wallace.

Self-organization: Process in which the internal organization of a system increases without being instructed or directed by external sources. An example of a methodical, mathematical representation of self-organization in biology is a ⇒ Turing system. The interacting participants (elements, system components, ⇒ agents, ⇒ cells) act according to simple rules, creating order out of chaos without needing to have knowledge of the overall evolution. Self-organized systems, in addition to their complex structures, also have new properties and capabilities that elements do not have. The concept of self-organization can be found in various fields of science besides biology, for example, in chemistry, astronomy, or sociology.

Signaling molecule: ⇒ Morphogen

Signal transduction: Cells perform numerous functions under different conditions and are interdependent in many ways. Therefore, it is essential that they communicate with each other. One way cells communicate is by transmitting signals via extracellular messengers. These include not only hormones but also morphogens. When such a signal substance reaches a target ⇒ cell, the signal must bind externally to suitable cell receptors on the cell surface (lock-and-key principle) and be transmitted across the plasma membrane to trigger the respective reaction on the inner membrane surface or in the cytoplasm. This process is called signal transduction. In ⇒ development, the corresponding external signals are often proteins that activate the expression of genes (⇒ gene expression) in the form of ⇒ morphogens. In the context of signal transduction, a whole signal cascade of individual signaling steps can be executed; some known signal pathways are the Sonic hedgehog (SHH) and the Wnt signaling pathway. Signal transduction is essential in ⇒ development and in ⇒ evo-devo.

Signaling pathway: ⇒ Signal transduction

Singularity: ⇒Technological singularity

Somatic cells: Body cells that, unlike ⇒ germ cells, cannot give rise to sperm or oocytes. They develop through differentiation during development, for example, into skin cells, muscle cells, or blood cells.

Species: Basic unit of biological systematics. A general definition of species that satisfies the theoretical and practical requirements of all biological sub-disciplines equally does not exist. Rather, there are various species concepts in biology that lead to overlapping, but not identical, classifications.

Speciation: ⇒ Allopatric speciation

Splicing: Step in the ⇒ translation of ⇒ RNA that takes place in the cell ⇒ nucleus of ⇒ eukaryotes. Different exons are excised from the RNA and, in the case of alternative splicing, even different ⇒ proteins are generated from the same genetic starting material.

Stasis: Standstill in the evolution of a particular species over a long period of time.

Stem cells: Body cells that can differentiate into different cell types or tissues. Depending on the type of stem cells and how they are manipulated, they have the potential to develop into any tissue (embryonic stem cell) or into certain specified tissue types (adult stem cells). Pluripotent stem cells can differentiate into any cell type because they are not yet committed to any particular tissue type. However, unlike totipotent stem cells, they are no longer capable of forming an entire organism. Induced pluripotent stem cells (iPS) are generated by artificial reprogramming of non-pluripotent somatic cells, e.g., from adult skin cells.

Superintelligence: System with intelligence superior to that of humans in many or all areas. The term is used especially in ⇒ transhumanism and in the field of ⇒ artificial intelligence. An intellectually superior system that fulfills the criteria of a superintelligent system is not known at present.

Superorganism: Living community of several, usually numerous, independent organisms that jointly develop abilities or properties that supersede the sum of the abilities of the individual members of the community. The classic example of a superorganism is the ant colony: each ant is theoretically capable of surviving on its own because it has all the organs that independent insects need to survive, but in reality, ants are specialized in such a way that they can survive only in a community. Human social systems can also be considered as superorganisms. Such systems can be analyzed using complex dynamic systems theory in computer models.

Survival of the Fittest: The survival of the best adapted individuals according to Darwin's theory. The term was coined by the British philosopher H. Spencer. Darwin adopted it in a later edition of the Origin of Species. Survival of the Fittest is no longer considered mandatory for evolution in the context of the ⇒ Extended Evolutionary Synthesis.

Symbiont: The smaller of the two species involved in a symbiotic relationship. The symbiotic partner with the larger body is also called the host.

Symbiosis: Association of individuals of two different species that is evolutionarily advantageous for both partners (⇒ symbiont).

Synthetic biology: Field at the interface of molecular biology, organic chemistry, engineering sciences, nanobiotechnology, and information technology. It can be considered the latest development of modern biology. In the field of synthetic biology, biologists, chemists, and engineers work together to create biological systems that do not occur in nature. The biologist thus becomes a designer of individual molecules, cells, and organisms, with the goal of creating biological systems with new properties.

Synthetic evolutionary theory: ⇒ Modern Synthesis

Systems biology: Branch of biological science that attempts to understand organisms or biological functions in the context of interactions between interconnected levels.

Taxon, taxa: (pl.) Group of living organisms recognized as a systematic unit. Usually, the systematic designation is also expressed by a separate name for the group, e.g., invertebrates, protozoa, mammals, etc. The main taxa of living things are domain, kingdom, phylum, class, order, family, genus, and species. Each one is further subdivided, for example, subphylum, subspecies, etc.

Technological singularity: Point in time from which machines rapidly improve themselves by means of artificial intelligence and thus accelerate technological progress to such an extent that the future of mankind beyond this event is no longer foreseeable (intelligence explosion). A replacement or displacement of mankind is conceivable. The earliest formulations of such a singularity date back to the mathematician John von Neumann in 1958. An early, famous article on it was authored by the American mathematician Vernor Vinge (1993).

Threshold effect: Phenomenon in which a target product (⇒ protein) no longer behaves linearly above a certain level (threshold), e.g., of an ⇒ enzyme. Developmental processes can be subject to threshold effects.

Toolkit: ⇒ Genetic toolkit

Transcription: (lat. *transcribere*, "to transcribe") in genetics, the synthesis of ⇒ RNA using ⇒ DNA as a template. Transcription is followed by protein synthesis (⇒ translation).

Transcription factor, gene regulator: Regulatory ⇒ protein that binds to ⇒ DNA and controls ⇒ gene expression.

Transhumanism: International philosophical school of thought that seeks to extend the limits of human capabilities intellectually, physically, and/or psychologically using biological or technological processes.

Translation: Synthesis of proteins in cells using genetic information contained in an ⇒ RNA molecule. Translation represents a shift from the language of nucleic acids to ⇒ that of amino acids. Translation follows ⇒ transcription.

Transposon, jumping gene: A gene capable of moving from a particular locus in ⇒ DNA to another locus. First described by the US Nobel laureate B. McClintock in the 1960s.

Turing model, Turing system: Self-organizing system of partial differential equations described by A. Turing in 1952. A Turing model is able to form self-organizing structures from diffusing substances (⇒ morphogens), which do not initially show any system organization. According to current understanding, these are not genetically predetermined in detail in the system. Turing models are gaining importance for explaining organismic form-finding in ⇒ evo-devo, such as in limb development. In addition to the biochemical-molecular level involving morphogens, Turing processes may exhibit diffusion-like behavior with pattern formation at the cellular level.

Unmasking: Revealing alternative, cryptic (hidden) developmental pathways that are hidden from selection. Unmasking occurs through a genetic or environmental change.

Vertebrates: Subphylum in the systematics of biology. The vertebrate subphylum includes five major groups: fish (bony and cartilaginous), amphibians, reptiles, birds, and mammals.

Weismann barrier: Doctrine of A. Weismann, according to which there is no way in an individual for properties to pass from a ⇒ somatic cell to a ⇒ germ cell. When considering epigenetic developmental processes, the Weismann barrier is no longer valid. Retroviruses can also enter the germline and alter evolutionary traits.

Zygote: ⇒ Diploid cell formed by the fusion of two ⇒ haploid gametes (⇒ germ cells), usually from an egg (female) and a sperm (male).

INDEX

23andMe.com 196

Abert's squirrel 19
accommodation
    genetic 79, 80, 89, 104, 151, 157
adaptation 77, 133, 162
    A. Wagner 45
    actively generated 159
    beetles 154
    best adapted 29
    bichirs 103
    catfish 24
    cichlids 105
    costal wolf 23
    Endler experiment 22
    explanatory value 26
    goat on two legs 78
    human chin 23
    ice frog 22
    lactose tolerance 142
    macaques 24
    morphological 22, 57
    not the best solution 24
    of the evolutionary process 109
    reciprocal process 135
    vertebrate extremity 56
adaptation process 11, 24, 105, 126, 184
adaptationism 23, 26
    critique of 25, 29
adaptationistic dogma 131, 221
African bichir (*Polypterus senegalus*) 103, 127
African clawed frog (*Xenopus laevis*) 186
African wildcat (*Felis sylvestris lybica*) 121
agencies in development 220
agency, instances of 73, 74, 94, 155, 159, 160, 161
aging process 190, 191, 192
agouti mice 63, 64
agriculture 27, 64, 138, 208, 223
AI
    algorithm 199, 200, 204, 213
    strong 202

system 198, 199, 200, 201, 202, 203, 204, 213
system, meta learning 213
system, superior to humans 207
weak 201, 202
Alberch, Pere 50, 120
    portrait 225
Alexa 202
algae 131, 132, 151
allopatric speciation 19, 221
AlphaGo 201
alternative splicing 55
Amur leopards 11
annelid 6
Anolis (*Anolis carolinensis*) 154
Anolis lizards 153, 154
Aristotle 23, 66, 117, 175
Arrival of the Fittest 39, 43, 46
artificial general intelligence 201, 202
artificial intelligence 67, 137, 175, 183, 193, 197, 209, 213
    risks 211
artificial life 137, 185, 186
artificial selection 1, 3, 106, 194, 195, 196, 212
Atlas, humanoid robot 201
Avicenna Latinus 117

Bach, Johann Sebastian 89
Baedke, Jan 222
Banting, Frederick 184
basipharyngeal joint 104
bat wing 56
Bateson, Patrick, portrait 225
Bateson, William 1, 7, 118, 125
Bauer, Joachim 174
bcd (*Bicoid*) 242
Beatty, John 221
beaver dam 132, 133, 135, 162
beavers 131, 132, 134, 135
bees 8, 158
beetle horn 88, 91, 93
behaviorism 168
Benz, Carl 196

Best, Charles 184
bias *see* developmental bias
biased development 105, 109, 126, 149, 162, 220
biased variation 127
Bicyclus anynana, African butterfly 106, 107, 108, 109, 116, 154
biodiversity 38, 131, 154, 212, 235
biosphere 61, 209
biotechnology 187, 189
bird feather 90, 110
bird wing 4, 42, 115
bird's nest 138, 154
bluehead wrasse (*Thalassoma bifasciatum*) 77
BMP4 102, 103
Boden, Margaret 199, 201
Böhme, Madelaine 115
Bonner, John Tyler 40
bonobo (*Pan paniscus*) 6, 7, 19
Bostrom, Nick 137, 202, 204, 205, 206, 207
bowerbird 226
brain 116, 139, 174, 175, 178, 195, 205
brain cells 57, 188
brain emulation 202
brain evolution 27, 66, 168, 171, 174, 181, 207
brain implant 192, 207
brain receptors 115
brain size 168, 169, 170, 171
brain structure of birds 179
brain upload 193
brain, finger representation 114
brain–computer interface 193
Brakefield, Paul 108
    portrait 226
Brenner, Sidney 70
Brosius, Jürgen 188, 189
brown bear (*Ursus arctos*) 28
buckler-mustard (*Biscutella laevigata*) 8
butterfly eyespot *see* eyespot, butterfly
butterfly wing 84, 101, 105, 108, 127
Bycyclus, butterfly genus 107

261

Caledonian crow (*Corvus moneduloides*) 177
Callebaut, Werner 52
Cambrian explosion 91, 157
canalization 18, 32, 33, 35, 42, 44, 48, 80, 81, 93, 94
cancer cells 137
capitalism 208
Carroll, Sean B. 41, 42, 43, 46, 47, 59, 81, 105, 106
    portrait 226
Cassirer, Ernst 159
causality 142, 154
    biological 208
    R. Riedl 231
    reciprocal 147, 148, 162, 220
causality, reciprocal *see* reciprocal causality
causal-mechanistic results 96
cell cytoskeleton 54
cell differentiation 59, 96, 122, 125, 151
cell division (mitosis) 186, 190
cell repair 191
cell signaling 57, 59, 123, 126
centipede 112, 113, 127
centipedes (Chilopoda) 112
central cogma of molecular biology 71
central nervous system 59
cesarean birth 141, 188
chaperone 81
chiasm 17
chicken wing bud doubling 119
chimera 184, 189
chimpanzee 6, 7, 19, 55, 172, 177, 178, 189
China 11, 112, 195, 196, 201
Chlamydomonas green alga 151
Chopin, Fréderic 89
chromatin 12, 63, 93
chromosome 16, 17, 21, 36, 125, 190, 196
cichlids 101, 104, 127
cis-element 89
cleaner wrasse (*Labroides dimidiatus*) 177
cloning 96, 183
closed-loop system 67, 203
coevolution 153
cognitive science 85, 199, 207, 213
Collingridge dilemma 209
Collingridge, David 209
comparative genome sequencing 41
compartmentation 53, 55, 58
complex variation 39, 43, 48, 123, 126
complexity 66, 67, 69, 84, 207
    cell 71
    cellular interactions 187
    developmental system 43
    discontinuity 9
    of evolution 95, 152
    heart pacemaker 74
    levels 41
    socio-technical society 213
Comte, Auguste 116

conjunction fallacy 206
connectome 193
conscience 176
consciousness 158, 167, 193, 198
    AI systems 66, 198, 202, 207
    bee 158
    evolution of 174
constructive development 86, 220
continuous variation 12, 90
cooperation 27, 58, 168, 171, 173, 175
    cells 59
    concept 172, 209
    global 209, 213
cooption 111
coral reef 132, 134, 154, 162
core process, preserved 53, 54, 55, 57, 58, 59
costal wolf 23
crayfish 42
creationism 177
Crick, Francis 11, 13, 35, 71
CRISPR 158, 187, 188, 189
cross vein 31
crossing over 16, 17
cultural adaptation 173
cultural evolution 27, 115, 173, 174, 208
cultural inheritance 27, 60, 137, 157, 180, 208, 220
culture 157, 168, 179, 183, 184, 189, 194, 195, 196, 208
    cooperation 168
    definition 222
    and evolution 183
    is evolution 208
    group-oriented 173
    and maladaptations 212
cyanobacteria 131, 134, 151, 152
cyborg 192, 202
cytoplasm 63, 186

Dahlem Conference Evolution and Development 40
dairy farming 137, 139, 140, 141, 142
Danuvius guggenmosi 115
Daphnia magna water flea 152
Darwin, Charles 1
    criticism 28
    earth worms 131
    embryology 35
    an epilogue 223
    fiew of thinking 167
    inheritance of acquired characteristics 73
    innovations 84
    maladaptation 212
    own criticism 223
    polydactyly 117, 118
    possible mutation 59
    speciation 20
Darwin's finches 19, 32, 80, 101, 102, 127
Darwin-Wallace doctrine 20
Dawkins, Richard 20, 66, 72, 73, 74, 133, 135, 231

de Beer, Gavyn Rylands 35
decanalization 122
deep learning 199, 203
deep-sea octopus (*Graneledone boreopacifica*) 112
Delbrück, Max 21, 77
descent with modification 3, 84
developmental bias 45, 46, 101, 150
    active character 160
    EES 51, 147, 148
    eye spots 106, 109
    innovation 151
    and natural selection 109
    plasticity 86
    polydactyly 123
developmental biology 36, 49, 50, 85, 106, 110, 225, 226
developmental constraints 39, 41, 47, 106, 108, 112, 116, 127, 160
developmental genes 36, 41, 42, 95, 223
developmental genetics 41, 86, 89, 95
developmental mechanism 39, 40, 49, 61, 82, 93, 105, 126, 220
developmental niche construction 137, 162
developmental plasticity 60, 77, 82, 84, 85, 103, 109, 147, 148, 149, 152, 220
devo-evo 47, 79
Devonian 115
diabetes 63, 67, 184, 185, 188, 191, 195, 196
Diamond, Jared 134
Dick, Philip K. 190
dimorphism 9
dinosaurs 3, 4, 20, 28
Diplodocus 116
diprosopus 120
discontinuity 8, 9, 10, 127, 172
discontinuous variation 8, 9, 10, 12, 39, 126
Distal-less (*Dll*) 105, 242
DNA double helix 11, 12, 180
DNA methylation 63, 64
DNA repair 61, 81, 191
DNA replication 72
DNA screening 195
DNA sequencing 75, 196
DNA structure 13, 35
DNA transcription 68
DNA, noncoding elements 119
Dobzhansky, Theodosius 18, 20, 228
dolphins 66
domestication of animals and plants 138
domestication of plants 134
Doudna, Jennifer 189
Drexler, Eric 206
Drosophila 55
    cross veins 33
    heat shock proteins 80
    homeobox 36
    homologous genes 37

Shh 38
    study of T. Hunt Morgan 15
    wingless mutants 105
dualism 175, 176, 181
dualisms of Western culture 184
Duboule, Denis 123
Dunbar, Robin 170
Dunbar's number 170
Dupré, John 68

E. coli 53, 62, 77
earth lover (*Geophilomorpha*) 112
eco-evo-devo 47, 75, 76, 84, 87, 93
ecological inheritance 144, 150
ecosystem engineering 136, 155
ectopic wings 89
egg cell 10, 63, 68
ego-consciousness 167, 177
ego-consciousness of birds 179
Ehrenreich, Ian 76
Einstein, Albert 171, 222
Eldredge, Niles 25, 29, 227
    portrait 226
Ellington, Duke 89
embryo
    avoided disintegration 39
    early research 34
    formation 93, 117, 180
    intrinsic mechanisms 143
    Modern Synthesis 35
    rediscovery 36
    role in evolution 223
    synthetic 194
embryology
    Aristotle 117
    insights into evolution 10
    Modern Synthesis 35
    19th century 35
embryonic development 10, 35, 36, 38, 40, 49, 160, 223
    evolution of 46
    genes 61
    Modern Synthesis 35
    and niche construction 144
    and phylogeny 31
    physical forces 104
    S.J. Gould's role 26
    self-organization 93
    transition to 6
    turtle shell 111
    vertebrate limb 75
Endler, John 21, 22
    portrait 226
environmental changes 48, 65, 80, 103, 131, 138, 150, 161, 162
environmental conditions 139, 151
    changing 28, 44, 48, 76, 77, 94, 138, 142, 190, 234
    constant 94
    convergent 150
    Darwin 8, 11
    different 152, 161
    human-health 87
    inheritance 152

    new 77
    newly created 132
    novelties 91
    physical and chemical 132
    similar 105
environmental stressor 33, 34, 109
enzymes 12, 17, 45, 63, 78, 185, 191
Epel, David 76
epigenesis 93
epigenetic inheritance 60, 61, 64
    DNA methylation 63, 64, 65
    EES 51, 148
    Jablonka&Lamb 156
    learned knowledge 208
    learning 65, 73
    Waddington 31
epigenetic marker 63, 65
epigenetics 31, 34, 63, 65, 85, 93, 136, 148, 156, 221
eugenics 156
eukaryote 81, 87
Eunotosaurus africanus, extinct turtle 112
Eurasian jay (*Garrulus glandarius*) 178
European catfish (*Silurus glanis*) 24
European Society for Evolutionary Developmental Biology (EED) 96, 148
evo-devo mechanisms 75, 121
evo-devo questions 46
evo-devo theory 31
evo-devo, research results 101
evolutionary biology 20, 50, 52, 85, 86, 87, 147, 149, 213, 226
    emergence of evo-devo 50
    enduring problems 84
    enduring questions 89
    evolutionary factors 51
    focus 90
    J. Huxley 17
    a one-way causal chain of genes 68
    a population science 11, 84
    questions failed to asnwer 87
evolvability 39, 105, 106, 107, 149, 150, 153, 160, 162, 214
exaptation 110
explorative behavior 56
Extended Evolutionary Synthesis 147
eyespot, butterfly 106, 107, 108, 116, 154, 226

facilitated variation 53, 54, 55, 58, 59, 79, 101, 102, 127, 150, 160, 161
fecundity 2
Feldman, Marcus W., portrait 226
female orgasm 23
fertility 64
Feyerabend, Paul 221
fire, use of 137, 139, 142, 171
firefly 83, 84, 90
Fisher, Ronald Aylmer 16, 17, 18, 20
fitness 2, 9, 23, 28, 62, 78, 112, 132, 134, 135, 149, 150, 174, 212

fitness advantage 23, 26, 78
fitness curve 105
fitness of a population 61
Fleck, Ludwik 222
fly wing 31, 33, 108, 109
foraging 171, 173
Franklin, Rosalind 180
free will 175, 181, 209
Frey, Carl 200
frontosa cichlid (*Cyphotilapia frontosa*) 104

game theory 209
Gates, Bill 206
Gehring, Walter 36, 38
gene centrism 69, 220, 221
gene-culture coevolution 137
gene expression 21, 35, 54, 68, 124, 125
    cascade of 36, 37
    changes 55, 81
    comparative analysis 153
    developmental 37
    ectopic 40, 120
    environmentally regulated 82
    epigenetic regulation 63
    extent of 55
    interplay with environment 80
    location of 40
    mechanisms of 54
    patterns 70, 188, 195
    robustness 82
gene frequency 53, 139
gene knockout 36
gene pool 184
gene regulation 7, 41, 42, 45, 46, 75, 101, 125
    chains of 36
    evolutionary changes in 89
    mechanisms 90
    new studies and findings 219
    robustness 44
    theories 221
gene regulation networks 36, 43, 67
gene regulatory network 82, 88, 91, 94, 95, 96, 159, 161, 219
gene switch 41, 42
gene-culture coevolution 137, 139, 142, 145, 183, 222
gene-culture niche construction 184
gene–culture–microbiome 153
general intelligence 171, 201
Genes are followers 75, 80, 94, 153, 234
genetic assimilation 75, 76, 80, 81, 82, 94, 127
    cross veins 31
    tobacco hornworm 109
genetic code 3, 6, 7, 54, 87
genetic determinism 78, 187, 231
genetic drift 17, 18, 113, 133, 149, 221
genetic engineering 187, 193, 196
genetic inheritance 60, 104, 136, 144, 161, 183, 220, 221
genetic toolkit 41, 47, 81

genetics 80, 148, 226
    ambiguities 61
    and botany 20
    early works 15
    and epigenetics 65
    of eyespots 106
    gaining momentum 11
    importance 195
    introduction by Bateson 7
    Mendelian 17
    original question 68, 73
    of polydactyly 119
    study of Morgan 16
genome 41, 65, 78, 92, 94, 96, 109, 122, 153, 196
    arrangement of genes 36
    biased response 62
    blueprint 41, 57
    as the book of life 68
    determines the phenotype 78
    expanding the coding capacity 42
    feedback from cells 56
    genetic redundancy 47
    human 53, 55, 62, 187, 188, 189, 195
    hypermutation 62
    intervention 158
    maternal 68
    Modern Synthesis 53
    new role 12
    not a program 70
    not only a read-only memory 62
    as a program 72
    read by the phenotype 68
    redundancy of genes 44
    reorganization 74
    use of genetic toolkit 41
genome editing 66, 183, 187, 188, 195, 196, 214
genome-wide association research 71, 195
genotype 31, 34, 45, 48, 69, 77, 78, 105, 134, 160
    does not determine the phenotype 234
genotype-phenotype relation 29, 31, 34, 35, 61, 96, 105, 149
Geospiza, genus of Darwin's finches 102
Gerhart, John C. 53, 54, 55, 56, 57, 58, 59, 62, 68, 69, 75, 79, 91, 92, 102, 142, 160
germ cell 7, 10, 61, 74, 189, 194, 196
germline 187, 189
    theory 10
Gierer, Alfred 175, 180
Gilbert, Scott F. 76, 80, 111
    portrait 227
Ginsburg, Simona 158
glucagon 58
goat on two legs 78
golden rice 185
Goldschmidt, Richard 9, 10, 113
Good, Irving John 205
Goodall, Jane 178
Google 191, 201, 202, 203, 233

Google Brain 202
gorilla 7
Gould, Stephen J. 25, 26, 29, 39, 41, 122
    portrait 227
gradualism 2, 15, 25, 28, 29, 110, 123, 125
Grant, Peter and Rosemary 32
great apes 172, 173
Grey, Aubrey de 190
group adaptation 151, 181
group selection 27, 174, 232
Güntürkün, Onur 179
guppy (*Poecilia reticulata*) 21

Haff, Peter K. 209
Haldane, J.B.S. 18
Hall, Brian K., portrait 227
Haraway, Donna 222
Harmony, sex robot 199
Hartmann, Nicolai 49
Hawaiian crow (*Corvus hawaiiensis*) 177, 178
Hawking, Stephen 188, 200, 206
heart chambers 7, 9
heart development 85
heart model 71
heart models 71
heart rhythm 71, 73
heartbeat 22, 67, 70, 71, 125, 180
heat shock 31, 32, 33, 109, 110
heat shock protein HSP90 55, 80
Heisenberg, Werner 222
Hemingway model 123, 124
Hemingway mutant 121
Hemingway, Ernest 120
hemoglobin 40, 70
heterochrony 39, 40, 47, 225
hexadactyly 118
hidden mutation 34, 161
hidden variation 81
histones 12
Hölldobler, Bert 27, 230, 235
homeobox genes 36, 38, 106
Homo habilis 115
Homo sapiens 66, 183
Homo synchronicus 178
homology 3, 38, 84, 85, 88, 89, 90, 91, 94
homonomy (serial homology) 85
homunculus 117, 118
hopeful monsters 9, 113
horned beetles (*Onthophagus*) 76, 82, 83, 84, 88, 116, 136, 153, 230
Houle, David 108
Hox gene cluster 37, 42, 116
Hox gene expression 47
Hox genes 36, 37, 38, 41, 42, 53, 54, 59, 123, 125
HSP90 protein 53, 55, 80, 81, 94
human brain 167
human chin 23
human mind 174

humankind 44, 66, 75, 138, 139, 140, 141, 177, 178, 179, 180, 183, 195
human–machine behavior 207, 213, 214
human–machine coevolution 183, 214
human–machine combination 192, 214
human–machine society 213
humanoid robot 178, 197
hump-head mouthbreeder (*Cyrtocara moorii*) 104
Huntington, Samuel P. 140
Huntington's disease 188
Huxley, Aldous 17
Huxley, Julian 16, 17, 18, 35, 185
Huxley, Thomas 16, 17
hypermutation 62

ice frog (*Lithobates sylvaticus*) 22
imitation 24, 66, 167
immortality 190, 191
immune system 23, 27, 139, 188
Imperative of Responsibility 211
in vitro fertilization (IVF) 194, 196
inclusive inheritance 31, 60, 77, 96, 127, 131, 147, 149, 152, 161, 220
infertility 194
inheritance *see* cultural i.; discontinuous i.; epigenetic i.; genetic i.; inclusive i.; symbolic i.
innovation 44, 155, 161, 220, 223
    A. Moczek 83, 84, 85
    A. Wagner 43, 44, 45, 46
    agents 159, 161
    bird's nest 137
    cichlids 104
    developmental contribution 48
    developmental emergence 127
    duplication of genes 42
    EES project 147, 149
    environmental stimulus 80
    examples 76, 84
    G.B. Müller 48, 87
    hidden mutations 28
    intelligence 171
    J. Monod 61
    J.A. Shapiro 62
    Kirschner&Gerhart 59
    Ontophagus 151
    plasticity 79, 152
    role of the environment 101
    shared intentionality 174
    technology assessment 209
    theory 46
    turtle shell 110
insect wing 16, 37, 42, 43, 48, 84, 88, 89, 90, 91
insulin 67, 70, 184
    synthetic 185
insulin pump 67, 192, 203
insulin secretion 58, 70
intelligence 27, 195, 196, 212

child 178
components 171
sequential 178, 179
intelligence explosion 205
intelligence quotient (IQ) 196
intentionality 169
collective 173, 174
individual 172
joint 173
orders of 169, 170
shared 172, 173

Jablonka, Eva 34, 50, 60, 66, 68, 92, 156, 160, 208
EES 219
portrait 227
Jacob, Francois 35
Janich, Peter 179
jellyfish *Turritopsis dohrnii* 190, 191
Johannsen, Wilhelm 72
Johanson, Donald 6, 179
John Templeton Foundation 147
Jonas, Hans 211
jumping gene 62, 156
jumping spider 116

Kahneman, Daniel 133, 145, 206, 207, 221
Kaibab squirrel 19
Kant, Immanuel 117
Karafyllis, Nicole C. 192
kea (*Nestor notabilis*) 177
Kenny, Anthony 73
Kimura, Motoo 27, 44, 148
Kipling, Rudyard 26
Kirschner, Marc, portrait 227
Kirschner, Marc W. 53, 54, 55, 56, 57, 58, 59, 62, 68, 69, 75, 79, 91, 92, 102, 142, 160
complex developmental apparatus 46
kiwi 19
Konrad Lorenz Institute for Evolutionary and Cognitive Research 51
Kuhn, Thomas 149
Kurzweil, Ray 191, 207

lactase persistence 139
lactase, enzyme 139
Laland, Kevin N. 138, 171, 220
EES 219
portrait 228
Lamarck, Jean-Baptiste 10, 60, 63, 66, 73, 74, 137, 189, 208, 227
Lamarckism 10, 73
Lamb, Marion J. 34, 60, 61, 62, 66, 68, 92, 157, 160, 208
language 87, 137, 167, 168, 169, 172, 173, 174, 193
evolution 115, 139, 157, 171, 172
inheritance 66
machines 198
robustness 44
technical 120

Laubichler, Manfred D. 95
portrait 228
Laughter 169
Layer, Paul 41
Leakey, Louis 178
Leakey, Mary 179
Leakey, Richard 179
learning 65, 150, 157, 161, 193, 198, 214, 225
evolutionary origins 134
learning systems 214
Leeuwenhoek, Anthonie van 117
Lem, Stanislaw 190
Lettice, Laura 119
level of selection 27
Levy, David 198
Lewis, Edward B. 38
Lewontin, Richard C. 39, 50, 131, 135
portrait 228
Libratus, AI system 198
limb 2, 4, 43, 93, 111, 115, 116, 119, 121, 153, 154
limb bud 93, 120, 122, 123
limb development 59, 116, 230
Lorenz, Konrad 49, 50, 66, 157, 230
Lucy, *Australopithecus* 179
Luria, Salvador E. 21
Luria-Delbrück experiment 77
Lyell, Charles 2
Lysenkoism 156
Lyson, Tylor 110

macaque 24
machine learning 195, 197, 198, 201, 203, 204, 213, 214
macroevolution 42, 154, 155
macromutation 9, 10, 110
Maine Coon cat 56, 121
Mainzer, Klaus 198, 201, 207
Making of the Fittest 39
maladaptation 212
Malaria resistance 184
male nipples 22
Malthus, Robert 5
Margulis, Lynn 74
portrait 229
Mars Rover 180
Marx, Karl 140
Marzolo, Francesco 118
mass extinction 28
master gene 38, 54
Maupertuis, Pierre-Louis Moreau de 118
Maynard Smith, John 50, 87
Mayr, Ernst 18, 50
McClintock, Barbara 62, 156
portrait 229
Meinhardt, Hans 38
meiosis 16, 194
meme 66
Mendel, Gregor 3, 7, 10, 15, 18, 60, 72, 76, 118, 136, 208
Mendelian genetics 17
Mendelian inheritance 16, 18, 47, 221
Mendelism 1

mentalization 169
Menzel, Randolf 158, 179
mesenchyme 122, 191
messenger RNA 55
metabolite 45
metamorphosis 78, 230
metformin 191
microbiome 84, 85, 133, 138, 152, 153, 188
microbiota 83, 152, 153
microevolution 28, 134
middle-out pathway 69, 70
millipede (*Illacme plenipes*) 112
millipedes (*Myriapoda*) 112
mind 175
mind-body problem 176
mind-brain relationship 175
Minelli, Allessandro 95
portrait 229
mirror test 177
missing link 6
Mitchell, Sandra 221, 222
Moczek, Armin 31, 68, 155
complex developmental apparatus 46
degrees of freedom in evolution 40
EES 219
genetic assimilation 82
heredity and development 142
homology 88
horned beetles 76, 79, 82, 89, 116
innovations 87, 89, 91, 94, 95
instances of agency 155
interview 82
new eye tissue 40
niche construction 133
phenotypic plasticity 76
portrait 230
Science magazine 88
serial homology 93
model organism 15, 96, 125, 151, 188
Modern Synthesis 15
Modular Prosthetic Limb 192
molecular biology 6, 28, 71, 228
molecular developmental biology 50, 69
molecular genetics 6, 11, 68
molly (*Poecilica sphenops*) 3
Mongols 139
monism 175, 176, 181, 221
Monod, Jacques 35, 49, 61
morality 179, 202
Moravec, Hans 206
Morgan, Thomas Hunt 15, 16
morphogen 106, 120
morphogenesis 34, 230
mouse gene 3
mRNA 55
Müller, Gerd B. 31, 34, 68, 75, 158, 160
complex developmental apparatus 46
EES 219
evo-devo theory 46

Müller (cont.)
    evo-Devo-questions 75, 87
    Genes are followers 94
    heredity and development 142
    innovations 91
    interview 48
    portrait 230
Müller-Wille, Staffan 222
multicellular organism 21, 35, 73, 74, 151, 152, 155
multicellularity 151, 152
multilevel causation 73
multilevel selection 26, 27, 208
MultiSense 199
Musk, Elon 193, 200
mutagenesis 96
mutation
    cryptic 80
    hidden 33
    masked 94, 214
    neutral 27, 44
    nonrandom 60
    random 1, 32, 57, 61, 65, 80, 102, 160, 220, 234
mutation process 80
mutation rate 18, 62
mutational hotspot 62
mutation-selection mechanism 22
Mycalesis, Asian butterfly 106, 107, 108
Mycoplasma bacterium 185

Nagel, Thomas 158, 176, 198
nanobot 190
nanobot technologies 190
narwhal (*Monodon monoceros*) 91
Natura non facit saltum 5
natural genetic engineering 62
naturalism 175, 176
Nature (magazine) 213
Neanderthal 169
neighbor proteins 44
nematode 42, 190
neocortex 167, 170, 179
neo-Darwinism 15, 73, 74, 219, 220, 231
Nesbit, Jeff 200
nest building 152, 154
nest building, birds 137
neural crest 59
neural crest cells 102
Neuralink, company 193
neurogenesis 38
neutral theory of molecular evolution 27
Newman, Stuart 50, 95, 158, 187, 188, 189
    portrait 230
niche construction theory 131
Niche Inheritance 136
Nida-Rümelin, Julian 198
Nietzsche, Friedrich Wilhelm 185
night heron (*Nycticorax nycticorax*) 178
nightingale 66
Nijhout, Fred 76, 83, 109, 110
    portrait 230

Noble, Denis 67, 68, 92, 122, 125, 142, 158, 160, 175
    biological function 68, 70
    causality 69
    chance 126
    complex developmental apparatus 46
    genetic program as an illusion 68
    inheritance 68
    interview 71
    portrait 231
    reductionism 69
non-adaptation 24, 41
non-adaptationism 25
noncausality 140
novelty, evolutionary 83, 84, 85, 88, 89, 90
nucleosome 12
nucleotide 17
nucleus 53, 68, 186, 190
Nüsslein-Volhard, Christiane 37, 38, 187

Odling-Smee, John 131, 132, 133, 212
    EES 219
    interview 134
    portrait 231
Odontochelys semitestacea 112
Online Mendelian Inheritance in Man (OMIM) 125
Onthophagus, dung beetle 94, 151
ontogenesis 31, 39, 175
Orca 66
organismic systems biology 67, 75, 142
Osborne, Michael 200
ostrich 33
oxygenation of the atmosphere 131

Pan American Society for Evolutionary Developmental Biology (EvoDevo PanAm) 96
panda, thumb 116
paraplegia 193
parental care 60, 112
Parisi, Luciana 208, 209
pattern formation 38, 43, 46, 116, 122, 123, 124, 126, 230, 233
Pavlicev, Mihaela 23
Pax6 36, 38
pentachlorophenol 45
pentadactyly 115, 116
Pepper, robot 203
pest control 152
pesticide 45, 209
Peterson, Tim 104
Pfennig, David W. 76
phenotype plasticity 28, 75, 77, 79, 82
phenotypic accommodation 79, 80, 150, 157, 161
Piechocki, Reinhard 178
Pigliucci, Massimo 50, 51, 94, 230
    portrait 231
placenta 6, 84

point mutation 40, 95, 113, 188
    polydactyly 114, 120, 121, 122
polar bear (*Ursus maritimus*) 28
polydactyly 8, 39, 48, 57, 75, 101, 108, 113, 114, 127
polyphenism 109
polypheny 122
Popper, Karl 143, 157
population genetics 11, 15, 17, 18, 20, 51, 72, 84, 134, 148, 228
positivism 116
posthumanism 184, 201
Precis octavia, African butterfly 78
preformation theory 118
preformationism 117
preimplantation diagnostics (PID) 194
preimplantation genetic testing 194
preimplantation screening (PGS) 194
programming languages 203
protein synthesis 54, 55
proteome 68, 75, 152
protozoa 152
PubMed, database 199
punctualism 25, 29, 227
punctuated equilibrium 25, 26

quantitative gene regulation 79

Raff, Rudolf, portrait 231
RAS2 190
rat 63
ratchet effect 173, 174, 208
reaction norm 78
reaction-diffusion processes 123, 233
reciprocal causality 86, 133, 140, 145, 147
reciprocal causation 147
reciprocity 133, 137, 142
recombination 55
reductionism 69, 221
retrovirus 7
Rheinberger, Hans-Jörg 222
Riedl, Rupert 49, 50, 70
    portrait 231
ringed tadpole (*Siphonops annulatus*) 112
RNA 12, 55, 63, 65, 68, 72
RNA translation 68
robotic hand 193, 202
robotic skin 192
robotics 183, 203, 209, 214
robotics industry 202
robustness 42, 43, 44, 45, 46, 48, 74, 80, 94, 117, 161, 214
rock pigeon (*Columba livia*) 24
Roux, Wilhelm 35

saltation 8, 9, 110, 111, 113
saltationism 9
SCH9 190
Scharmer, Claus Otto 211
Schmidhuber, Jürgen 198, 206
Schrödinger, Erwin 72, 75
Schummer, Joachim 186
Schwille, Petra 186
Science (magazine) 64, 88, 106

Scolopendromorpha (centipedes) 82, 113
selection and adaptation 21, 24, 62, 93, 103, 142, 143, 148, 168, 171
selection paradox 84, 90, 109
selection pressure 27, 133, 137, 142, 154, 157, 168, 220
self-organization 46, 89, 93
    on the cellular level 43, 101
serial homology 89, 91, 93, 94
sex robot 197, 198
sexual selection 2, 3, 21, 73, 74, 167
sexuality 10
Shakespeare, William 73
Shapiro, James A. 62
    portrait 232
shark 7
Shh see Sonic Hedgehog
SHH signaling pathway 120
SHH, morphogen 120
shore leave 103, 127
sickle cell anemia 184, 188
silk bowerbird 65
Simpson, John Gaylord 20
single-celled organism 3
single-celled organism, artificial 185
singularity 205
sirtuin 191
skinks (*Scincidae*) 77
Skinner, Michael K. 65
social brain theory 168, 170, 171, 174, 181, 209
social intelligence 171, 212
social learning 139, 152, 171
sociobiology 24, 27
Sonic Hedgehog (Shh) 37, 120, 122
Sophia, humanoid robot 200
Sorgner, Stefan Lorenz 201
speciation 19, 21, 221, 234
speech 178
speech recognition 201
Speman, Hans 106
Spengler, Oswald 140
Sphingobium chlorophenolicum, bacterium 45
spider web 154, 162
Standen, Emily 103
stasis 25, 59
Stebbins, George Ledyard 20
stem cells 186
    human-induced pluripotent (iPS) 184
    primordial 194
stem cells, human-induced pluripotent (iPS) 194
Sterelny, Kim 170, 171
    portrait 232
stochasticity 74, 157, 160
strawberry poison frog (*Oophaga pumilio*) 60, 112
stress resilience 63, 64
stress resistance 208
Sultan, Sonia 155
    portrait 232

superintelligence 201, 205, 206, 207
superorganism 27, 208, 219
    insect societies 27
Survival of the Fittest 2, 11, 12, 39, 168, 172, 184
Suzuki, Yuichiro 109
swordtail (*Xiphophorus hellerii*) 3
symbiogenesis 74, 157
symbolic inheritance 60, 174, 208
synthetic biology 184, 185, 207
syrinx 66
systems biology 12, 67, 69, 71, 75, 207
Szathmáry, Eörs 50, 87
    portrait 233

Tabin, Clifford 116
technology assessment 209, 213
technosphere 197, 207, 209, 210, 211, 212, 214
Tegmark, Max 200, 201
telomerase 190, 191
telomere 190
telomere shortening 191
temperature stressor 109, 110
termites 152
Theißen, Günter 9
Theory of mind 169
theory pluralism 221
theory structure 143, 220, 221
theory-induced blindness 221
thinking, evolution of 167
threshold effect 81, 89, 92, 93, 119, 122, 123, 124, 127, 147, 161
threshold mechanism 93, 126
threshold value 123, 124
threshold, intelligence explosion 205
tiger salamander (*Ambystoma tigrinum*) 78
Tiktaalik 103
Tinbergen, Nikolaas 213
tobacco hornworm caterpillar (*Manduca sexta*) 32, 109, 110, 127
Tomasello, Michael
    portrait 233
tool use 171, 178
transcription 12, 55, 63, 68
transcription factor 81
translation 55, 68
transposable element (transposon) 62
treehopper 83
trunk, elephant 115
Tuck, Jay 198
Turing system 123, 124, 233
Turing, Alan 73, 197
    portrait 233
turtle shell 90, 109, 111, 127

Uller, Tobias 109
    portrait 233
unitarianism 222
United Nations 200, 210
upright walk 115, 178
urea cycle disorder 185

Use it or lose it! 22, 42

variation *see* biased v.; complex v.; discontinuous v.; facilitated v.; hidden v.
Venter, Craig 185, 186
vertebrate extremity 56
Voland, Eckart und Renate 176
Volvox algae 151
Vries, Hugo de 10

Waddington, Conrad Hal 31, 32, 35, 44, 46, 61, 75, 80, 81, 93, 94, 109, 122, 135, 142
    portrait 233
Wagner, Andreas 41, 42, 43, 81, 93
    portrait 234
Wagner, Günter 23, 50, 89
    portrait 234
Wallace, Alfred Russel 5, 6, 11, 15, 17, 18, 20, 21, 28, 168
Walsh, Denis 159
Walsh, Toby 200, 204, 207
Wang, Yuja 115
Warwick, Kevin 193
Watson, James 11, 13
Watson, Richard A. 214
weak regulatory couplings 53, 55, 57
Weddell Seal (*Leptonychotes weddellii*) 108
Weismann barrier 10, 15, 61, 73, 74
Weismann, August 10, 15, 16, 73, 74, 118
West-Eberhard, Mary Jane 75, 76, 77, 79, 80, 82, 91, 94, 109, 142, 157
    complex developmental apparatus 46
    portrait 234
Whitelaw, Emma 63
Wieser, Wolfgang 57, 58
Wilson, David Sloan 26, 27, 138, 143, 208
    portrait 234
Wilson, Edward Osborne 27, 50
    portrait 235
wing pattern 84, 90, 91, 106
Wischaus, Eric 38
wisdom teeth 184
Wittgenstein, Ludwig 176
Wnt signaling pathway 111
Wozniak, Steve 206
Wright, Sewall 16, 18, 118, 119

xenobot, biological mini-robot 186, 190

Y chromosome 16, 22

Zalasiewicz, Jan 209
zone of polarizing activity (ZPA) 119, 120
ZRS, cis element 120
Zwenger, Thomas 141
zygote 68, 189

For Product Safety Concerns and Information please contact our EU representative GPSR@taylorandfrancis.com Taylor & Francis Verlag GmbH, Kaufingerstraße 24, 80331 München, Germany